MICROBIOLOGY RESEARCH ADVANCES

# ALIMENTARY MICROBIOME

# A PMEU APPROACH

# MICROBIOLOGY RESEARCH ADVANCES

Additional books in this series can be found on Nova's website
under the Series tab.

Additional e-books in this series can be found on Nova's website
under the e-book tab.

MICROBIOLOGY RESEARCH ADVANCES

# ALIMENTARY MICROBIOME

# A PMEU APPROACH

ELIAS HAKALEHTO
EDITOR

Nova Science Publishers, Inc.
*New York*

Copyright © 2012 by Nova Science Publishers, Inc.

**All rights reserved.** No part of this book may be reproduced, stored in a retrieval system or transmitted in any form or by any means: electronic, electrostatic, magnetic, tape, mechanical photocopying, recording or otherwise without the written permission of the Publisher.

For permission to use material from this book please contact us:
Telephone 631-231-7269; Fax 631-231-8175
Web Site: http://www.novapublishers.com

**NOTICE TO THE READER**

The Publisher has taken reasonable care in the preparation of this book, but makes no expressed or implied warranty of any kind and assumes no responsibility for any errors or omissions. No liability is assumed for incidental or consequential damages in connection with or arising out of information contained in this book. The Publisher shall not be liable for any special, consequential, or exemplary damages resulting, in whole or in part, from the readers' use of, or reliance upon, this material. Any parts of this book based on government reports are so indicated and copyright is claimed for those parts to the extent applicable to compilations of such works.

Independent verification should be sought for any data, advice or recommendations contained in this book. In addition, no responsibility is assumed by the publisher for any injury and/or damage to persons or property arising from any methods, products, instructions, ideas or otherwise contained in this publication.

This publication is designed to provide accurate and authoritative information with regard to the subject matter covered herein. It is sold with the clear understanding that the Publisher is not engaged in rendering legal or any other professional services. If legal or any other expert assistance is required, the services of a competent person should be sought. FROM A DECLARATION OF PARTICIPANTS JOINTLY ADOPTED BY A COMMITTEE OF THE AMERICAN BAR ASSOCIATION AND A COMMITTEE OF PUBLISHERS.

Additional color graphics may be available in the e-book version of this book.

**Library of Congress Cataloging-in-Publication Data**

**Library of Congress Control Number: 2011946298**

ISBN: 978-1-61942-692-4

*Published by Nova Science Publishers, Inc. † New York*

# Contents

# Preface

Over a decade ago Professor Joshua Lederberg introduced the concept "Alimentary Microbiome". This term includes all the micro-organisms constituting the microbial ecosystems in the alimentary tract. Their actions are so closely interrelated with the host body functions that the microbial communities could be considered as an independent organ of the human body system. Neurological, hormonal and immunological regulation is extended to the microbial populations, which in turn secrete molecular messages influencing their host. These microbes also live in closely balanced communities, where the competition is often a secondary goal, if it is striven for at all by the micro-organisms in many alimentary tract surroundings.

The members of the microbiome base their existence on interactions. Mutual benefit of the microbes and the human host contribute to our nutrient uptake and the overall health. The PMEU (Portable Microbe Enrichment Unit) technology makes it possible to study the microscopic and molecular interactions as well as metabolic developments in the cultures. This technology has revealed lactic acid bacteria surviving on the gastric epithelia, and symbiotic associations between various facultative coliforms in the duodenum. The strive for the balance between members of intestinal ecosystems, improved recovery of pathogenic and indicator strains in the microbiological samples, as well as microflora regulation by gaseous compounds are among the numerous striking features of the alimentary microbiome also studied with the PMEU. By maintaining the BIB (Bacterial Intestinal Balance) the microbiome promotes our health.

This book describes the microflora in mouth and on the teeth, in the gastric areas with low pH, in the upper small intestines where most nutrients are taken up from food, and in the fecal samples as well as in the environment. It surveys reasons behind microbial diseases and imbalances of the gastrointestinal tract. One of the severe situations is dysbiosis. Some exceptional conditions, such as the colonization of the gut of immunocompromised patients, small children, and patients treated with extensive antibiotic medications, have been taken into account. These could lead to new emerging infections, like to the overgrowth of *Clostridium difficile*, or the antibiotic resistant bacterial strains, or the yeasts. The ecological succession, and the composition of the microflora of the neonates is presented. Digestive enzymes and their role in the microbiome is considered with some information on the uptake of medicines from the GI tract. The developments of microbial ecosystems and the probiotic usage are discussed with respect to human health and its protection. The metabolic variations among micro-organisms provide excellent basis for interactions between man and his microbiome. Continuous research on these issues is a prerequisite for the future development of medicine.

I wish to thank the colleques for their contribution to this book. Let us hope that the endeavouring atmosphere of the team will prevail and flourish.

Elias Hakalehto
Adjunct Professor in Biotechnical Microbe Analytics
Department of Biosciences
University of Eastern Finland
Kuopio, Finland
email: elias.hakalehto@gmail.com

In: Alimentary Microbiome: A PMEU Approach
Editor: Elias Hakalehto
ISBN: 978-1-61942-692-4
© 2012 Nova Science Publishers, Inc.

*Chapter I*

# Introduction of the Alimentary Tract with Its Microbes

***Elias Hakalehto***
Department of Biosciences, University of Eastern Finland,
FI, Kuopio, Finland

## Abstract

The huge metabolic potential of the human body is a joint result of our alimentary tract functioning together with the microbiome. In order to investigate this multitude of activities, which ultimately makes our physical appearance what it is, we need to examine the interactions between man as a host and its microbiota. The latter one is composed of hundreds of species, and many times more of strains, which have an individual relationship with our body while at the same time belonging to the several gastrointestinal ecosystems in the alimentary tract.

Many structures of the alimentary tract are complicated and challenging in the microbiological investigation. All parts of the digestive system are interdependent, and so are the microbial communities in various parts. Even though the flow of food and chyme is downwards, many regulatory functions are exercised by the entire body system via nervous or hormonal or immunological control. Human and microbial enzymes continuously process the materials in order 1. to provide adequate nutrition for the host, and 2. to maintain the balanced function of the microbiome. Tissue hormones are secreted locally in order to regulate the function of the digestive organs. All changes in the system are interrelated with the microbiome.

The concept "molecular communication" came across the author for the first time some thirty years ago. Then it was used by a Finnish microbiologist and numerical taxonomist, Prof. Helge Gyllenberg, who was also the chairman of the Central Scientific Committee of the Academy of Finland. This concept includes both the interactions between various intestinal strains and communities, as well as the physicochemical signaling between the microbes and the host. All these biological activities could be studied in the PMEU (Portable Microbe Enrichment System).

# Introduction

The human body carries even kilograms of micro-organisms, which mainly locate in the large intestine. In addition they occupy our skin and all mucosal membranes. The microbiota includes hundreds of species. These taxonomical units (taxons) are further divided into strains. The strains have their specific genetic and physiological characteristics [Shenderov, 2011] as well as surface structures [Foroni *et al.*, 2011]. In the human gastrointestinal tract they constitute the alimentary microbiome [Lederberg, 2000]. It is a kind of superorganism within the body system. The intestinal flora is a continuum of the microflora in the upper areas of our nutrient uptake. These microbial communities are restricted by many host defense systems. Occasionally, during the meals, the ecosystem starts an accelerated pace [Hakalehto *et al.*, 2008]. The microbes interact with each other and with the human host. Therefore the concept microbiome is justified; it is a series of ecosystems, which strive for balance. Ideas reflecting the supposed competition between various micro-organisms have turned out to be collaboration. If a particular microbe continues its existence, the best approach would definitely be a symbiotic action in order to:

1. avoid excessive host defenses
2. get all nutrients readily in an exploitable form
3. manipulate the environment into a favorable direction
4. set up biofilms for continued survival OF THE ENTIRE COMMUNITY
5. get distributed within the current host and beyond

From this point of view, pathogenesis is an opportunistic strategy. In fact, it is often an indication of inadequate balance within the alimentary microbial ecosystems. The host actually interacts strongly with its microbiome by neurological, hormonal and immunological signals [Lyte, 2010]. There are also many mechanisms for molecular communication between the microbial strains. Physiologically, they exploit the nutrients by joint metabolic functions. For example, the *Escherichia coli* and *Klebsiella mobilis* strains were shown to contribute positively to each other′s growth in mixed PMEU (Portable Microbe Enrichment Unit) simulation of the intestinal conditions [Hakalehto *et al.*, 2008]. They were growing in a more equal manner in a mixed culture than in separate ones. Similarly, the lactobacilli were shown to have a booster effect on anaerobic butyric acid bacteria [Hakalehto and Hänninen, 2012]. These fascinating windows into the microbiological world should be opened further in order to understand the interactions on the epithelia, between cells, and on the molecular level.

# Alimentary Flow of Substances

The human body is capable of exploiting numerous variable food sources. The alimentary tract takes care of the acquisition of energy and raw materials. This is the basis for metabolism, which is traditionally divided into catabolism and anabolism. The microbial metabolism is connected with that of the host. This combination has given the ground for the concept "Alimentary Microbiome" [Lederberg, 2000]. The overall functions of the body

WITH its microbes need to be evaluated. What are then the consequences, if something goes wrong?

The energy metabolism liberates both thermal energy and ATP molecules which dissipate energy for metabolism. Wastes are secreted into the colon, or via the urinary tract or skin or the respiratory system, in order to be delivered out of the body. Microbes live in parallel, but they also participate in the uptake of the nutrients into our body. In these conditions of an incoming and occasionally overwhelming flow of food substances, they have to use a third kind of metabolism mode, the overflow metabolism [Tempest *et al.,* 1983]. As a matter of fact, the nutrient uptake constitutes the basis of our physical existence. For the maintenance and daily activities, we need the catabolic functions that are keeping our cells in an active state. Microbes take part in the generation of resources for that part of our body mechanisms. The interactions need to be studied effectively also *in vitro*. The PMEU provides tools for studying the microbial growth and metabolism with the actual speed that reactions proceed in the GI tract either on the epithelial surfaces, or in the chyme.

Anabolic metabolism, in turn, requires various minerals and trace elements (that is minerals whose demand is limited to small amounts). By the circulation of all these substances, besides the organic molecules, our cells are interconnected with nature, and our environment. The minerals take part in enzymatic reactions. These reactions, in turn, catalyze both the energy metabolism and the construction of our cells and tissues. In the anabolism, the organic molecules form building blocks for the cells. For example, fatty acids are necessary in the cell membrane synthesis, and proteins serve as enzymes and structural molecules in all cells.Nutritional substances are taken up in the gastrointestinal column that has a length of several meters [Garcia Luna and Lopez Gallardo, 2007]. Salivary glands, stomach, pancreas, liver and the bile secretions contribute to the digestion which continues in the mouth, pharynx, esophagus, stomach, the small intestine (duodenum, jejunum and ileum), ascending and descending colon and the rectum. Bacteriologically important regions in the GI tract are:

- - the hypopharynx (laryngopharynx), which is the bottom part of the pharynx connecting it with the esophagus
- - the gate between the pyloric part of ventricle and duodenum, which is the narrowest point in the GI tract
- - the ileocaecal valve between the small and large intestines

Food digestion starts in the mouth where teeth begin the mechanical treatment of the various materials. The saliva moistens the mass making swallowing easier. Salivary amylase starts to degrade carbohydrates [Nunes *et al.*, 2011]. Food is swallowed into the stomach, which stores and mixes it, and thermostats it to the body temperature. The glands in the gastric walls secrete inactive pepsinogen, a preform of the pepsin protease [Neubert *et al.*, 2010]. This is activated by the low pH of the hydrochloric acid secreted by the same glands as pepsinogen itself. The pepsin molecules cut the peptide bonds in the proteins, producing smaller peptides. Besides the pepsinogen, the stomach glands produce gastrin [Goo *et al.*, 2010]. The gastrin regulates small intestinal functions, and the amount of mucin slime that is protecting the stomach walls against the acid. The stomach epithelial cells are regenerated fast since they stay alive only for a couple of days. Similarly, it is expectable that a fraction of the bacteria delivered into the stomach together with food, or from the oral communities, could develop some kind of a temporary survival strategy inside the acidic stomach surroundings [Hakalehto *et al.*, 2011]. Such strains include the strains of various LAB (Lactic Acid Bacteria). The low pH in the stomach is necessary to activate the pepsin and for destroying a majority of the incoming microbes. Nevertheless, helicobacteria protected by mucus and ammonium ionscan cause problems [El-Eshmawy *et al.*, 2011]. Its colonies are protected by carbohydrates, and can resist the acidic challenge in the gastric areas. If these microcolonies get rooted too deeply into the epithelium, problems will arise from the host point of view. Therefore, it is of crucial importance for the host body to clean up and flush away any permanent microbial colony formations from the upper GI tract. After meals the muscle relax and increase the stomach volume by up to 1-3 litres. The food stays in the stomach for an average 3-4 hours.

The cholecystokinin (CKK) is produced in the duodenal mucosal cells . It causes the release of digestive enzymes and bile substances from the pancreas and the gallbladder, respectively [Temler *et al.*, 1984]. Such tissue hormones as the gastrin and the CKK have a key role in the regulation of food digestion in our alimentary tract. Besides these host-made regulators, the alimentary microbes control some parts of the degradation of organic molecules in the food. They also seem to regulate the conditions in the duodenum, for their own advantage, and for the benefit of the host [Hakalehto *et al.*, 2008].

The small bowel of 3-5 meters in length, receives the chyme in small portions by the closing muscles in the bottom of the pylorus [Brizzee, 1990]. Thus the small bowel enzymes are able to digest even large meals. The hydrolysis of food substances continuesto monomers which are absorbed from the small intestines. Proteases and peptidases secreted from the intestinal walls and by the pancreascomplete the digestion. Besides the lipases from the pancreas, also the bile tract liberates fat hydrolyzing bile acids produced in the liver [Hebanowska, 2010]. The bile consisting of water, bicarbonates, cholesterol, bile salts and bilirubin, dissociates the lipid droplets into smaller ones in a way that the lipases are able to degrade them.

Amylases from the pancreas hydrolyze the starch into sugars. Trypsin and other proteinases produce amino acids and peptides, and nucleases dissociate nucleic acid. The pancreas secretes also hormones from its endocrine part, of which most important are insulin and glucagone participating in the regulation of the blood sugar levels.

The absorption takes place in the *villi intestinalis* whose epithelial cells contain many small appendices, *microvilli* [Chopra *et al.*, 2007]. Each *villus intestinalis* contains capillary veins and lymphatic veins, into which the nutrients are transferred either via the epithelial cells or from the space in between the cells. Amino acids and monosaccharides are taken up by active transport mechanisms, and further to blood circulation by diffusion. The dissociation products of various lipids, fatty acids, are not water soluble, and they form, together with the bile salts, cholesterol and other substances, the small droplets which can approach the cells to where the fatty acids and monoglycerides are transferred by diffusion [Gangl, 1975]. This is one example of the necessity of cholesterol in human tissues, because the fatty acids are a prerequisite for anabolism. Inside the cells the fatty acids and monoglycerides join together and form triglycerides, which are surrounded by a protein membrane making them water soluble and absorbable to the lymphatic veins which could be more easily penetrated than the capillary veins. Water and fat-soluble vitamins are absorbed differently, the latter in a somewhat similar fashion to the lipids, whereas the water solubles are transferred into the blood circulation by active transport or diffusion. Trace elements are electrically charged ions, which pass the cell membrane through the ion pump or ion channels formed by protons.

The released hydrolysis products are fastly absorbed and avoid bacterial uptake. Small intestine contains only limited numbers of bacteria although conditions are favourable for them because the gut motility moves the masses towards colon.

The unabsorbed material is stored into the colon to wait for disposal. In the colon, some vitamins are produced by bacteria, such as vitamin K. The feces consists of desquamated human cells and microbes with the fibers and other nonhydrolysed material. Little digestive enzymes is lost.

## PMEU as a Research Tool to Simulate the Gut Conditions

All microbiological culture methods are based on the multiplication of the cells. By division they form communities and colonies, which more or less reflect the status of an individual cell. This metabolic condition becomes thus demonstrated on population level. In the PMEU the cultures are usually homogenous due to mixing, and cultures are studied within a few hours from the onset of the cultivation [Hakalehto, 2010]. Instead of an overnight culture only (or a cultivation of several days), the results are accumulated hour by hour during the course of the incubation in highly controllable conditions. Any effects of additions or environmental manipulations become easily documented, and the metabolic behavior and interactions can be easily comprehended. Single parameters, such as gas composition and flow, or antibiotic influences, can be followed up in circumstances parallel to the ones faced by a single cell on the gastrointestinal epithelium, or in the intestinal fluid [Hakalehto, 2011; Hakalehto, 2012]. The reactions of the microflora, as a whole, on nutrition or on medications, could also be screened with different PMEU versions [Pesola *et al.*, 2009; Hakalehto *et al.*, 2011; Pesola and Hakalehto, 2011].

The idea of planning a cultivator where microbial enrichment starts on the spot (at the sampling site), produced the Portable Microbe Enrichment Unit (PMEU). In order to avoid losses in time and in sample quality, a new tool for microbiology research both in laboratory and field, the PMEU was developed. During the ten years of applying the PMEU for microbiological research, we have developed an entire ideology, the "PMEU Technology".

It has helped us to transform microscopic metabolic events into larger and better observable scales. It has also assisted us in enhancing the recovery and growth of all kinds of bacteria and other microbes. The detection processes now last only hours instead of days and days instead of weeks. If we follow microbes from the real start and practically every minute after that, we can see that the microbiological reactions often take place faster in natural or original settings compared to observations in the laboratory. The microbe detection is also seen as a process, which consists of separate unit tasks:

1. Sampling
2. Sample transport/storage
3. Pre-enrichment

4. Enrichment
5. Detection
6. Recognition/Identification
7. Characterisation/Analysis

Paradoxically, even though providing means mainly for the submerged cultivations, the PMEU has turned out to be an excellent instrument in researching the microscale behaviour of individual cells on the surfaces. Because it is a tool for overcoming the diffusion limitation on the culture level, it reflects the optimal circumstances for the cells, and their true action in short time. This gives an idea of the reactions of the populations on the very surfaces that they have attached to. When the biofilms are further structured, they exhibit other features. These could also be studied with the PMEU system by installing the surfaces within the cultivation syringes thus making room for actual attachment [Mentu *et al.*, 2009]. However, the speed of metabolic reactions could be understood best when the diffusion barriers are lowered using the conditions in the PMEU for that purpose.

In the PMEU one can simultaneously cultivate the bacteria and study the interactions of different bacteria in different culture media and atmospheres and also examine their responses to different antimicrobial agents.

## Probiotics and Human Efforts to Improve the Alimentary Microflora

Nowadays, the common understanding on the importance of the alimentary microbes on our health has rapidly increased. The author was at 1986 the first scientist in a joint project between Helsinki University and the Finnish Dairy company Valio (then a cooperative, now Valio Oyj, the largest dairy corporation in Finland). At that time, 25 years ago, the aim of the investigation was to make a recommendation on ways for Valio to get involved in the probiotics business. The conclusion of the preliminary work was to suggest the acquisition of a microbial strain that was as profoundly researched as possible. The intention was to find a probiotic bacterium whose beneficial influences on human health had been documented widely. On this basis Valio then purchased the rights for the GG (*Lactobacillus rhamnosus*) strain, whose trade name was to be Gefilus™ in Finland [Yli-Knuuttila *et al.*, 2006].

The research behind, or on the basis of the Gefilus™, "good bacteria", has been further accumulated since then. Nevertheless, the research has mostly focused on the microbiota in the colon (and thus the fecal microbes) representing the most numerous populations of the human normal flora. However, even more important ecosystems could exist within our alimentary tract.

The microbes of the large intestines take a benefit of our waste materials, but the microbiota in the duodenum, for example, takes part in the nutrient intake of individuals. In researching these microbial communities, we also have to make a clear distinction between the biofilms attached onto the surfaces and the microbial growth in the fluidic contents of the alimentary tract. The latter one often originates, or gets inoculated, from the biofilm microflora. The mechanisms behind the population dynamics in these different niches need to be better understood, as well as their interactions.

# Human Regulation of the Microbiome and Its Interactions

The several ecosystems in the alimentary tract form the alimentary microbiome. – It is a series of ecosystems dependent on each other .The material flow goes mostly downwards, but gases are emitted, and nutrients circulate on the basis of microbial functions. In order to remain in good health, we need to take care of the biodiversity of our alimentary ecosystems. These ecosystems are regulated by the host neurological, hormonal and immunological systems, and the information in this network is distributed to every direction.

This book arises from a fountain of scientific laboratory works. The PMEU is opening up new windows into the hidden lives of the intestinal microbes. Being a simple-to-use instrument, it can rapidly give extensive information about the ecological phenomena in the microenvironments, which also safeguard our health. Although the ideas presented in this book are well based on sound scientific research and countless experiments, at the same time we wish to "extend our vision behind the horizon". This is made possible by the simple adjustments in the PMEU, and by choosing the appropriate conditions for specific pure and mixed cultures.

In order to understand the conditions behind diseases, and their etiology, it is of great importance to get a view on the actual mechanisms of microbial behavior in the small scale. We need to test the reactions of individual strains towards environmental changes, as well as the impact of the same circumstances on co-cultures. Our body system is in a central role in producing the niches into the alimentary tract. The PMEU simulates the conditions in both cavities of the mouth and the intestinal fluids or the mucosal membranes. The body exhibits protective functions and symbiotic actions at the same time. These reflect the sensitive balances in the alimentary tract and its microbiome, and are also the factors behind our health and disease. The latter could be called "a systematic error", or a dysfunction on the ecosystem level within our body system.

# References


Brizzee, KR. Mechanics of vomiting: a minireview. *Can. J. Physiol. Pharmacol.*, 1990; 68, 221-229.

Chopra, S; Saini, RK; Sanyal, SN. Intestinal toxicity of non-steroidal anti-inflammatory drugs with differential cyclooxygenase inhibition selectivity. *Nutr. Hosp.,* 2007; 22, 528-537.

El-Eshmawy, MM; El-Hawary, AK; Abdel Gawad, SS; El-Baiomy, AA. *Helicobacter pylori* infection might be responsible for the interconnection between type 1 diabetes and autoimmune thyroiditis. *Diabetol. Metab. Syndr.,* 2011; 3, 28.

Foroni, E; Serafini, F; Amidani, D; Turroni, F; He, F; Bottacini, F; O'Connell Motherway, M; Viappiani, A; Zhang, Z; Rivetti, C; van Sinderen, D; Ventura, M. Genetic analysis and morphological identification of pilus-like structures in members of the genus *Bifidobacterium*. *Microb. Cell Fact.,* 2011; 10 Suppl 1, S16.

Gangl, A. The lipid metabolism of the small intestine and its correlation to the lipid and lipoprotein metabolism of the total organism. *Acta Med. Austriaca Suppl.,* 1975; 2, 1-49.

Garcia Luna, PP; Lopez Gallardo, G. Study on intestinal absorption, metabolism, and adaptation. *Nutr. Hosp.,* 2007; 22 Suppl 2, 5-13.

Goo, T; Akiba, Y; Kaunitz, JD. Mechanisms of intragastric pH sensing. *Curr. Gastroenterol. Rep.,* 2010; 12, 465-470.

Hakalehto E, Hänninen, O. Lactobacillic $CO_2$ signal initiate growth of butyric acid bacteria in mixed PMEU cultures. Manuscript in preparation. 2012.

Hakalehto, E. Antibiotic resistant traits of facultative *Enterobacter cloacae* strain studied with the PMEU (Portable Microbe Enrichment Unit). In Press. In: Méndez-Vilas A, editor. *Science against microbial pathogens: communicating current research and technological advances. Microbiology book series Nr. 3.* Badajoz, Spain: Formatex Research Center; 2012.

Hakalehto, E. Hygiene monitoring with the Portable Microbe Enrichment Unit (PMEU). *41st R3 -Nordic Symposium. Cleanroom technology, contamination control and cleaning. VTT Publications 266.* Espoo, Finland: VTT (State Research Centre of Finland); 2010.

Hakalehto, E. Simulation of enhanced growth and metabolism of intestinal *Escherichia coli* in the Portable Microbe Enrichment Unit (PMEU). In: Rogers MC, Peterson ND, editors. *E. coli infections: causes, treatment and prevention.* New York, USA: Nova Publishers; 2011.

Hakalehto, E; Humppi, T; Paakkanen, H. Dualistic acidic and neutral glucose fermentation balance in small intestine: Simulation *in vitro*. *Pathophysiology,* 2008; 15, 211-220.

Hakalehto, E; Vilpponen-Salmela, T; Kinnunen, K; von Wright, A. Lactic acid bacteria enriched from human gastric biopsies. *ISRN Gastroenterol,* 2011; 109-183.

Hebanowska, A. Bile acid biosynthesis and its regulation. *Postepy Hig. Med. Dosw. (Online),* 2010; 64, 544-554.

Lederberg, J. Infectious history. *Science,* 2000; 288, 287-293.

Lyte, M. The microbial organ in the gut as driver of homeostasis and disease. *Medical Hypotheses,* 2010; 74, 634-638.

Neubert, H; Gale, J; Muirhead, D. Online high-flow peptide immunoaffinity enrichment and nanoflow LC-MS/MS: assay development for total salivary pepsin/pepsinogen. *Clin. Chem.,* 2010; 56, 1413-1423.

Nunes, LA; Brenzikofer, R; Macedo, DV. Reference intervals for saliva analytes collected by a standardized method in a physically active population. *Clin. Biochem.,* 2011; Dec; 44(17-18), 1440-1444.

Pesola, J; Hakalehto, E. Enterobacterial microflora in infancy - a case study with enhanced enrichment. *Indian J. Pediatr.,* 2011; 78, 562-568.

Pesola, J; Vaarala, O; Heitto, A; Hakalehto, E. Use of portable enrichment unit in rapid characterization of infantile intestinal enterobacterial microbiota. *Microb. Ecol. Health Dis.,* 2009; 21, 203-210.

Shenderov, BA. Probiotic (symbiotic) bacterial languages. *Anaerobe,* 2011; 17, 490-495.

Temler, RS; Dormond, CA; Simon, E; Morel, B. The effect of feeding soybean trypsin inhibitor and repeated injections of cholecystokinin on rat pancreas. *J. Nutr.,* 1984; 114, 1083-1091.

Tempest, DW; Neijssel, OM; Teixeira De Mattos, MJ. Regulation of metabolite overproduction in *Klebsiella aerogenes*. *Riv. Biol.,* 1983; 76, 263-274.

Yli-Knuuttila, H; Snall, J; Kari, K; Meurman, JH. Colonization of *Lactobacillus rhamnosus* GG in the oral cavity. *Oral Microbiol. Immunol.,* 2006; 21, 129-131.

In: Alimentary Microbiome: A PMEU Approach
Editor: Elias Hakalehto

ISBN: 978-1-61942-692-4
© 2012 Nova Science Publishers, Inc.

*Chapter II*

# Research on the Normal Microflora and Simulations of Its Interactions with the PMEU

***Elias Hakalehto***
Department of Biosciences, University of Eastern Finland,
FI, Kuopio, Finland

## Abstract

The various interactions between the microbial strains consist of metabolic co-operation, sensing, and molecular messages, as well as antagonistic effects. When the PMEU (Portable Microbe Enrichment Unit) has been used for simulating these effects it has revealed a trend within the mixed population for balancing their coexistence. This kind of organization seems to be possible and achievable to prevail in the microbial communities. In case of the gastrointestinal tract we have designated it as BIB (Bacterial Intestinal Balance). It also has beneficial effects on the host body system, which, in turn, has means for controlling the above-mentioned balance *in vivo*. In the PMEU we have studied the effects of pH, bile substances, gases, antibiotics and defensins on various bacterial populations. This chapter is an introduction to the possibilities of enhanced enrichment in research on microbial ecology in the alimentary tract with the PMEU.

The possibilities to monitor each and every interaction between the cells are limited. However, the PMEU culture being in a homogenous, highly active state, multipliesthe effect of *e.g.* strain-to-strain interactions onto population level. Any swift changes are also detectable. This "amplification" of the culture activities gives new views into the microbial ecosystems. The adjustment of conditions in the PMEU could mimic the host functions, as well as addition so substances, such as bile acids. An example of the follow-up of the interactions of the adjacent microbial cells in the body system, is the linkage of the outgoing gas flow from one PMEU syringe into the other one The flow of volatiles can well regulate the environment on the mucosal membranes and in the intestinal fluids. It is also carrying information on cell-to-cell and species-to-species basis.

# Introduction

As the PMEU method is producing advantageous growth conditions for individual bacterial cells, monitoring of the bacterial populations and detecting the presence of potential contaminants is optimised in terms of speed and accuracy. As long as the limits of the culture environment are not met, all different strains in the sample could all grow optimally, in fact being enriched at the same time.

Any sample taken during these first hours of the mixed cultures thus represents the collective response of the population to the prevailing conditions (Hakalehto, 2011).

Isolates of various bacterial strains could be cultivated aerobically, microaerobically or anaerobically using standard broth media by standard PMEU enhanced enrichment technology, or by PMEU Spectrion® units equipped with optical, UV or IR sensors or PMEU Scentrion® gas sensored units. The equipment were planned and designed by the author in his company, Finnoflag Oy, and produced by Samplion Oy in Siilinjärvi, Finland according to the ISO 9001 standard protocol.

The reference cultivations were carried out using standard microbiological procedures. The PMEU Spectrion® is being validated by the VTT of Finland for water hygiene measurements (Wirtanen and Salo, 2010). Similarly, the PMEU Scentrion® is in the process of hospital validation (Hakalehto *et al.*, 2009; Pesola and Hakalehto, 2011).

The PMEU versions facilitated ultra-fast detection of coliformic bacteria, *Bacillus cereus*, salmonellas, yersinia, staphylococci, streptococci, campylobacteria, pseudomonads and other groups. The PMEU Scentrion® equipped with gas sensors for volatile organic compounds detected the contaminants at concentrations of bacterial levels around 10-1000 cfu/ml in 2-5 hours (Hakalehto *et al.*, 2009). The PMEU has also been used for helping the environmentally stressed cells to recover and become viable in the enrichment cultures in clinical studies, or water monitoring (Pesola *et al.*, 2009; Hakalehto *et al.*, 2011). Hygiene sampling with the PMEU is carried out using a specific sampling syringe serving also as an incubator container in the PMEU.

# Studies on Bacterial Pure Cultures Using PMEU Spectrion® Device with an Option of Altering Incoming Gas

The various bacteria and other microbes behave in an individual way in the PMEU. This unit can be used for cultivation of aerobic, microaerobic and anaerobic bacteria equally well. The effects of the gas flow and its composition can be effectively studied.

In Figure 1 a-j, we have compared the impact of sterile air with the gas flow containing slightly elevated levels of $CO_2$. These elevated levels could prevent the growth, and alternatively also provoke its onset (Hakalehto and Hänninen, 2012). The strains are derived from the laboratory collection of Finnoflag Oy. The shapes of the various growth curves in the PMEU Spectrion® are presented in Chapter 6 of this book.

Figure 1. PMEU Spectrion® unit with holder for 10 sampling and enrichment syringes. The syringe size is changeable. Some models can take 20 samples. The central control unit with cable and mobile connections is situated on the right half of the device, gas adjustment valves and batteries on the right. This unit is operable both within the laboratories and in the field, and in hospital units.

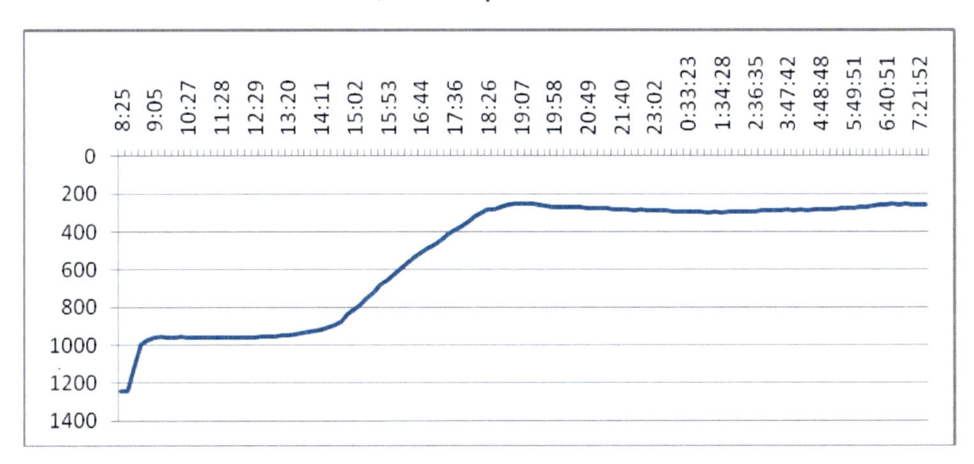


Figure 2a. The growth of *Escherichia coli* in the TYG (Tryptone Yeast Extract) medium in the PMEU Spectrion® at body temperature with sterile air as the gas flow. The x axis represents time flow in real time (during a 8 h working day in most cases). The y axis is relative loss of IR light penetration. Studies with the *E. coli* cultures in the PMEU have been reviewed earlier (Hakalehto, 2011).

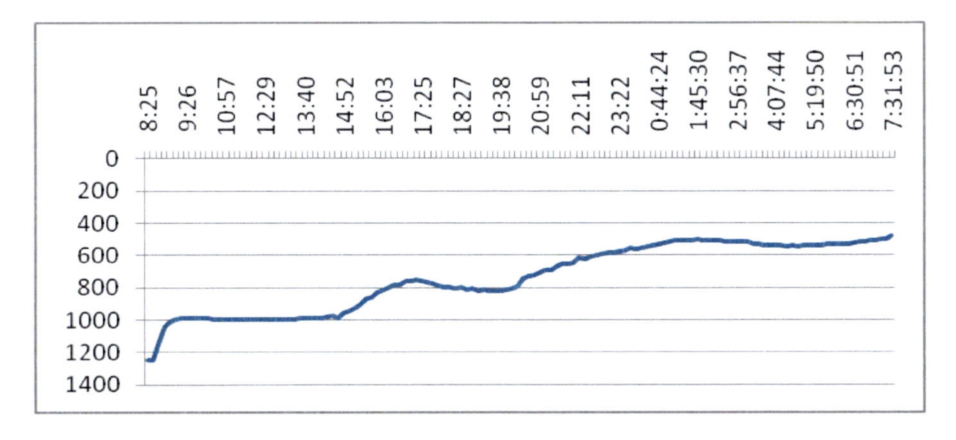


Figure 2b. Same as 2a. but with sterile 5% $CO_2$ as the cultivation gas.

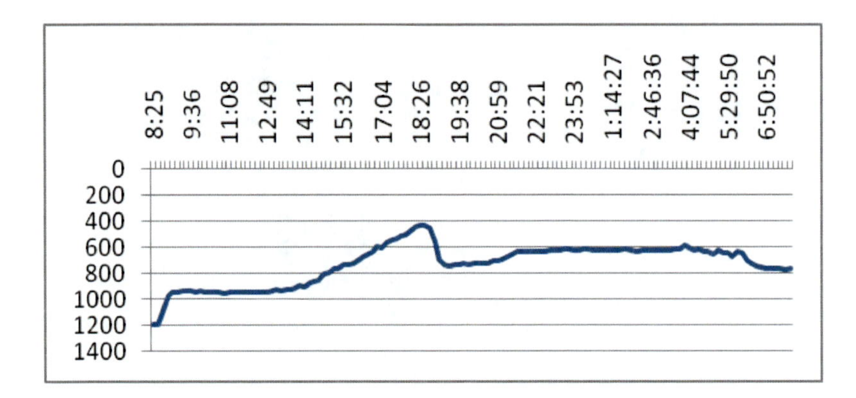


Figure 2c. The growth of *Klebsiella mobilis* in the TYG (Tryptone Yeast Extract) medium in PMEU Spectrion® at body temperature with sterile air as the gas flow. The x axis represents time flow in real time (during a 8 h working day in most cases). The y axis is relative loss of IR light penetration in the PMEU enrichment syringe. The studies regarding the important pH and microflora balancing function of the *Klebsiella* / *Enterobacter* group of facultative bacteria in the duodenal flora is published by Hakalehto *et al.* (2008 and 2010).

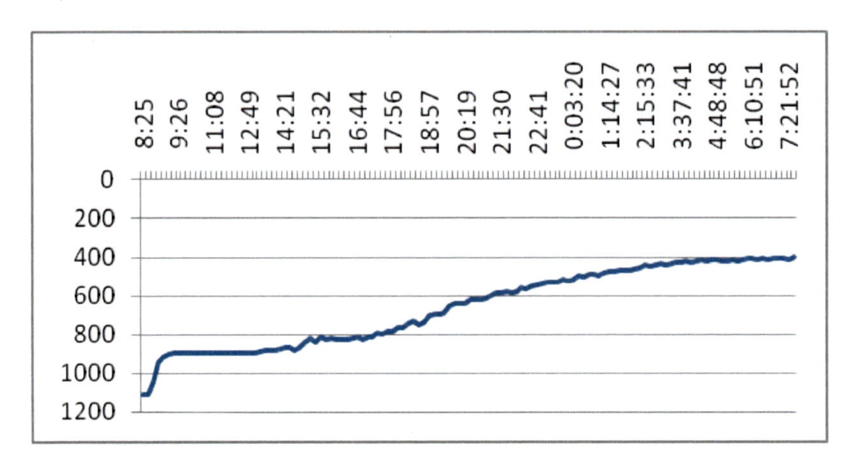


Figure 2d. Same as 2c. but with sterile 5% $CO_2$ as the cultivation gas.

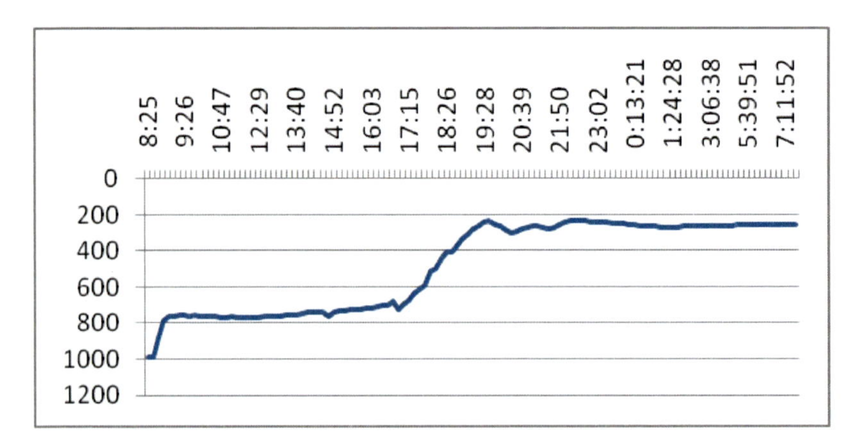


Figure 2e. The growth of *Pseudomonas aeruginosa* in the TYG (Tryptone Yeast Extract) medium in PMEU Spectrion® at body temperature with sterile air as the gas flow. The x axis represents time flow in real time (during a 8 h working day in most cases). The y axis is relative loss of IR light penetration in the PMEU enrichment syringe.

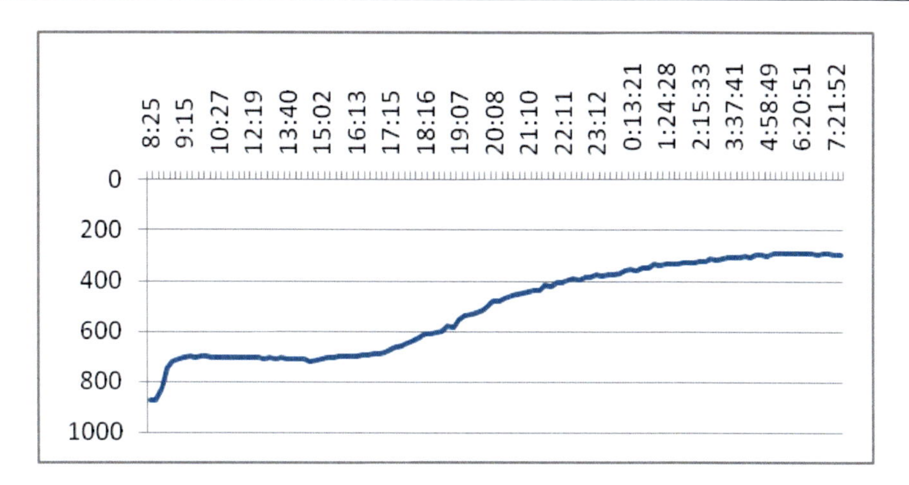


Figure 2f. Same as 2e. but with sterile 5% $CO_2$ as the cultivation gas.

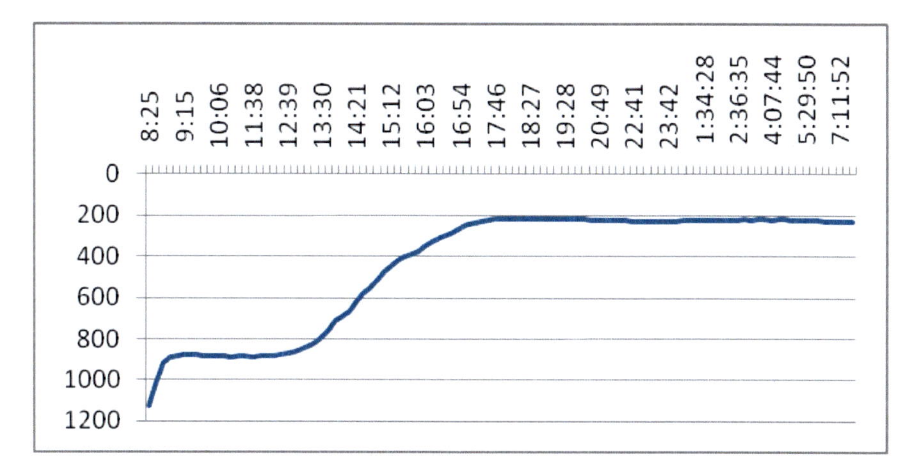


Figure 2g. The growth of *Enterobacter cloacae* in the TYG (Tryptone Yeast Extract) medium in PMEU Spectrion® at body temperature with sterile air as the gas flow. The x axis represents time flow in real time (during a 8 h working day in most cases). The y axis is relative loss of IR light penetration in the PMEU enrichment syringe. The studies with the PMEU regarding multiresistant *E. cloacae* strain is published before by Hakalehto (2011).

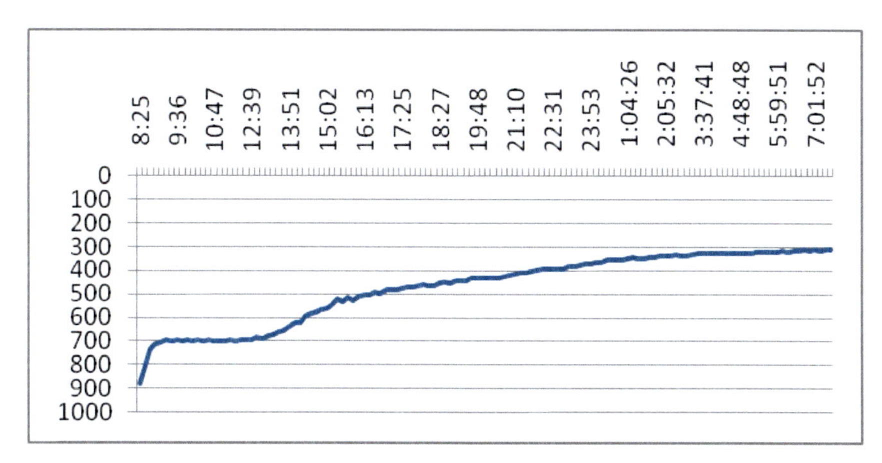


Figure 2h. Same as 2g. but with sterile 5% $CO_2$ as the cultivation gas.

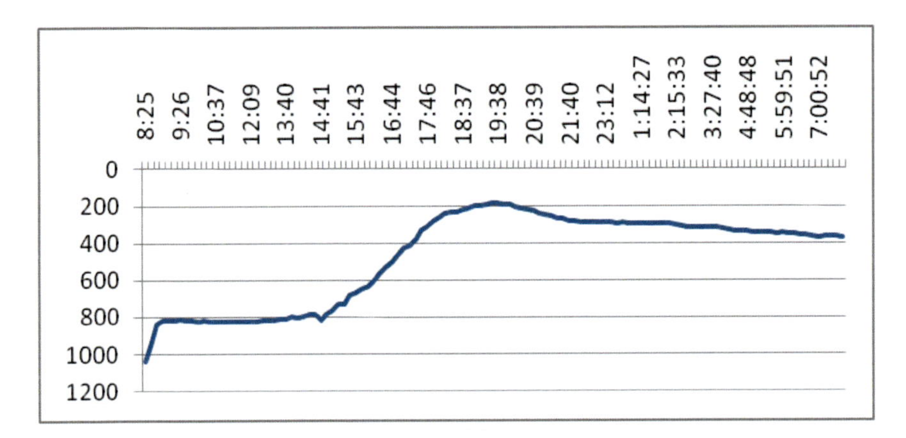


Figure 2i. The growth of *Salmonella enterica* Serovar *typhimurium* in the TYG (Tryptone Yeast Extract) medium in PMEU Spectrion® at body temperature with sterile air as the gas flow in the PMEU enrichment syringe. The detection of different salmonellas with the PMEU have been presented earlier (Hakalehto *et al.*, 2007; Hakalehto, 2011). In the latter publication the tests were carried out with the PMEU Scentrion® version which monitors the volatiles emitted from the bacterial cultures (Hakalehto *et al.*, 2009).

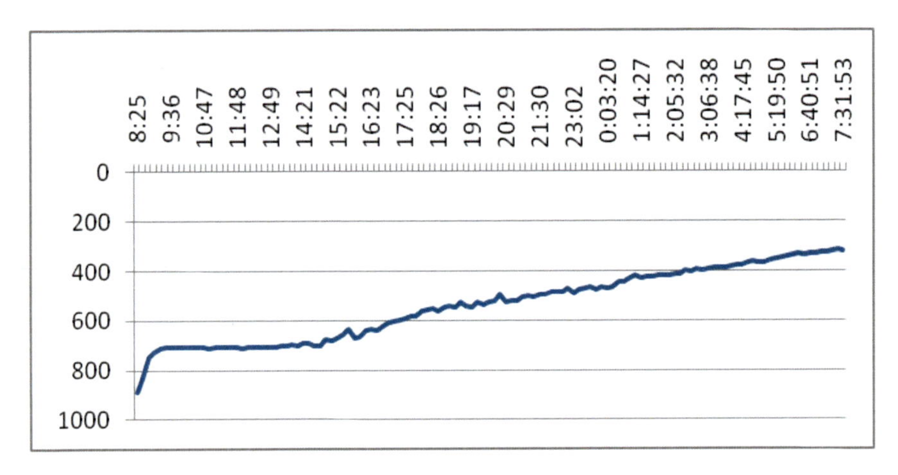


Figure 2j. Same as 2i. but with sterile 5% $CO_2$ as the cultivation gas.

# Quantification with the PMEU Method

In several series of experiments we have used the standard plate count techniques for the determination of the cell counts (cfu's, colony forming units). This is also in line with the traditional views. However, if and when the PMEU cultivation allows more germs to get culturable, this means an increase in the detected cells or cfu´s. Correspondingly, we can ask questions regarding the reliability of any results in this respect. The correlation of *E. coli* cultures in the PMEU with the initial inocula concentration has been illustrated by growth curves earlier (Hakalehto, 2011). These results indicated that one decade dilutions in the inoculum size cause a delay of one hour in the onset of growth of the PMEU cultures. In the Figure 3 we demonstrate this effect of diluted inoculum (or "reverse correlation") in the case of *Pseudomonas fluorescens* bacterium (Figure 3.). The illustration of the bacterial cultures on BD CHROMagar™ Orientation plates (Becton, Dickinson & Co., Sparks, MD, USA) for enumeration is seen in Figure 4. and the effect of various cultivations in the PMEU on the final pH in Table 1.

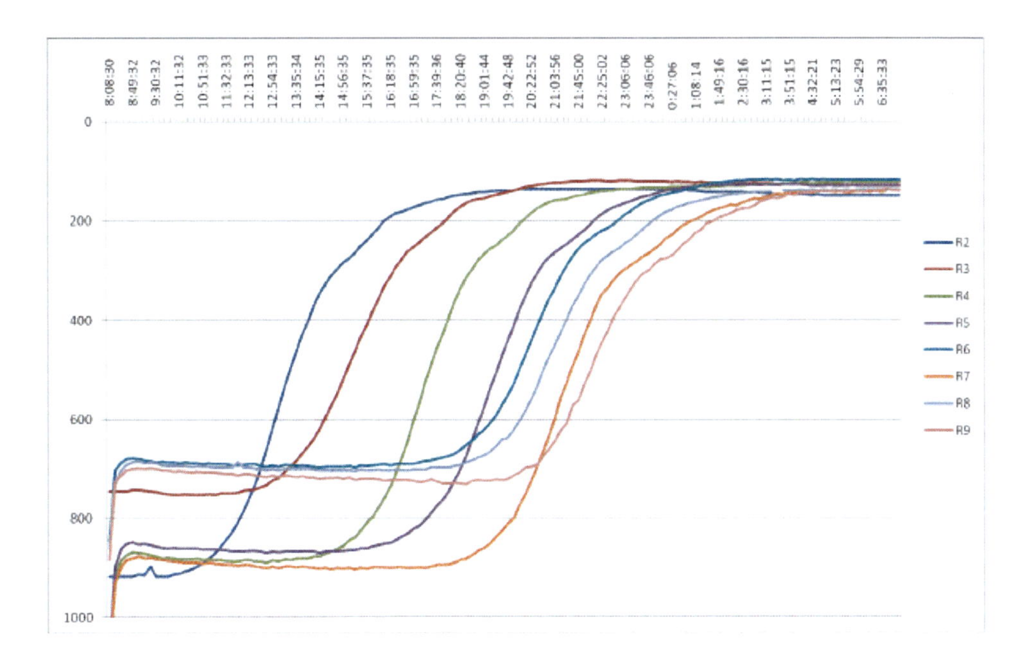


Figure 3. The quantitative PMEU Spectrion® cultivation of *Pseudomonas fluorescens* at 37 °C with sterile air flow. For dilutions of the inocula: the curves from left (1:100) to the right (1:1 000 000 000) correspond to ten fold dilutions. In case of more diluted samples (more than 1:100 000) the effect of inoculum dilution is surprisingly not delaying much the onset of the culture. Otherwise, the culture growth seems to follow the same pattern as with *E. coli* where every ten fold dilution means ONE HOUR DELAY IN THE ONSET OF THE ACTIVE GROWTH.

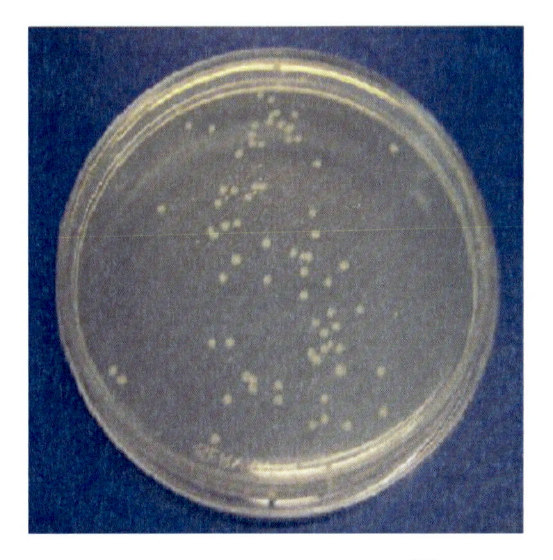

Figure 4. *Pseudomonas fluorescens* colonies on the BD CHROMagar™ Orientation plates (Becton, Dickinson & Co.).

**Table 1. Cultivation of *Pseudomonas fluorescens* in the PMEU Spectrion® at 37 °C with sterile air flow. The final pH values in the TYG medium after around 24 hours of aerobic culture time**

| Syringes and dilution of the inoculum | Final pH |
|---|---|
| 1:100 | 7,90 |
| 1:100 | 7,86 |
| 1:1000 | 7,98 |
| 1:10 000 | 7,87 |
| 1:100 000 | 8,04 |
| 1:1 000 0000 | 7,95 |
| 1:10 000 000 | 7,94 |
| 1:100 000 000 | 8,05 |
| 1:1000 000 000 | 8,04 |
| 1:10 000 000 000 | 8,14 |

## Microbial Interactions Studied in the PMEU

In these stressing conditions the bacterial species are together formulating their surroundings for the common benefit of the microbial communities. The interaction is important for all participants, and is also maintaining healthy conditions in the host body. The recent results from the PMEU studies emphasize the need for investigating the entire microbe populations and their interactions in the human body, and in the host – microbe ecosystem established therein.

Microbiological research is based on isolating individual strains as pure cultures starting from single colonies. The isolates are then inoculated to further cultivation on new Petri plates (solid media), for instance, or into liquid broth media. However, the conditions often restrict the microbial action on the cellular level. The nutrients are available but the speed of metabolism is limited by several factors inside the culture broth or on the plate. The nutrients are not diffusing to the cell surfaces fast enough and, on the other hand, also the waste materials from the metabolic reactions form gradients, which set up limits for growth and microbial interactions.

These factors may strongly interfere the investigations on the symbiotic relations between different microbes, as well as mislead our idea on their metabolism. The PMEU is a tool, which assists to overcome the methodological limitations. It acts as a simulator for the microbial growth conditions on cellular level.

Any general or selective liquid media can be used in the PMEU. It can also be applied for the cultivation of native samples without the addition of growth media. The close to natural conditions surrounding the cells in the PMEU make it possible to find answers to questions regarding the molecular communication between different bacterial cells or strains including the follow up of antibiotic influences.

The PMEU technologies have been developed in order to

1. begin the microbial cultivation immediately after sampling,
2. speed up the enrichment of the desired strains,

3. investigate the events on cellular level in a liquid bacterial culture, and
4. monitor the interactions between various microbes, as well as microbes and their hosts.

Any droplet of the culture broth inside the PMEU is identical to the others illustrating the bacterial behavior in specific temperature and growth medium, and with respect to gas composition and flow bubbling through the culture. The improved diffusion of gaseous compounds, nutrients and waste materials enhance the microbial metabolism. Therefore it is possible to gain an improved recovery of the strains, as well as more active growth and metabolism. These advantages make the PMEU the most effective way to cultivate the microbes for research or detection purposes. It is possible to usethe PMEU Scentrion® equipment to monitor the volatiles emitted from the microbial culture (Hakalehto *et al.*, 2009). This method is an extremely fast way to detect the microbes and to research their metabolism. Some examples are presented in the Figure 5, where clinical isolates of *Staphylococcus aureus* have been monitored by two various sensors of the PMEU unit, the ScC probe making in this case the early VOC (Volatile Organic Compounds) emissions detectable, and the MOS sensor tracking the later volatiles production. Other formats for presenting the PMEU Scentrion® results are also possible. Another alternative can be seen in the Chapter 6 of this book.

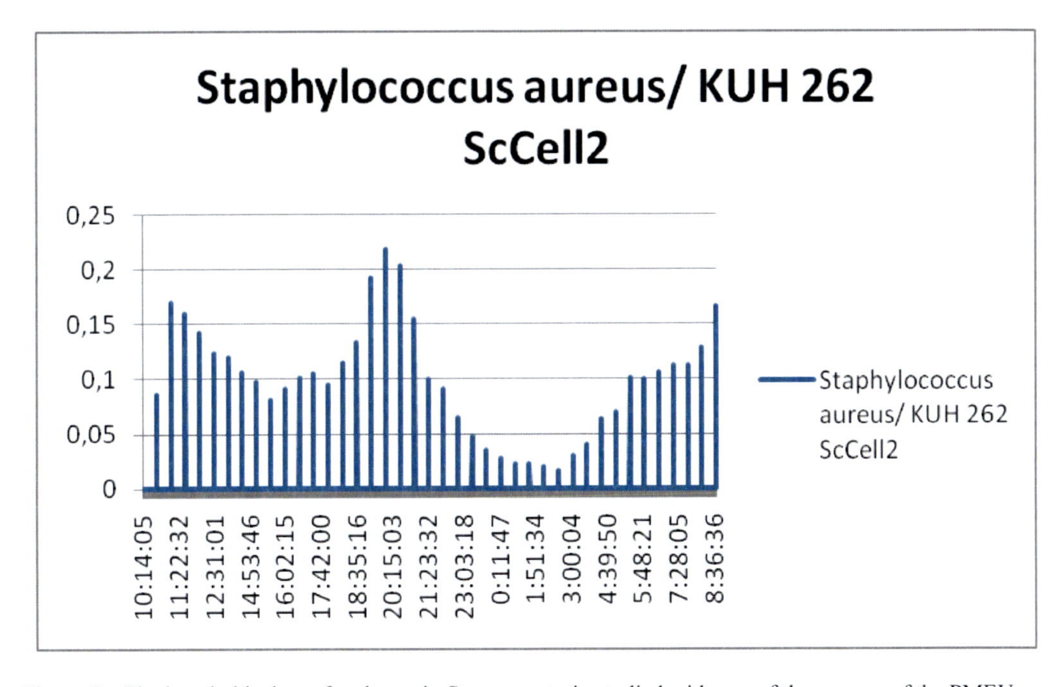


Figure 5a. The hospital isolate of pathogenic *S. aureus* strain studied with one of the sensors of the PMEU Scentrion® designed for the verification of volatile compounds. The ScC (Semiconductive Cell) recorded the gaseous compounds almost immediately after the onset of the culture (on the TYG medium at body temperature).

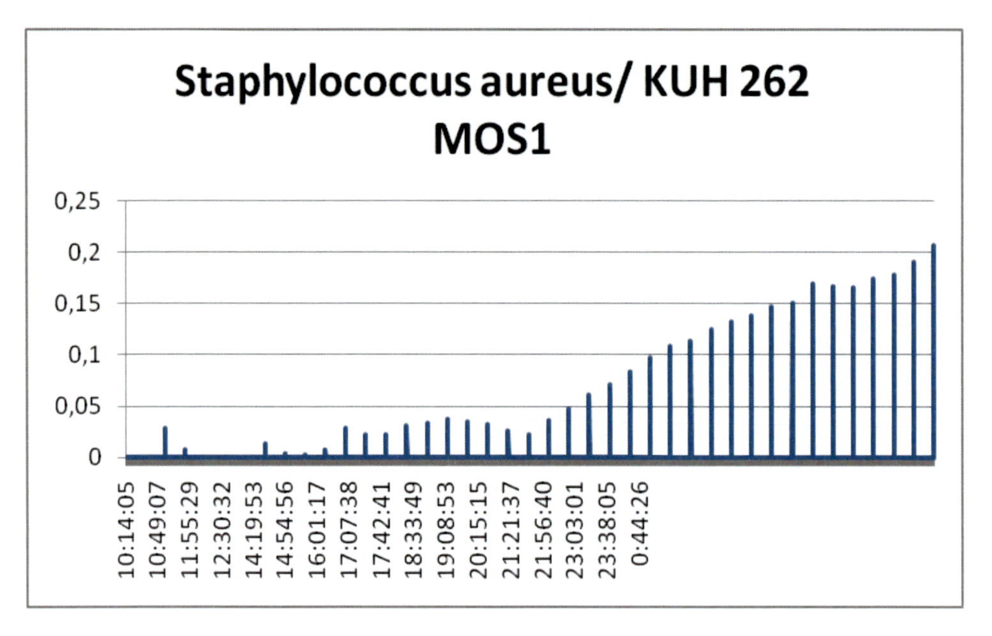


Figure 5b. The volatiles liberated by the hospital isolates of pathogenic *S. aureus* strain recorded by one of the MOS (Metal Oxide Sensor) probes of the PMEU Scentrion®. In this case the gaseous compounds emitted on the later stages of the culture were observed.

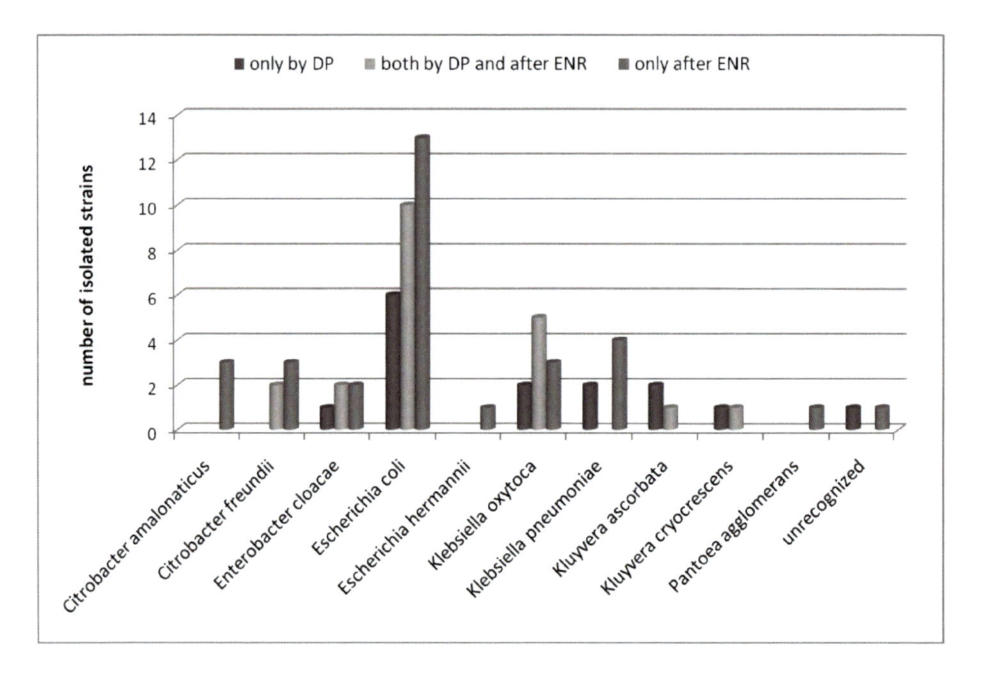


Figure 6. Enterobacterial strains isolated from fecal samples collected from healthy infants by only direct plating (DP) to BD CHROMagar™ Orientation plates (Becton, Dickinson & Co., Sparks, MD, USA), both by direct plating and after 8 hours of enrichment (ENR) in MacConkey broth (Becton, Dickinson and Co.) and only after enrichment. Modified from Pesola *et al.*, 2009.

The PMEU method has been used for screening the microflora of clinical samples for monitoring the prevailing strains and their interactions (Hakalehto, 2009; Hakalehto *et al.*, 2009; Pesola *et al.*, 2009; Hakalehto *et al.*, 2010; Pesola and Hakalehto, 2010). In the Figure

6 the potential of the PMEU approach in improving the results by recovering more cultivable strains, is illustrated.

In the PMEU studies we have been able to find and document symbiotic relationships between different bacteria. These interactions could take place *in vivo* in the midst of several host body functions. For example, there is likely to exist a BIB (Bacterial Intestinal Balance) maintaining the pH and other conditions in the small intestines, and consequently in the entire intestinal tract. In these studies it was demonstrated by the PMEU method that the so called mixed-acid producing *Escherichia coli* strains could co-operate with the more neutral substances producing *Klebsiella/Enterobacter* group of strains, which produce 2,3-butanediol besides ethanol, some organic acids, hydrogen and carbon dioxide gases. Together these differently metabolising coliforms produce equal growth in the PMEU as in pure cultures indicating a symbiotic relationship (Hakalehto *et al.*, 2008). The co-operation between these facultatively anaerobic bacteria is maintaining a dualistic balance which could be the foundation of the BIB, this balance being capable of sustaining other intruding species, which could not alter the proportions between the two facultative groups (Hakalehto *et al.*, 2010).

## Conclusion

Portable Microbe Enrichment Unit techniques have been developed in Finland by Finnoflag Oy in Kuopio and Siilinjärvi. These methods provide the researcher a quick look at the microscale events in the microbial culture. This view is opening up the multiplied effect of all homogenous cells in the enhanced enrichment cultivation. According to many research documentations we have a full reason to think that the results and ideas produced by the PMEU research get closer to the actual modes of life of the intestinal microbes than the present methodology. In this book the concept of BIB (Bacterial Intestinal Balance) is highlighted on the basis of these findings.

## References

Hakalehto E, Hänninen, O. Lactobacillic $CO_2$ signal initiate growth of butyric acid bacteria in mixed PMEU cultures. Manuscript in preparation. 2012.

Hakalehto, E, inventor. Anonymous method and apparatus for concentrating and searching of microbiological specimens. US Patent No. 7,517,665. 2009.

Hakalehto, E. Simulation of enhanced growth and metabolism of intestinal *Escherichia coli* in the Portable Microbe Enrichment Unit (PMEU). In: Rogers MC, Peterson ND, editors. *E. coli infections: causes, treatment and prevention.* New York, USA: Nova Publishers; 2011.

Hakalehto, E; Heitto, L; Heitto, A; Humppi, T; Rissanen, K; Jääskeläinen, A; Paakkanen, H; Hänninen, O. Fast monitoring of water distribution system with portable enrichment unit – Measurement of volatile compounds of coliforms and *Salmonella* sp. in tap water. *JTEHS,* 2011; 3, 223-233.

Hakalehto, E; Hell, M; Bernhofer, C; Heitto, A; Pesola, J; Humppi, T; Paakkanen, H. Growth and gaseous emissions of pure and mixed small intestinal bacterial cultures: Effects of bile and vancomycin. *Pathophysiology,* 2010; 17, 45-53.

Hakalehto, E; Humppi, T; Paakkanen, H. Dualistic acidic and neutral glucose fermentation balance in small intestine: Simulation *in vitro*. *Pathophysiology,* 2008; 15, 211-220.

Hakalehto, E; Pesola, J; Heitto, A; Bhanj Deo, B; Rissanen, K; Sankilampi, U; Humppi, T; Paakkanen, H. Fast detection of bacterial growth by using Portable Microbe Enrichment Unit (PMEU) and ChemPro100i((R)) gas sensor. *Pathophysiology,* 2009; 16, 57-62.

Hakalehto, E; Pesola, J; Heitto, L; Närvänen, A; Heitto, A. Aerobic and anaerobic growth modes and expression of type 1 fimbriae in *Salmonella*. *Pathophysiology,* 2007; 14, 61-69.

Pesola, J; Hakalehto, E. Enterobacterial microflora in infancy - a case study with enhanced enrichment. *Indian J. Pediatr.,* 2011; 78, 562-568.

Pesola, J; Vaarala, O; Heitto, A; Hakalehto, E. Use of portable enrichment unit in rapid characterization of infantile intestinal enterobacterial microbiota. *Microb. Ecol. Health Dis.,* 2009; 21, 203-210.

Wirtanen, G, Salo, S. PMEU-laitteen validointi koliformeilla (Validation of the PMEU equipment with coliforms). 2010;Report VTT-S-01705-10, Statement VTT-S-02231-10.

In: Alimentary Microbiome: A PMEU Approach
Editor: Elias Hakalehto
ISBN: 978-1-61942-692-4
© 2012 Nova Science Publishers, Inc.

*Chapter III*

# Oral Microbiology and Human Health

***Ilkka Pesola*[1], *Jouni Pesola*[2], *Hannu Kokki*[3], *and Elias Hakalehto*[4]**
[1] Clinic of Oral and Maxillofacial Diseases,
Kuopio University Hospital, FI, Kuopio, Finland
[2] Clinic of Children and Adolescents,
Kuopio University Hospital, FI, Kuopio, Finland
[3] Anaesthesia and Operative Services,
Kuopio University Hospital, FI, Kuopio, Finland
[4] Department of Biosciences, University of Eastern Finland,
FI, Kuopio, Finland

## Abstract

The micro-organisms of the oral cavity have a significant impact not only on the gastrointestinal health but also on overall well-being in the human body. A good oral health is a prerequisite for satisfactory general health. The disturbances in the mouth biofilm may have significant impacts on the progress of several diseases.

Newborn babies have a sterile mouth, but soon after birth the sterile mouth surfaces are covered with microbial strains from the mother and the surroundings. Already at the age of two years, the children usually possess a well-developed oral microflora. This is then further developed and stabilized until the age of 3-4 years. This mouth microbiota is rather constant in nature during the entire life-span of the individual.

In early life streptococci are a common pioneer species in our mouth. These strains are not involved in the pathogenesis of mouth bacterial diseases but the IgA-1 protease producing streptococcal strains (*e.g. Streptococcus oralis, S. sanguis, S. mitis*) are supposed to take part in the development of diabetes mellitus type 1. It is supposed that the oral biofilm may contribute to the development of other autoimmune diseases as well.

During the first months besides streptococci one of the main colonizers of the mouth mucosal membranes is *Fusobacterium nucleatum*, which species is important in the formation of the diverse microbial community in the oral surfaces and for the development of host defense system. The establishment of the mouth flora and its interactions have a key role in the built up of the gastrointestinal and respiratory tract microflora, as well as in constituting of the oral tolerance. These functions lay the **foundation for the body's defense against intruding microbial pathogens** and other harmful agents. The microbes on the mucosal surfaces of the mouth, pharynx and the intestines take part in the initiation of the inflammatory responses, antibody formation in lymphatic tissues, cytokine response, and contribute to their progressing on the epithelial cells.

Proper host defense system is important since pathogens in the oral mucosa may cause endocarditis and affect the development of other cardiovascular diseases. Pathogens in the mouth mucosa are also a known risk for the infections on the endoprotheses and other implants. Therefore, the knowledge on the qualities and quantities of the normal flora in the mouth cavity is important for understanding how we can improve the health of the patient as a whole.

An example of the complex relationships between the alimentary microbiota and the host health in general are oral cariogenic bacteria. These species (*e.g. S. mutans)* are not usually pathogenic in the perspective of general health but many less cariogenic bacteria like *Streptococcus viridans* strains (*e.g. S. sanguis*) may have significant impact on the body functions as a whole.

Oral cariogenic bacteria especially (*e.g. S. mutans*) are not usually pathogenic in general health status unlike many less cariogenic bacteria like *Streptococcus viridans* strains.

For better understanding of the complex interactions in the oral cavity, and its impacts on the entire alimentary biome, there is a need for improved tools and methods to assess the function and consistence of oral microbes. In this chapter the correlation between microbial communities in the oral tract and the rest of the gastrointestinal flora as well as their contribution to human health are reviewed.

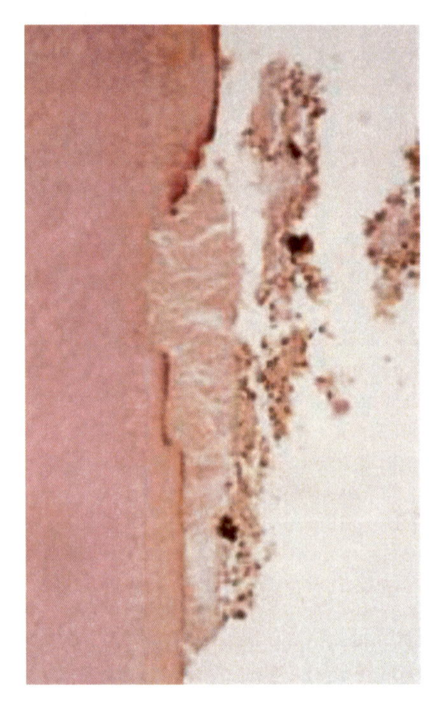

Figure 1. "*I believe they exceed the number of men in a Kingdom*", noted Antonie van Leeuwenhoek in 1683, when he viewed under his microscope a microbial sample taken from tooth surface.

# Introduction

The microflora of the mouth, as like that of the entire alimentary tract, starts to develop soon after birth, originating from sterile conditions in the womb. The skin, nostrils and oral cavity of newborn child act as a raceway for colonizing microbes. Oral cavity is perhaps the main microbial route in perspective of the whole body (Könönen, 2000). Salivary contamination occurs when child gets microbes and antibodies from his mother via saliva *e.g.* with spoon or pacifier contact. A child is in contact with outer world mainly via his mouth. *E.g.* oral bacteria *Prevotella melaninogenica* in mother's mouth correlates positively with child's mucosal finding (Könönen *et al.*, 1992). Besides the bacteria that participate in biofilm formation, a child gets maternal specific s-IgA –class antibodies via the mother's saliva.

The oral stage of development (0-18 months) is an important period in the developing of oral biofilm and biofilms of whole gastrointestinal and respiratory tracts. During the oral stage of development the child is in contact to environment especially with touching it by lips and tongue. Already during first two months after birth infant starts to develop an individual bacterial flora onto the oral surfaces.

First species in the oral cavity are aerobic viridans-group streptococci and anaerobic *Fusobacterium nucleatum, Veillonella sp.* and *Prevotella melaninogenica* (Könönen, 1999). Particularly *F. nucleatum* has been shown to facilitate the early stages of the biofilm formation (Könönen, 2005).

The eruption of deciduous teeth significantly effects the formation of oral flora, when the amount of microbes increase and bacteria adhere to hard tissues. As pioneers of bacteria in formation of tooth surface biofilm are Actinomyces species and streptococci, which in turn create a favorable ground to other bacteria to adhere (Mishra *et al.*, 2010).

The significance of adhesion in biofilm formation is of great importance. Local preparations that inhibit bacterial adhesion to underlying structures are under active research. If such microbe-specific products could be produced, it is possible in the future to guide biofilm formation into favorable direction.

Antibodies start to develop on the structure of oral biofilm soon after birth. A newborn child has IgG-class antibodies received via the mother's placenta and s-IgA antibodies obtained via breast milk. Even at the age of 12 years children have been shown to possess same total antibody titers of s-IgA than adults in the salivary samples (Mellander *et al.*, 1984).

The interactions between the biofilms on the teeth and the soft tissues have been less evaluated, but researching the mechanisms of adhesion could give some clues into the issue of the developing parallel biofilm structures elsewhere in the body. *E.g.For example, Fusobacterium nucleatum* is important and apparently a basic microbe also in the intestinal biofilm and its constitution (Dharmani *et al.*, 2011). It is obvious that there are several other similar types of key microbes that participate to biofilm construction.

Certain bacterial species are able to adhere exceptionally well to the oral mucosal surfaces *e.g.* with fimbria (Mishra *et al.*, 2010). The type 1 fimbriae have a role in the bacterial attachment on the small intestinal surfaces, for example (Hakalehto *et al.*, 2007).

Especially important is the communication between microbes and the cells of the epithelial surface. Disturbances in this communication may cause clinical disorders. It has been shown that oral *S. mutans* and *S. sobrinus*, specific cariogenic bacteria, can adhere and invade host cells both from planktonic and biofilm origin even if not a single tooth, *i.e.* a hard tissue, has erupted in the mouth (Berlutti *et al.*, 2010). Many bacteria and fungi adhere to the oral mucosa by lectin-like interactions. On the other hand some probiotics can inhibit the adhesion of other bacteria to mucosal cells (Cannon *et al.*, 2010).

In recent studies regarding the etiology and mechanisms of psoriasis, significant knowledge about consequences of altered communication between biofilm and epithelial cells has been emerged. Chronic inflammative reaction involving plenty of T-lymphocytes and cytokines (*e.g.* TNF-alfa) near ceratinocytes of the skin contributes to the chronic changes in atherosclerosis, diabetes, inflammatory bowel disease and parodontium (Davidovici *et al.*, 2010).

On the other hand, it has been shown that pathogens of chronic parodontitis may have a protective role against allergies or asthma of sensitive children. Even few months after birth there have been detected antibodies of parodontopathogens in serum. This kind of early immunization against these bacteria has an impact in the onset of allergic symptoms (Arbes and Matsui, 2011).

The interactions between microbes inside biofilm have often a remarkable contribution in pathogenesis. As a clinic example is *e.g.* that person having mainly parodontopathogenic biofilm around their gums and teeth have usually less caries incidences and *vice versa*.

It is a major issue to find out what is the influence of bacterial overgrowth on subclinical disorders in host body. *E.g. Candida albicans* remains usually in mucosa with no disorders, but in case of antibiotic usage, cortisone treatment and many other medications, and even during HIV-infection, yeast multiplies, and host defense/biofilm balance wavers. What happens in the intestines when bacteriota overgrows? How epithelial cells can cope with this phenomenon on? It is known that the microbial growth could be extremely rapid. How effectively microbes can resist antibiotics and also bowel enzymes that split the bolus along the intestinal tract? The distinction between subclinical and clinical significant disorders is not always straightforward.

Microbial or food antigens, which start the immune responses in oral and pharyngeal lymphatic tissue, pass deeper to the lymphatic tissue via epithelial crypts of tonsils with saliva. The concentration of antigens has a key role in the process of immune response in cryptal cells.

Bacterial cell accumulation and adhesion on surfaces are interactive with host mechanisms. The important contribution of the mucosal antibodies in setting up the matrix for the biofilm formation has been suggested in the case of the intestinal mucosal flora (Fagarasan and Honjo,. 2004). Mucins of outer mucosa have a significant impact how mucosal bacteriota develops and survives (Johansson *et al.*, 2011).

## Development or Oral Biofilm

During the first months of life the mucosal colonization evolves species by species including both aerobic and anaerobic microbes. Already at the age of six months the microflora exhibits individual characteristics between various infants and has found unique features of homeostasis (Könönen, 1999; Könönen *et al.*, 1999). Data from clinical studies has shown that even after prolonged antibiotic treatment the structure of biofilm returns to normal for each individual and keeps the balance achieved during the early months of life.

This is consistent with the results obtained with the PMEU (Portable Microbe Enrichment Unit) in clinical trials evaluating the effects of antibiotic treatment on the developing neonatal intestinal microflora (Pesola and Hakalehto, 2011).

Earlier it was believed that some specific bacteria remain in the mouth only if there are hard tissue growth surfaces such as teeth. Recently it has been shown that those bacteria adherent to hard tissues commonly exist also in other soft tissue clusters even after all teeth have been extracted (Sachdeo *et al.*, 2008). This is supposed to be related with the immunological memory. Some type of immunological memory seems to be apparently abundant, but the hypothesis is difficult to be exactly defined or conceptualized in a measureable way. This phenomenon, immunological memory is connected with the microflora development in several ways. It is supposed to be related to the antibody production of pharyngeal, intestinal, spleen and other lymphatic tissues (Brandtzaeg, 2011a). Adenotonsillar tissue in dorsal tongue and pharynx is the location of entry of different spectrum of antigens to the oral, respiratory and alimentary tracts. Lymphatic tissues are prominent in childhood and diminished to rudimentary in most adults. The relatively large size of lymphatic tissue could be the explanation for the developing maturation of immune responses during the early childhood.

One aim of this chapter is to highlight the consequences of bacterial overgrowth in biofilms to the human body and homeostasis. It is known that an accumulation of parodontal bacteria can create a clinically visible gingivitis within 3-4 days without tooth brushing. Gum infection with redness, soreness and bleeding are the easily observed signs of acute gingivitis in these situations. In case of caries clinic consequences appear within few months when cariogenic bacteria dissolve dental enamel in conjunction with acids to carious lesions. Treatment of both the common diseases, gingivitis and caries, is based on mechanically breaking down the biofilm. When biofilm is crashed its capability to infect host cells subsides and the amount of bacterial load to dental tissues declines.

## Divergence of Biofilm in Different Oral Surfaces

Different oral cavity tissues are favorable for the growth of various bacterial populations, and saliva forms ideal habitat to some parts of the microbiota. Consequently, the saliva, oral mucosa, rough surface of tongue, gingival pockets and different surfaces of erupting teeth form their own, distinct habitats for different prevailing microbes (Mager *et al.*, 2003).

Salivary proteins form an organic pellicle to tooth surfaces. The construction of pellicles has a major role in determining the kind of biofilm that colonizes the dental hard tissue. Pellicles have important contribution to protein adsorption, mineralization / demineralization-balance, bacterial adhesion and colonization and function of the entire biofilm.

Saliva can affect the adherence of bacteria to hard surfaces. For example, among streptococcal species there is a significant divergence on adherence to tooth enamel determined by whether the teeth are saliva coated or not. If there is a similar divergence in pellicle proteins preferences of cariogenic microbiota, this could have significant consequences on colonization. High expression of those bacteria produces cariogenity of the whole biofilm (Bardow *et al.*, 2005).

**Table 1. Frequency and location of bacteria in the oral cavity (Adapted from Zilberstein B *et al.*, 2007)**

| Microorganism | Saliva (%) | Tongue (%) | Upper buccal biofilm (%) | Lower buccal biofilm (%) |
|---|---|---|---|---|
| *Actinomyces* sp. | 40 | 30 | 30 | 20 |
| Anaerobic rods | 30 | 10 | 50 | 10 |
| *Candida albicans* | 20 | 30 | 10 | 0 |
| *Fusobacterium* sp. | 60 | 0 | 40 | 0 |
| *Lactobacillus* sp. | 90 | 50 | 30 | 0 |
| *Peptostreptococcus* sp. | 70 | 50 | 50 | 20 |
| *Staphylococcus* sp. | 50 | 30 | 10 | 10 |
| *Streptococcus* sp. (alpha-/hemolytic) | 100 | 90 | 90 | 80 |
| *Streptococcus* sp. (gamma-/hemolytic) | 20 | 10 | 0 | 10 |
| *Veillonella* sp. | 100 | 60 | 80 | 70 |
| *Escherichia coli* | 10 | 20 | 20 | 0 |

Oral microflora consists of a huge number of microbial strains. In early 1980's it was estimated that the oral cavity contains circa 300 different bacterial species (Evaldson *et al.*, 1982). This was based on the fact that approximately 300 oral bacterial species have been identified by traditional culture methods, but more recently, with molecular methods 600 species and phylotypes have been identified in the mouth (Dewhirst *et al.*, 2010).

However, it is estimated that in the oral cavity there are even more than 700 different species, majority of which belong to the normal flora. Because significant number of species in the oral microflora could not be identified with traditional culture procedures, new methods, such as the PMEU, are warranted in order to better elucidate the microbiological niches in different oral cavity tissues in variable circumstances.

Depending on balance between biofilm and oral defense mechanisms, it is determined how and when oral microbes are able to create infection and clinical symptoms. Consequently, the pathogenesis of any particular infectious agent or mixture of microflora is dependent on the status of the biofilm and the host oral defense mechanisms. In other words, the better the conditions of the diverse microflora and its reciprocal molecular communication with the host are, the more resistant both the individual and his or her microbial ecosystem are toward external threats or disintegration.

The strength of the balance in this ecosystem is at least partially determining the susceptibility of the mouth mucosa to ulceration and other stresses. If the infants developing immune system against pathogenic bacteria or nonpathogenic antigens from food and other sources are not working properly, there is an increased risk for recurrent infections, and even autoimmune diseases and allergies can develop for susceptible persons (Aureli *et al.*, 2011).

# Colonization of the Gastro-Intestinal Tract

Colonization of the gastro-intestinal track is more complex process than that of the oral cavity. In contrast to the oral microbiota where the microbial balance is quite stable already at the age of sixmonths, the evolving of the gastrointestinal flora takes few years. Altering of the intestinal flora with different microbial agents is under excited interest (Mitsuyama and Sata, 2008).

Gastro-intestinal track of fetus is sterile, and colonization begins soon after birth with swallowed saliva and its contents, which are carrying microbes. Several factors affect the colonization process. First of them is the mode of delivery. Distinct differences in the early gastrointestinal (GI) tract flora have been shown in infants after vaginal delivery and those born by cesarean section. In vaginal delivery babies are exposed to mother's vaginal and intestinal microflora. In contrast to that, in neonates born by cesarean section the first species of the GI microbiota are those in the medical environment. Over the last few years there have been distinct alterations in the colonization of the intestine reflecting changes in hygiene. Recent studies indicate that neonates are nowadays colonized with *E. coli* later than what was the case few decades ago, and as a whole the number of early colonizing species has decreased. Earlier coagulase-negative staphylococci and *S. aureus* were common early phase constituents in infants' gastorointestinal microflora, but the colonization with these species seems to be diminished (Parm *et al.*, 2011). This may indicate changes in the overall nutrition in the population.

Initiatation of the feeding with breast milk and later with solid food is the second exposure after saliva to intestinal microflora development. The immunoreactions are first generated in the oral cavity, Mother is the principal source of microbes during the first weeks after the birth and mother's diet during the late pregnancy and in early breastfeeding period affects the content of the infant's gastrointestinal microflora. The mode of feeding may also have some impact on the infant's gastrointestinal flora. However, it has not established how much modern formulas differ from mother's milk in their properties to affect the intestinal microflora. Some research on this issue has been carried out with the PMEU indicating that different milk formulas have a remarkable impact on the enterobacterial variation (Pesola *et al.*, 2009). Not only in the infancy but also in later growth the food that a child is eating affects maturating intestinal flora (Adlerberth and Wold, 2009). The structure of developing biofilm is guided by many signal molecules and in many stages. The purpose of those molecules is to affect microbial adhesion, colonization, cohesion and maturation on different surfaces in the mouth, esophagus and intestines in a certain order of strains and genus (Saito *et al.*, 2008). This procedure is dependent of competition, *i.e.* first colonizers find their locus in biofilm first. Maternal biofilm creates the fundamental basis for newborns biofilm structures in mucosa. In addition to it genetic base and inherited antibodies, different enzymes take part on the procedure.

The establishment of the intestinal flora is regulated by many host functions, such as bile acid formation (Hakalehto *et al.*, 2010). Due to the immature stomach circumstances maternal and food-derived microbiota of newborn can pass trough mildly acidic barrier of stomach further to intestine. Also the less mature immunological preparedness of the body is probably allowing some strains to get rooted into the GI tract.

In early infancy intestine is an aerobic environment and, thus, aerobic and facultative anaerobes are often the first species found in intestine. However, soon oxygen content in the gut is decreased and then anaerobes are the most common species in the intestine (Adlerberth and Wold., 2009). Many of the gastrointestinal track anaerobes are difficult to culture with traditional methods, and thus more widespread use of new methods, such as the PMEU, should facilitate our understanding of the complexity of the gastrointestinal track microflora establishment.

In the process of the GI microflora development non-pathogenic pioneering stem microbes have inhibitory influences on toxic pathogenic bacteria in intestines. The pioneers of intestinal flora are *e.g. Escherichia coli* and other coliforms, streptococcal and *Bifidobacterium sp.* strains. The latter becomes the major component of the neonatal intestinal microflora (Salvini *et al.*, 2011). The latter then diminish as a consequence of developing nutrition, but remains as an indicator of good intestinal bacterial balance in the later age. Even the diet of the mother effects the colonization of intestinal microbes in childhood (Mushi, *et al.*, 2010).

## About the Composition of Saliva

The parotid, submandibular and sublingual glands, as well as hundreds of small salivary glands in the oral mucosa and beneath it secret mucinous saliva. Oral health is strongly dependent of the saliva secretion. Severe complications are following if salivary flow rate is decreasing (Taji *et al.*, 2011).

Salivary secretion performs both as a host defense mechanism for oral health and as a source for nutrients for several micro-organisms. Saliva removes organism by physical means, it regulates the pH level in the mouth cavity, and its content, such as secretory-IgA (s-IgA), may have diverse effects on dental and periodontal health.

The saliva secreted by major and minor salivary glands contains rich of necessary and obligatory substances, inorganic salts, proteins and lipids. Salivary amylase initiates the digestion by breaking down starch to maltose. Mucins in saliva act as preservatives and facilitate food bolus swallowing, and also in speech articulation (Kawas *et al.*, 2012).

It has been estimated that 1 ml of saliva contains 100 million living bacterial cells, 2 mg proteins, 800 mg lipids and 100 mg antibodies (Tenovuo, 1998). The content of saliva affects significantly with the oral health. Persons with salivary glyco- and phospholipids have a high caries activity. An increase in the lipid content of dental surface biofilm seems to increase the adhesion of cariogenic bacteria.

Saliva has a high concentration of bicarbonate ions, and thus has a high buffer capacity. For most microbes in the mouth the optimum pH for growth is in the range between 6.5 and 7.5. The pH of stimulated saliva is neutral, pH 7.3. After meal bacterial metabolism decreases the pH of saliva to more acidic, but in a couple of minutes the buffering ability of saliva restores pH to normal. In most individuals the oral biofilm is adapted to slightly acidic level.

The pH level in esophagus is similar to that in the mouth, pH 6-8, but the gastric pH is significantly lower, pH 1-2 (Ranjitkar *et al.*, 2012). The low gastric pH is one of the main host defense mechanism in human body, as a result of acidic environment, only a small part of swallowed microbes are able to stay viable up to duodenum (Dicksved *et al.*, 2009).

Different defense factors in saliva resist pathogenity of microbes (Lendenmann *et al.*, 2000), and thus saliva has a major role in first line defense mechanism against foreign particles and antigens.

## Antibodies in Saliva

Age is an important factor for the development of the salivary immunoglobulin system. In early infancy the oral defencing is based on non-immunologic factors and antibodies of a mother. Antibody production inherited from mother is replaced gradually with developing own antibody production of an infant about by the age of six months (Rashkova and Toncheva,. 2010). The amount of salivary antibodies reaches the concentrations obtained in adults in early school age (de Farias and Bezerra, 2003).

The primary contact of salivary antigens with the oral mucosa is important for the development of mouth resistance. Antigens dissolved into saliva induce antibody production in the lymphatic tissue of tongue and tonsils that is an important oral cavity defense system for the entire body (Brandtzaeg, 2011b).

For the oral mucosal defense system development one of the most important events is the beginning of s-IgA antibody secretion. Children having plenty of saliva contacts with their mother in early life develop significantly more IgG–class antibodies *e.g.* against *S. mutans* than children with less salivary contacts to adults (Tenovuo and Aaltonen,. 1991).

Secretory IgA (s-IgA) and IgG are the two most important antibodies in the saliva. Secretory antibodies develop in major and minor salivary glands and in their plasma cells, while IgG is mostly derived from serum (Reynolds, 1999). From salivary glands secretory antibodies flow to salivary ducts changing at the same time from monomers to dimeric and polymeric forms. The secretions flow of salivary s-IgA maturates at the school age and after that the amount of s-IgA in saliva is quite stable throughout the entire lifespan.

Secretory IgA has several protective functions in the gastrointestinal mucosa too. Interesting *in vitro* observations in tissue culture have been made from the intestinal epithelial cells (Bollinger *et al.*, 2003). It is supposed that s-IgA may be involved not only in host defense alone, but it may also interact with the gut mucosa by limiting local inflammation induced by bacteria. s-IgA may also promote the adhesion of non-pathogenic biofilm to the gut mucosa with mucins, and may have an defensive activity by this mechanism. If it is the case on the oral mucosa, too, this should be evaluated in future studies (Fagarasan and Honjo, 2004).

In contrast to s-IgA that is secreted from salivary tissues, IgG and IgM-type antibodies are secreted from the bloodstream into the oral cavity through periodontal pockets (Awartani,. 2009). The increase of antibodies in saliva implicates the inflammation activity of periodontal tissue.

In addition to antibodies lysozyme, lactoferrin, peroxidase, and the defensins have an interactive antimicrobial activity in the mouth cavity. These compounds slow down the metabolism of bacteria in the mouth and have a host defense function also by the colonizing different oral surfaces.

The number of oral defensins (particularly human Beta-Defensins, hBD-2 and hBD-3) has been found to be elevated in those areas of the mouth, where there is evidence of inflammatory reaction (Vankeerberghen *et al*., 2005). The antimicrobial efficacy of defensins against most common oral aerobes is high. On the contrary, about half of anaerobes are resistant to defensins. It has been shown that cytokines and bacteria induce defensin secretion and that in particular, mucosal inflammation strongly induces hBD-2 secretion. In the PMEU studies the effect of hBD-2 on *E. coli* was shown to by far exceed the antibiotic impact on the bacterium (Hakalehto,. 2011).

hBD-2 has been shown to be highly effective in the prevention of *Candida albicans* infection (Meyer *et al*., 2004). Supporting the role of defensins in immunofunction, it has been recently shown in the PMEU, that the human defensins have a potent antimicrobial activity in low concentrations on *e.g. E. coli* strains (Hakalehto,. 2011).

The dominating antibody class to bacteria in saliva is s-IgA. The role of antibodies in saliva has been evaluated recently, and antigenic specific s-IgAs against pathogens is known to play a key role in effective barrier system on the oro-nasal mucosa (Fukuyama *et al*., 2010). It is shown that the presence of s-IgA is connected with immune response to antigens in oro-pharyngeal mucosa (Miranda *et al*., 2011).

A key finding of studying the systemic effects of parodontitis is that the two most important Gram-negative parodontopatogens, *Actinomyces actinomycetemcomitans and Porphyromonas gingivalis*, induced IgG antibody formation in the serum. The antibody formation induced by both bacteria could be detected easily and reliably by ELISA method. The researchers suggest that the method is suitable for epidemiological studies, when seeking to determine the severity of periodontal disease and connection to potential general diseases.

## Mucin Layers of Oral Mucosa

Recent studies on mucosal surface metabolism have increased knowledge about interaction between epithelial cells and antigens. Mucosal layer in the region of mouth, esophagus, stomach and intestines protects epithelial cells from outer physical and chemical stimulus and antigen load. Body system has developed several defense and elimination mechanisms especially against environmental microbes. During the last two decades our awareness has increased about mucins significance and especially about function of mucins on the duodenal epithelium (Deplancke and Gaskins,. 2001).

The interaction between epithelial mucin productive cells and extracellular mucins is under interesting study.

Mucins of the salivary and gastrointestinal tract create an important protective layer against pathogenic microbes and communicate at the same time with non-pathogenic microbiota. Intestinal mucin studies show that many microbes have both mucolytic and toxic effects and the ability to pass through the inner mucin layer (Johansson *et al*., 2010). As a result the effects on epithelial cells can be harmful as it is the case with *Helicobacter pylori*. This microbe can cause even mutagenic damage on gastric mucosa.

Further studies are needed which bacteria and in which way microbiota create the invasion through defense layers causing mucosal inflammation and infection (McGuckin *et al*., 2011). It has been shown for example that there is possible a connection between *Helicobacter pylori* and oral mucosal aphthous ulcers (Karaca,, *et al*., 2008).

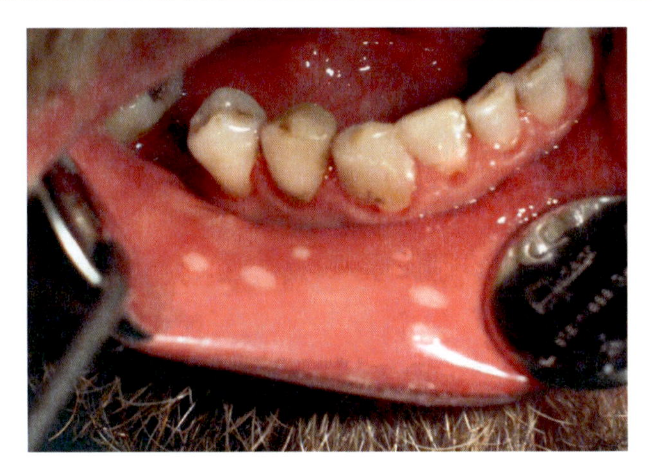

Figure 2. Aphthous ulcers in lower lip. Patient had symptoms also in stomach. During coeliacial diet most of the oral ulcers disappeared.

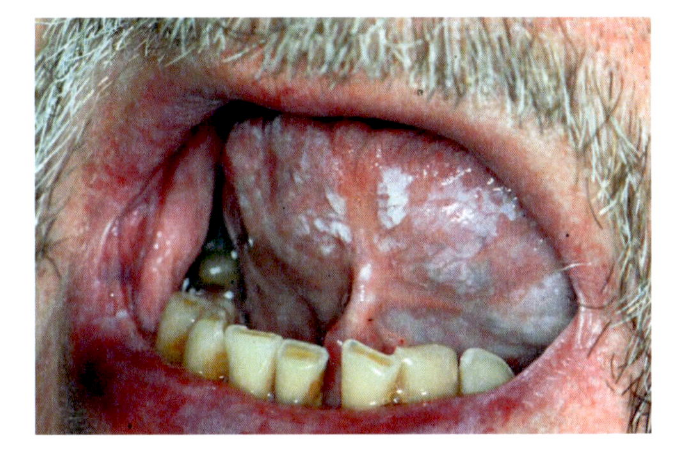

Figure 3. Mucosal lesions of tongue and oral base. Histological diagnosis was oral lichen planus.

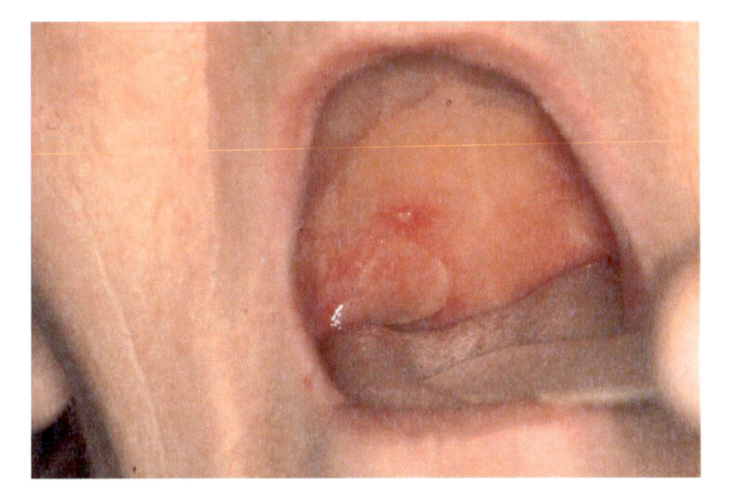

Figure 4. Bullous pemphigoid in dorsal palate.

# Parodontal Pathogens

The amount of parodontal pathogens is approximately 90/5/5 between anaerobic/aerobic/microaerophilic bacteria. Major proportion of anaerobic bacteria are Gram negative rods (for example *P. gingivalis, B. forsythus, P. intermedia, Fusobacterium sp.*) but some anaerobic bacteria, mostly Gram positive cocci (for example *Peptostreptococcus micros*), can also be detected (Maiden *et al.*, 1996).

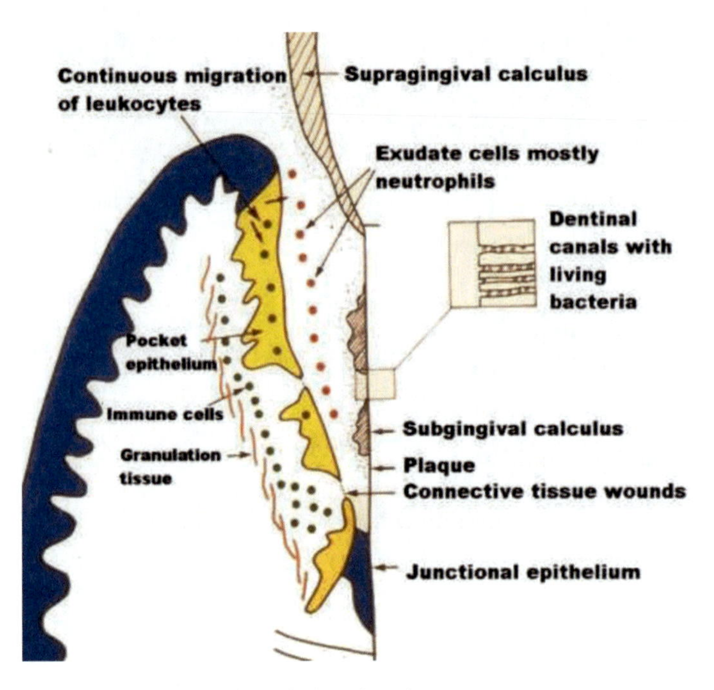


Figure 5. Inflammation with leucocytosis in periodontal pocket.

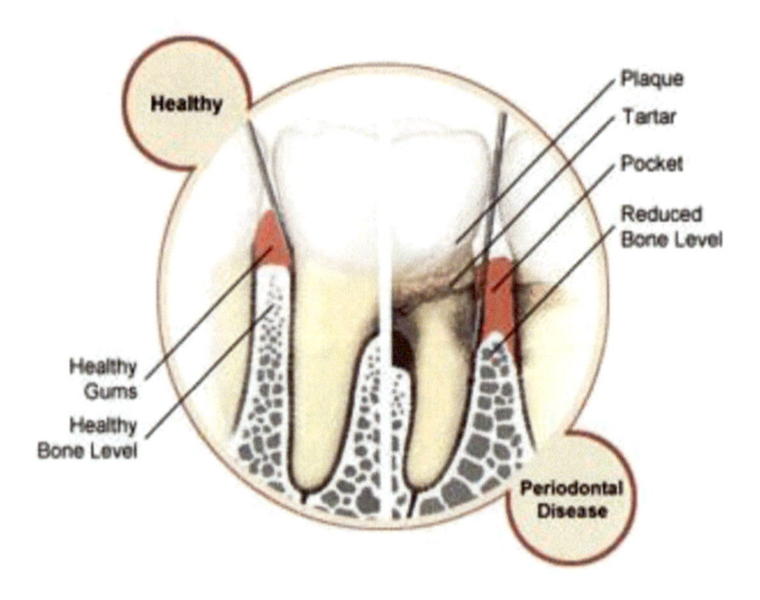


Figure 6. Supragingival calculus that offers a porous basis for microbial adhesion.

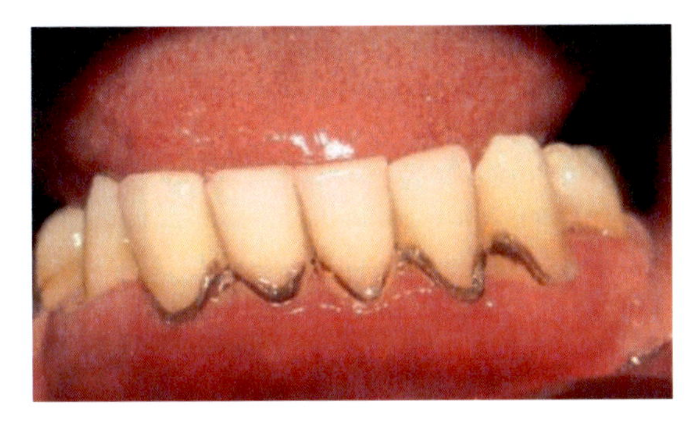

Figure 7. Clinic view of supragingical calculus with chronic inflammation around lower teeth.

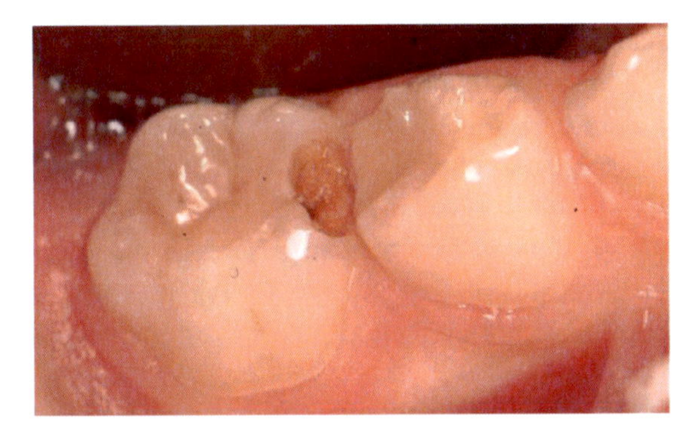

Figure 8. Dental caries lesion is a consequence of bacterial acid formation. Typical active cariogenic bacteriota in deep lesions consists of *e.g. Str. mutans, Lactobacilli, Actinomyces odontolyticus. Fusobacterium sp, Veillonella sp.*

Expanding evidence is published regarding the importance of chronic parodontitis to general health. In oral environment parodontium gives several opportunities to study the mechanisms of chronic infection locally and systemically. Cytokine secretion and specific antibody formation, especially s-IgA, reflect the local imbalance in microflora in parodontal pocket.

This condition spreads to whole body via blood stream and lymph veins. It is evident that this chronic inflammation may cause potential risk elsewhere in the body and organs. Presently researchers of different fields are studying the role of chronic inflammation with the pathogenesis of different diseases in the human body.

## Pathogens of Oral Infections with Local Symptoms

Serious oral infections appear roughly in the following propositions: ostitis periapicalis 70 %, parodontitis and pericoronitis 15-20%, others 10-15%. All these infections can create harmful symptoms and adverse effects also to general health, if microbes transfer to blood circulation. General symptoms of infection are for example local swelling, pain and fever, and as a laboratory parameters leucocytosis and increase in CRP (C reactive protein) are commonly detected.

Simple oral infections can lead to serious consequences if not treated appropriately. For example, oral gingivitis with mild bleeding and redness can lead untreated to deeper infection in parodontal pockets and severe bone loss, even tooth shedding. Changes in bacterial structures of biofilm during the life cycle of parodontal diseases are obvious.

Also more serious infections may occur. In Finland with a population of 5 million annually more than one hundred patients need to be hospitalized because of life-threatening oral infections, and out of them about one third requires an admission into an intensive care unit. Therefore early renovation of infective focuses and proper antimicrobial pharmacotherapy is essential. This is important especially in patients with general disease, whose microbiological resistance is compromised.

In the oral mucosa one quite common microbe causing general symptoms is the yeast *Candida albicans* (Sanita, *et al.* 2011). *C. albicans* exists in the oral mucosa in about 3-50% of adult population and in 45-65% among healthy children. Risk groups to develop severe oral and/or systemic *C. albicans* infections are elderly people, those who get antibiotics or corticosteroids, immunosuppressive drugs, or x-ray therapy to head and neck region. The risk for oral candidosis is significantly higher in HIV-patients than in others. Symptoms vary from mild irritation and mucosal redness to ulcers and systemic disorders that endanger general health. Several studies have shown that microbes have potential role in maintaining the defense system balance in mucosal epithelium and in restoring the microbiological defense when that balance breaks down.

Therefore it would be of high clinical importance to seek bacteria with beneficial effects on the development and stability of the healthy microflora. The oral probiotic strains could then extend their influence on maintaining the microbiological and immunological balances not only in the oral cavity but also in the entire body.

Herpes viruses, *e.g.* simplex and zoster, are typical viruses that cause infections in the oral region, too (Tugizov *et al.*, 2011). Clinical symptoms especially in children occur as painful aphthous like ulcers, vesicles, redness and fever. The occurrence of viral as well as yeast infections is often related with imbalance or impairment of microbiological resistance or with some systemic medication. These often opportunistic infections could originate from the lessened protection of the host mucosa by the normal microbial communities. In turn, the microfungi and viruses also cause further harm to the epithelial structures, which often prolongs the recovery from the illness, and in the worst cases opens up the tissues for other pathogens with still unknown mechanisms. Consequently, it could be postulated that the infection could have a kind of ecological succession. In the treatment, these pathological processes should be taken in account and effectively prevented or interrupted.

## Oral Microbiota and General Diseases

So far over 700 bacterial species have been isolated in the oral flora, of which roughly 60% have been cultivated. About one third of the human population suffers from different stages of gingival or parodontal infections. It has been calculated, that the area of oral mucosa is as large as the area of palmar skin. Therefore, the oral mucosa and especially inflammated mucosa and parodontal pockets (Tsakos *et al.*, 2010) are significant sources of contamination and hence risk for general infections.

**Table 2. Oral pathogenic bacteria and local / systemic infections**

| Cultivation finding | Importance / detection point |
|---|---|
| *Streptococcus viridans* sp. | Endocarditis |
| *Streptococcus mutans* | Dental caries |
| *Streptococcus salivarius* | Dental biofilm |
| *Streptococcus sanguis* | Tongue |
| *Streptococcus mitis* | Oral mucosa |
| *Streptococcus milleri* | Parodontal abscess |
| *Streptococcus anginosus* sp. | Serious odontogenic infections |
| *Lactobacillus* sp. | Caries indicators |
| *Aggregatibacter actinomycetemcomitans* | Parodontal pathogen |
| *Porphyromonas gingivalis* | -- " -- |
| *Prevotella intermedia* | -- " -- |
| *Tannerella forsythia* | -- " -- |
| *Peptostreptococcus micros* | -- " -- |
| *Treponema denticola* | -- " -- |
| *Fusobacterium* sp. | Odontogenic abscesses |
| *Bacteroides melaninogenicus* sp. | -- " -- |
| *Actinomyces* sp. | -- " -- |
| *Candida albicans* | Infection of oral mucosa |

The causes of these problems are the unregulated growth of bacteria on dental surfaces and infection in gingival pockets; bacteria that cover dental surfaces and gingival pocket epithelium. In different studies microbes in the oral biofilm and infections in the tooth and surrounding mucosa have proved to be connected with several general diseases in population by spreading through blood stream and lymph veins. Strong association exists, for example, between chronic parodontitis and cardio-vascular diseases (Bohnstedt *et al.*, 2010), and also between parorontitis and poor diabetes control (Pradeep *et al.*, 2011).

The increased risk between chronic parodontitis and cardiovascular diseases, stroke, rheumatoid arthritis and preterm birth has been well documented (Pussinen *et al.*, 2002). In addition to that we know that oral infections have a high potential to lead to general diseases (Zoellner, 2011): chronic intestinal diseases (*e.g.* ulcerative colitis, Mb. Crohn) (Dharmani *et al.*, 2011), diabetes, many skin diseases (*e.g.* psoriasis), lung diseases and ocular diseases (*e.g.* iritis).

*Fusobacterium nucleatum* is believed to be in a key role in the mouth and in the intestines not only in biofilm formation but also in developing of mucosal proinflammatoric reaction and in invasive behavior of microbes. It has been shown that biopsies from colonoscopy rich of *F. nucleatum* strains can be diagnostic for inflammatory bowel disease (Strauss *et al.*, 2011).

Normally the body develops together with the microbiota a symbiotic interaction. Biofilm on different surfaces depends on the environmental conditions (*e.g.* pH, quality of mucin layer, the composition of the liquid phase). The defense mechanisms of healthy body system are able to cope with pathogenic bacteria that have forced their way into blood circulation. However, if for some reason the resistance of body is impaired or there is some structural defect (*e.g.* in cordial valve) even quite apathogenic bacteria in the blood stream can cause serious consequences.

Septic infection is always serious risk to the whole body. For several years oral bacteria have been known to be associated with hematogenic endocardial infection. This is based on the fact, that often in the blood culture of endocarditis patients the growth of oral microbial strain is indicated. In addition in cardiac valves of endocarditis patients there has been a growth of same strains of bacteria that are common in the mouth. In addition to this oral microbes can spread via blood stream to endoprotheses. This can be a dangerous development and require even an extraction of endoprotheses. This may have a significant negative impact on the quality of life in patients with endoprotheses.

Recent studies have provided new insights on septic infections of oral origin. In the case of neonatal septic conditions studied with the PMEU, the causative agents were often strains belonging to the normal flora of healthy adults (Pesola *et al.*, 2009, Pesola and Hakalehto, 2011).

The causes of septic infections originate by up to 80% patient's own microflora. Infection could be based on the contamination of oral mucosa, gastrointestinal tract, urinary tract, perineum, or skin puncture cannular openings. Common bacteria from the buccal mucosa spreading via blood circulation (especially *Streptococcus viridans* group) may cause serious infections in heart structures and other organs. Oral microbes have been detected for example, in heart valves, inner membrane of the heart, atherosclerosis bar in veins, cardiac thrombus, brain abscesses, renal tubulus, joints and endoprotheses (Asikainen and Alaluusua,. 1993).

Nearly 300 bacterial species reported to cause endocarditis have been isolated from blood samples of septic cardiovascular disease patients. The most common types of bacteria in sepsis samples were *Streptococcus, Prevotella, Actinomyces, and Fusobacterium* (Lockhart *et al.*, 2008). In addition, it has been demonstrated that the arterial plaques and their inflammatory reactions are important in pathogenesis of blood-spread infections. From these plaques there have been isolated *e.g. Porphyromonas gingivalis, Fusobacterium nucleatum, Prevotella intermedia and Actinomyces actinomycetemcomitans* strains. These are typical pathogens in the parodontium (Ford *et al.*, 2006), (Saito *et al.*, 2008). The renovation of oral infections and treatment of parodontitis significantly reduces the body's infection load and reduces the risk for the development of the general diseases. It is also suggested the correlation between parodontal treatment and arterial welfare. The thickness of intima-media in carotid artery is reduced, and fibrinogen concentrations and leukocytosis is reduced when patients with periodontitis will be renovated (Piconi *et al.*, 2009). Serum IgG response is clearly linked to parodontal pathogen *Porphyromonas gingivalis* and the occurrence of cardiovascular events (Bohnstedt *et al.*, 2010).

Pregnant women, immunocompromised patients and the elderly have been studied with a focus for risk to develop septic conditions of oral origin. In these studies it has been shown that chronic mouth and jaw region infection may be one reason to preterm birth (Kloetzel *et al.*, 2011). The body's pro-inflammatory status is known to rise during chronic infection in parodontium. Pro-inflammatory cytokines can increase the risk for preterm birth. Pathogenesis for this may be as a increased risk of ruptures in fetal membranes (Straka, 2011). It has been shown in mice that the intravenously injected *Fusobacterium nucleatum* can increase the risk of preterm birth (Han *et al.*, 2004).

Similarly, the general status of immunocompromised patients and patients with multiple diseases as well as frail elderly people can get worse, or their healing slow down if they have oral infectious focuses that have not been renovated properly (Shimoe *et al.*, 2011).

A normal tooth brushing causes a short-term bacteremia of about 20 minutes to one third of persons. The risk for bacteremia is related on the degree of gingival inflammation. This is an essential risk, especially in persons with lowered infection resistance, pre-existing general disease (*e.g.* valvular heart disease), or prosthesis in joint (Lockhart *et al.*, 2008).

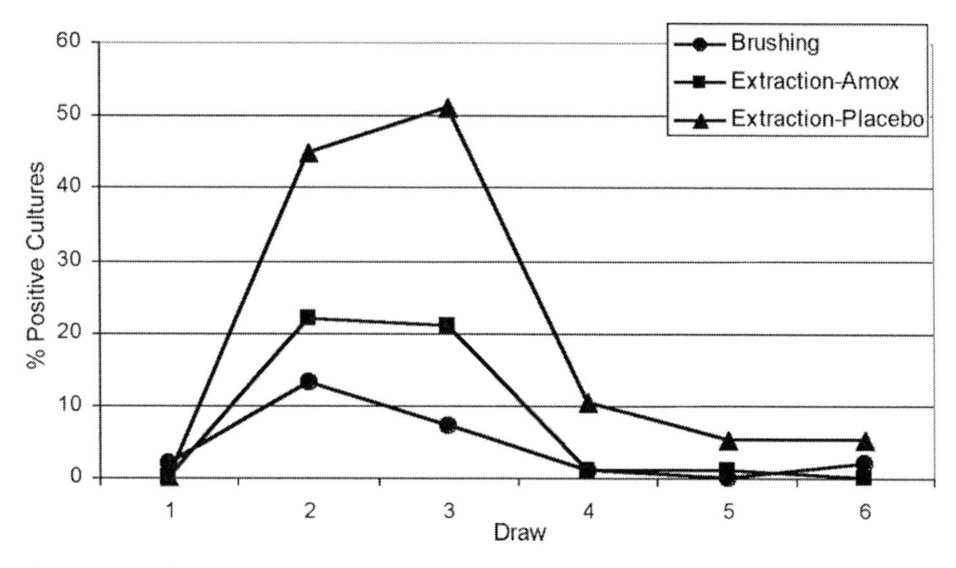


Figure 9. Bacteremia in blood samples after tooth brushing and tooth extraction (with preoperative antibiotics or placebo).

Oral health problems associated with diabetes are well known. Particularly infections in the supporting tissues around the teeth can worsen the diabetic balance, i.e. gingivitis and high blood sugar levels are positively correlated (Ben-Aryeh *et al.*, 1993). Hyperglycemia promotes the development of microinflammation around the teeth by causing bacterial growth in favorable conditions for microbiota (Lim,, *et al.*, 2007).

A positive response to rheumatoid arthritis activity and severity has been achieved by active parodontal treatment. The secretion of pro-inflammatory mediators is the combining factor in pathophysiology of parodontal disease and rheumatoid arthritis. The destruction of hard tissue and infection of connective tissue is a consequence of the secretion of these pro-inflammatory mediators. It has been suggested that patients with ankylosing spondylitis have a sevenfold risk of developing severe parodontitis compared to healthy subjects (Al-Katma *et al.*, 2007), (Loyola-Rodriguez *et al.*, 2010).

There is a positive correlation between oral infections and lung infections. Particularly, in the elderly and multi-sick people lung infections and poor oral hygiene create an elevated risk that infections become chronic and more severe. Oral microbes play a key role in the development of mucosal tolerance, including the airways. However, the full role of the mucosal microflora throughout the body's immune system as a key regulator for the human health has not been fully elucidated (Noverr and Huffnagle, 2004).

A large Japanese study found a significant association between tooth loss and increased risk of cancer. This was the case in particular in patients with head and neck, esophageal and lung cancer (Hiraki *et al.*, 2008). This finding supports the notions of the importance of oral hygiene and dental care in cancer prevention.

Researchers have shown that tobacco and alcohol in contact with the oral microbiota create a high carcinogenic risk to oro-pharyngeal and esophagus-gastric cancers (Salaspuro and Salaspuro, 2004). Oral bacteria have potential ability to transform alcohol to acetaldehyde. This is important because carsinogenic acetaldehyde is one of the most important factors to mucosal mutagenesis (Nieminen *et al.*, 2009; Chocolatewala *et al.*, 2010).

In Helsinki lymphoma study it was found that up to 46% chemotherapy-treated patients developed sepsis during one year follow-up period (Laine *et al.*, 1992). In a large proportion of these infections the pathogens were normal oral bacteria. This high risk of serious infections in the case of severely sick and immunocompromised patients indicates the importance of developing more effective and accurate methods for patient control and hygiene monitoring in hospitals (Hakalehto, 2006, Hakalehto *et al.*, 2010). The improved level of early-warning techniques for the risk of severe infections is proposed readily saving lives.

Chronic pancreatitis and pancreatic cancer are other examples of positive correlation of diseases with salivary microbiota. Especially salivary *Neisseria elongata* and *Streptococcus mitis* seemed to be as a non-invasive biomarker valuable for discovering systemic severe disease (Farrell *et al.*, 2011).

## Probiotic Bacteria and Mouth

The use of probiotics in regulation the microbiota of the GI tract has been studied years and based on that data we know that many probiotics have a favorable influence to intestinal functions and balance. A total amount of 1.5 kg of bacterial mass is permanently colonizing intestinal surfaces. Probiotic and antibiotic effects to that alimentary ecosystem are complex and diverse. Less is known on the effects of probiotics in the oral cavity. Thus, the effects of probiotics on to the oral mucosa should be in focus of future studies. Patients with immunosuppressive status in the oral mucosa or getting radiotherapy to head or neck region are examples of patient groups that could benefit from this preliminary research.

Long term effects of probiotic therapy of oral diseases have been evaluated poorly. Probiotic bacteria have possible favorable effect in parodontal infection, but when dosing of probiotic bacteria will be stopped, the oral biofilm returns back to its previous state in a couple of weeks. Reliable long term clinical knowledge and evidence of this issue is lacking (Teughels *et al.*, 2011).

*In vitro* data support the protective effects of the probiotics. When probiotics of yoghurt *(Lactobacillus bulgaricus, Streptococcus thermophilus, Lactobacillus acidophilus, and Bifidobacterium)* were inoculated with parodontopathogens those probiotics prevented the growth of most important parodontopathogens (Zhu *et al.*, 2010).

Against mainly Gram negative anaerobic parodontopathogens can combat, at least *in vitro*, with so called BALO-bacteria. Those bacteria isolated from intestinal or fecal samples have been used *in vitro* to predation of *e.g. Aggregatibacter actinomycetemcomitans, Porphyromonas gingivalis and Prevotella intermedia, Fusobacterium nucleatum.* These predator bacteria are promising treatment for parodontal diseases in dental clinics in the future (Van Essche *et al.*, 2011).

# Conclusion

Oral mucosa is one of the most significant routes for microbes to cause infections in the human body. Microbes can initiate infection when swallowed into and through the gastrointestinal canal or they can by pass through the mouth cavity mucosa and spread throughout the entire body by blood circulation. Normal oral flora consists of both bacteria and yeasts, which form a rather stable biofilm on the all soft and hard tissues in the mouth. Mechanical forces, eating or tooth brushing, crashes the biofilm, plaque is broken up and bacteria are released in the mouth cavity. The microbes released from biofilm can behave as predisposition factors for diseases and they have a potential to create new microbe colonies in the mouth cavity. On the other hand, to mechanically break up the biofilm around the teeth is a therapeutic method to cure parodontal infection. By scheduled tooth brushing the balance in the biofilm can be restored and as a clinical sign gums get better in a few days. This is an example regarding the importance of the balance and interactions between microbes in biofilm with respect to maintenance and eradication of infections.

Local infections in the mouth can be associated with the pathogenesis of several general diseases. The severity of mouth infections may also correlate with the severity and clinical picture of some systemic disease. Thus, a good oral health is of paramount importance for the human health.

Gene therapy is a rapidly evolving field of medicine, which potentially offers new treatments for different type of diseases. Gene therapy may have applications also in the oral health. Genetically altered microbes can be used for protein transport through mucosa of the intestines, mouth or nose. Gene therapy has been proven out to be a possible method to transport proteins through biofilms to mucosal cells. Using microbes of normal mucosa should be safe treatment to host cells (Steidler, 2005). In addition, in the future these genetically modified bacteria can obviously be exploited to transport different drugs to target organs. Recently, *Lactococcus lactis* bacteria has been applied as a vector for IL-10 in the treatment of Crohn`s disease (Bermudez-Humaran, 2009). However, the risk of developing genetically risky combinations should be considered carefully before any implementation of the technology in routine clinical practise (Pozzetto and Garraud, 2011).

As the salivary secretion is between half and one liter per day, it is evident that the swallowed amount of bacteria with saliva is significant. Saliva may not only carry potentially pathogenic microbes into the gastrointestinal canal but, at least hypothetically, may help to restore affected intestinal microflora. It is supposed that the oral microbiota works as a reserve for the entire gastrointestinal tract in maintaining microbial balance. This may be important to the host health for example after antimicrobial medication or after intestinal infections.

Intestinal dendritic cells are important in the antigen presentation and in the start-up of microbial immunity caused by antigens. There are dendritic cells also in oral mucosa. Dendritic cells are located in those regions where there is antigenic activity, and they are detected from clinically observed ulcers with inflammation (Santoro *et al.*, 2005).

Oral mucosal epithelial cells and mucins in biofilm have an etiological role in diseases of the oral mucosa. Oral aphthous ulcers, oral lichen planus, pemphigoid, erythema multiforme, celiacia, Crohn`s disease, Bechet disease, several drugs and x-ray therapy develop different degrees of ulcers and pseudomembranotic erosions in mucosa. Many of these are considered to be autoimmune diseases. Further evidence and studies are needed to evaluate what role the imbalance of normal microbiota and the occurrence of these diseases may have. Interesting clinical question is whether altered oral microflora is a cause or consequence to clinic mucosal symptoms. When the protective effect of mucin layer is disturbed, epithelium may also get damaged, and this can lead to clinic mucosal ulcer and even malignant transformations in mucosal cells.

The immune system in oro-pharynx and in intestines has a key role in maintaining peripheral tolerance against intestinal antigens. T-lymphocytes in the intestines' lymphatic tissue regulate the immune response against food antigens, against microbial structure of normal flora and against autoantigens.

Oral tolerance is a situation in which there exists no local or systemic immunologic reaction to orally dispensed antigens. The role of oral microbes in developing this tolerance during first years of life is obvious. Favorable load of microbial antigens creates in healthy host a comprehensive protection against pathogenic antigens. It remains to be seen if it is possible to regulate the formation of this tolerance in future to beneficial direction and result in an immunologically potent defense system. Thus, we need further expand our knowledge on the development and constituent of normal oral microbiological flora with proper diagnostic tools such as the PMEU.

## References


Adlerberth, I; Wold, AE. Establishment of the gut microbiota in Western infants. *Acta Paediatr.*, 2009; 98, 229-238.

Al-Katma, MK; Bissada, NF; Bordeaux, JM; Sue, J; Askari, AD. Control of periodontal infection reduces the severity of active rheumatoid arthritis. *J. Clin. Rheumatol.*, 2007; 13, 134-137.

Arbes, SJ Jr; Matsui, EC. Can oral pathogens influence allergic disease? *J. Allergy Clin. Immunol.*, 2011; 127, 1119-1127.

Asikainen, S; Alaluusua, S. Bacteriology of dental infections. *Eur. Heart J.*, 1993; 14 Suppl K, 43-50.

Aureli, P; Capurso, L; Castellazzi, AM; Clerici, M; Giovannini, M; Morelli, L; Poli, A; Pregliasco, F; Salvini, F; Zuccotti, GV. Probiotics and health: an evidence-based review. *Pharmacol. Res.*, 2011; 63, 366-376.

Awartani, F. Evaluation of the relationship between type 2 diabetes and periodontal disease. *Odontostomatol. Trop.*, 2009; 32, 33-39.

Bardow, A; Hofer, E; Nyvad, B; ten Cate, JM; Kirkeby, S; Moe, D; Nauntofte, B. Effect of saliva composition on experimental root caries. *Caries Res.*, 2005; 39, 71-77.

Ben-Aryeh, H; Serouya, R; Kanter, Y; Szargel, R; Laufer, D. Oral health and salivary composition in diabetic patients. *J. Diabetes Complications,* 1993; 7, 57-62.

Berlutti, F; Catizone, A; Ricci, G; Frioni, A; Natalizi, T; Valenti, P; Polimeni, A. *Streptococcus mutans* and *Streptococcus sobrinus* are able to adhere and invade human gingival fibroblast cell line. *Int. J. Immunopathol. Pharmacol.*, 2010; 23, 1253-1260.

Bermudez-Humaran, LG. *Lactococcus lactis* as a live vector for mucosal delivery of therapeutic proteins. *Hum. Vaccin.*, 2009; 5, 264-267.

Bohnstedt, S; Cullinan, MP; Ford, PJ; Palmer, JE; Leishman, SJ; Westerman, B; Marshall, RI; West, MJ; Seymour, GJ. High antibody levels to *P. gingivalis* in cardiovascular disease. *J. Dent. Res.,* 2010; 89, 938-942.

Bollinger, RR; Everett, ML; Palestrant, D; Love, SD; Lin, SS; Parker, W. Human secretory immunoglobulin A may contribute to biofilm formation in the gut. *Immunology,* 2003; 109, 580-587.

Brandtzaeg, P. Immune functions of nasopharyngeal lymphoid tissue. *Adv. Otorhinolaryngol.,* 2011a; 72, 20-24.

Brandtzaeg, P. Potential of nasopharynx-associated lymphoid tissue for vaccine responses in the airways. *Am. J. Respir. Crit. Care Med.,* 2011b; 183, 1595-1604.

Cannon, RD; Lyons, KM; Chong, K; Holmes, AR. Adhesion of yeast and bacteria to oral surfaces. *Methods Mol. Biol.,* 2010; 666, 103-124.

Chocolatewala, N; Chaturvedi, P; Desale, R. The role of bacteria in oral cancer. *Indian J. Med. Paediatr. Oncol.,* 2010; 31, 126-131.

Davidovici, BB; Sattar, N; Prinz, JC; Puig, L; Emery, P; Barker, JN; van de Kerkhof, P; Stahle, M; Nestle, FO; Girolomoni, G; Krueger, JG. Psoriasis and systemic inflammatory diseases: potential mechanistic links between skin disease and co-morbid conditions. *J. Invest. Dermatol.,* 2010; 130, 1785-1796.

de Farias, DG; Bezerra, AC. Salivary antibodies, amylase and protein from children with early childhood caries. *Clin. Oral Investig.,* 2003; 7, 154-157.

Deplancke, B; Gaskins, HR. Microbial modulation of innate defense: goblet cells and the intestinal mucus layer. *Am. J. Clin. Nutr.,* 2001; 73, 1131S-1141S.

Dewhirst, FE; Chen, T; Izard, J; Paster, BJ; Tanner, AC; Yu, WH; Lakshmanan, A; Wade, WG. The human oral microbiome. *J. Bacteriol.,* 2010; 192, 5002-5017.

Dharmani, P; Strauss, J; Ambrose, C; Allen-Vercoe, E; Chadee, K. *Fusobacterium nucleatum* infection of colonic cells stimulates MUC2 mucin and tumor necrosis factor alpha. *Infect. Immun.,* 2011; 79, 2597-2607.

Dicksved, J; Lindberg, M; Rosenquist, M; Enroth, H; Jansson, JK; Engstrand, L. Molecular characterization of the stomach microbiota in patients with gastric cancer and in controls. *J. Med. Microbiol.,* 2009; 58, 509-516.

Evaldson, G; Heimdahl, A; Kager, L; Nord, CE. The normal human anaerobic microflora. *Scand. J. Infect. Dis. Suppl.,* 1982; 35, 9-15.

Fagarasan, S; Honjo, T. Regulation of IgA synthesis at mucosal surfaces. *Curr. Opin. Immunol.,* 2004; 16, 277-283.

Farrell, JJ; Zhang, L; Zhou, H; Chia, D; Elashoff, D; Akin, D; Paster, BJ; Joshipura, K; Wong, DT. Variations of oral microbiota are associated with pancreatic diseases including pancreatic cancer. *Gut,* 2011; Oct 12. [Epub ahead of print].

Ford, PJ; Gemmell, E; Chan, A; Carter, CL; Walker, PJ; Bird, PS; West, MJ; Cullinan, MP; Seymour, GJ. Inflammation, heat shock proteins and periodontal pathogens in atherosclerosis: an immunohistologic study. *Oral Microbiol. Immunol.,* 2006; 21, 206-211.

Fukuyama, Y; King, JD; Kataoka, K; Kobayashi, R; Gilbert, RS; Oishi, K; Hollingshead, SK; Briles, DE; Fujihashi, K. Secretory-IgA antibodies play an important role in the immunity to *Streptococcus pneumoniae*. *J. Immunol.,* 2010; 185, 1755-1762.

Hakalehto, E. Semmelweis' present day follow-up: Updating bacterial sampling and enrichment in clinical hygiene. *Pathophysiology,* 2006; 13, 257-267.

Hakalehto, E. Simulation of enhanced growth and metabolism of intestinal *Escherichia coli* in the Portable Microbe Enrichment Unit (PMEU). In: Rogers MC, Peterson ND, editors. *E. coli infections: causes, treatment and prevention.* New York, USA: Nova Publishers; 2011.

Hakalehto, E; Hell, M; Bernhofer, C; Heitto, A; Pesola, J; Humppi, T; Paakkanen, H. Growth and gaseous emissions of pure and mixed small intestinal bacterial cultures: Effects of bile and vancomycin. *Pathophysiology,* 2010; 17, 45-53.

Hakalehto, E; Pesola, J; Heitto, L; Narvanen, A; Heitto, A. Aerobic and anaerobic growth modes and expression of type 1 fimbriae in *Salmonella. Pathophysiology,* 2007; 14, 61-69.

Han, YW; Redline, RW; Li, M; Yin, L; Hill, GB; McCormick, TS. *Fusobacterium nucleatum* induces premature and term stillbirths in pregnant mice: implication of oral bacteria in preterm birth. *Infect. Immun.,* 2004; 72, 2272-2279.

Hiraki, A; Matsuo, K; Suzuki, T; Kawase, T; Tajima, K. Teeth loss and risk of cancer at 14 common sites in Japanese. *Cancer Epidemiol Biomarkers Prev,* 2008; 17, 1222-1227.

Johansson, ME; Ambort, D; Pelaseyed, T; Schütte, A; Gustafsson, JK; Ermund, A; Subramani, DB; Holmén-Larsson, JM; Thomsson, KA; Bergström, JH; van der Post, S; Rodriguez-Piñeiro, AM; Sjövall, H; Bäckström, M; Hansson, GC. Composition and functional role of the mucus layers in the intestine. *Cell Mol. Life Sci.,* 2011; 68, 3635-3641.

Johansson, ME; Gustafsson, JK; Sjoberg, KE; Petersson, J; Holm, L; Sjovall, H; Hansson, GC. Bacteria penetrate the inner mucus layer before inflammation in the dextran sulfate colitis model. *PLoS One,* 2010; 5, e12238.

Karaca, S; Seyhan, M; Senol, M; Harputluoglu, MM; Ozcan, A. The effect of gastric *Helicobacter pylori* eradication on recurrent aphthous stomatitis. *Int. J. Dermatol.,* 2008; 47, 615-617.

Kawas, SA; Rahim, ZH; Ferguson, DB. Potential uses of human salivary protein and peptide analysis in the diagnosis of disease. *Arch. Oral Biol.,* 2012; 57, 1-9.

Kloetzel, MK; Huebner, CE; Milgrom, P. Referrals for dental care during pregnancy. *J. Midwifery Womens Health,* 2011; 56, 110-117.

Könönen, E. Anaerobes in the upper respiratory tract in infancy. *Anaerobe,* 2005; 11, 131-136.

Könönen, E. Development of oral bacterial flora in young children. *Ann. Med.,* 2000; 32, 107-112.

Könönen, E. Oral colonization by anaerobic bacteria during childhood: role in health and disease. *Oral Dis.,* 1999; 5, 278-285.

Könönen, E; Jousimies-Somer, H; Asikainen, S. Relationship between oral gram-negative anaerobic bacteria in saliva of the mother and the colonization of her edentulous infant. *Oral. Microbiol. Immunol.,* 1992; 7, 273-276.

Könönen, E; Kanervo, A; Takala, A; Asikainen, S; Jousimies-Somer, H. Establishment of oral anaerobes during the first year of life. *J. Dent. Res.,* 1999; 78, 1634-1639.

Laine, PO; Lindqvist, JC; Pyrhönen, SO; Strand-Pettinen, IM; Teerenhovi, LM; Meurman, JH. Oral infection as a reason for febrile episodes in lymphoma patients receiving cytostatic drugs. *Eur. J. Cancer B. Oral. Oncol.,* 1992; 28B, 103-107.

Lendenmann, U; Grogan, J; Oppenheim, FG. Saliva and dental pellicle--a review. *Adv. Dent. Res.,* 2000; 14, 22-28.

Lim, LP; Tay, FB; Sum, CF; Thai, AC. Relationship between markers of metabolic control and inflammation on severity of periodontal disease in patients with diabetes mellitus. *J. Clin. Periodontol.,* 2007; 34, 118-123.

Lockhart, PB; Brennan, MT; Sasser, HC; Fox, PC; Paster, BJ; Bahrani-Mougeot, FK. Bacteremia associated with toothbrushing and dental extraction. *Circulation,* 2008; 117, 3118-3125.

Loyola-Rodriguez, JP; Martinez-Martinez, RE; Abud-Mendoza, C; Patino-Marin, N; Seymour, GJ. Rheumatoid arthritis and the role of oral bacteria. *J. Oral Microbiol.,* 2010; 2, 10.3402/jom.v2i0.5784.

Mager, DL; Ximenez-Fyvie, LA; Haffajee, AD; Socransky, SS. Distribution of selected bacterial species on intraoral surfaces. *J. Clin. Periodontol.,* 2003; 30, 644-654.

Maiden, MF; Tanner, A; Macuch, PJ. Rapid characterization of periodontal bacterial isolates by using fluorogenic substrate tests. *J. Clin. Microbiol.,* 1996; 34, 376-384.

McGuckin, MA; Linden, SK; Sutton, P; Florin, TH. Mucin dynamics and enteric pathogens. *Nat. Rev. Microbiol.,* 2011; 9, 265-278.

Mellander, L; Carlsson, B; Hanson, LA. Appearance of secretory IgM and IgA antibodies to *Escherichia coli* in saliva during early infancy and childhood. *J. Pediatr.,* 1984; 104, 564-568.

Meyer, JE; Harder, J; Gorogh, T; Weise, JB; Schubert, S; Janssen, D; Maune, S. Human beta-defensin-2 in oral cancer with opportunistic *Candida* infection. *Anticancer Res.,* 2004; 24, 1025-1030.

Miranda, DO; Silva, DA; Fernandes, JF; Queiros, MG; Chiba, HF; Ynoue, LH; Resende, RO; Pena, JD; Sung, SS; Segundo, GR; Taketomi, EA. Serum and Salivary IgE, IgA, and IgG(4) Antibodies to *Dermatophagoides pteronyssinus* and its major allergens, Der p1 and Der p2, in Allergic and Nonallergic Children. *Clin. Dev. Immunol.,* 2011; 2011, 302739.

Mishra, A; Wu, C; Yang, J; Cisar, JO; Das, A; Ton-That, H. The *Actinomyces oris* type 2 fimbrial shaft FimA mediates co-aggregation with oral streptococci, adherence to red blood cells and biofilm development. *Mol. Microbiol.,* 2010; Jun 10. [Epub ahead of print],

Mitsuyama, K; Sata, M. Gut microflora: a new target for therapeutic approaches in inflammatory bowel disease. *Expert. Opin. Ther. Targets,* 2008; 12, 301-312.

Mushi, D; Byamukama, D; Kivaisi, AK; Mach, RL; Farnleitner, AH. Sorbitol-fermenting *Bifidobacteria* are indicators of very recent human faecal pollution in streams and groundwater habitats in urban tropical lowlands. *J. Water Health,* 2010; 8, 466-478.

Nieminen, MT; Uittamo, J; Salaspuro, M; Rautemaa, R. Acetaldehyde production from ethanol and glucose by non-*Candida albicans* yeasts *in vitro*. *Oral. Oncol.,* 2009; 45, e245-8.

Noverr, MC; Huffnagle, GB. Does the microbiota regulate immune responses outside the gut? *Trends Microbiol.,* 2004; 12, 562-568.

Parm, U; Metsvaht, T; Sepp, E; Ilmoja, ML; Pisarev, H; Pauskar, M; Lutsar, I. Mucosal surveillance cultures in predicting Gram-negative late-onset sepsis in neonatal intensive care units. *J. Hosp. Infect.,* 2011; 78, 327-332.

Pesola, J; Hakalehto, E. Enterobacterial microflora in infancy - a case study with enhanced enrichment. *Indian J. Pediatr.,* 2011; 78, 562-568.

Pesola, J; Vaarala, O; Heitto, A; Hakalehto, E. Use of portable enrichment unit in rapid characterization of infantile intestinal enterobacterial microbiota. *Microb. Ecol. Health Dis.,* 2009; 21, 203-210.

Piconi, S; Trabattoni, D; Luraghi, C; Perilli, E; Borelli, M; Pacei, M; Rizzardini, G; Lattuada, A; Bray, DH; Catalano, M; Sparaco, A; Clerici, M. Treatment of periodontal disease results in improvements in endothelial dysfunction and reduction of the carotid intima-media thickness. *FASEB J.,* 2009; 23, 1196-1204.

Pozzetto, B; Garraud, O. Emergent viral threats in blood transfusion. *Transfus. Clin. Biol.,* 2011; 18, 174-183.

Pradeep, A; Raghavendra, N; Kathariya, R; Pradeep Patel, S; Arjun Raju, P; Sharma, A. association of serum and crevicular visfatin levels in periodontal health and disease with type 2 diabetes mellitus. *J. Periodontol.,* 2011; Oct 3. [Epub ahead of print]

Pussinen, PJ; Vilkuna-Rautiainen, T; Alfthan, G; Mattila, K; Asikainen, S. Multiserotype enzyme-linked immunosorbent assay as a diagnostic aid for periodontitis in large-scale studies. *J. Clin. Microbiol.,* 2002; 40, 512-518.

Ranjitkar, S; Smales, RJ; Kaidonis, JA. Oral manifestations of gastroesophageal reflux disease. *J. Gastroenterol. Hepatol.,* 2012;27, 21-27.

Rashkova, MP; Toncheva, AA. Gingival disease and secretory immunoglobulin a in non-stimulated saliva in children. *Folia Med. (Plovdiv),* 2010; 52, 48-55.

Reynolds, HY. Defense mechanisms against infections. *Curr. Opin. Pulm. Med.,* 1999; 5, 136-142.

Sachdeo, A; Haffajee, AD; Socransky, SS. Biofilms in the edentulous oral cavity. *J. Prosthodont.,* 2008; 17, 348-356.

Saito, Y; Fujii, R; Nakagawa, KI; Kuramitsu, HK; Okuda, K; Ishihara, K. Stimulation of Fusobacterium nucleatum biofilm formation by Porphyromonas gingivalis. *Oral Microbiol. Immunol.,* 2008; 23, 1-6.

Salaspuro, V; Salaspuro, M. Synergistic effect of alcohol drinking and smoking on *in vivo* acetaldehyde concentration in saliva. *Int. J. Cancer,* 2004; 111, 480-483.

Salvini, F; Riva, E; Salvatici, E; Boehm, G; Jelinek, J; Banderali, G; Giovannini, M. A specific prebiotic mixture added to starting infant formula has long-lasting bifidogenic effects. *J. Nutr.,* 2011; 141, 1335-1339.

Sanita, PV; Pavarina, AC; Giampaolo, ET; Silva, MM; de Oliveira Mima, EG; Ribeiro, DG; Vergani, CE. Candida spp. prevalence in well controlled type 2 diabetic patients with denture stomatitis. *Oral Surg. Oral Med. Oral Pathol.. Oral Radiol. Endod.,* 2011; 111, 726-733.

Santoro, A; Majorana, A; Roversi, L; Gentili, F; Marrelli, S; Vermi, W; Bardellini, E; Sapelli, P; Facchetti, F. Recruitment of dendritic cells in oral lichen planus. *J. Pathol.,* 2005; 205, 426-434.

Shimoe, M; Yamamoto, T; Iwamoto, Y; Shiomi, N; Maeda, H; Nishimura, F; Takashiba, S. Chronic periodontitis with multiple risk factor syndrome: a case report. *J. Int. Acad. Periodontol.,* 2011; 13, 40-47.

Steidler, L. Delivery of therapeutic proteins to the mucosa using genetically modified microflora. *Expert Opin. Drug Deliv.,* 2005; 2, 737-746.

Straka, M. Pregnancy and periodontal tissues. *Neuro Endocrinol. Lett.,* 2011; 32, 34-38.

Strauss, J; Kaplan, GG; Beck, PL; Rioux, K; Panaccione, R; Devinney, R; Lynch, T; Allen-Vercoe, E. Invasive potential of gut mucosa-derived *Fusobacterium nucleatum* positively correlates with IBD status of the host. *Inflamm. Bowel. Dis.,* 2011; 17, 1971-1978.

Taji, SS; Savage, N; Holcombe, T; Khan, F; Seow, WK. Congenital aplasia of the major salivary glands: literature review and case report. *Pediatr. Dent.,* 2011; 33, 113-118.

Tenovuo, J. Antimicrobial function of human saliva--how important is it for oral health? *Acta Odontol. Scand.,* 1998; 56, 250-256.

Tenovuo, J; Aaltonen, AS. Antibody responses to mutans streptococci in children. *Proc. Finn. Dent. Soc.,* 1991; 87, 449-461.

Teughels, W; Loozen, G; Quirynen, M. Do probiotics offer opportunities to manipulate the periodontal oral microbiota? *J. Clin. Periodontol.,* 2011; 38 Suppl 11, 159-177.

Tsakos, G; Sabbah, W; Hingorani, AD; Netuveli, G; Donos, N; Watt, RG; D'Aiuto, F. Is periodontal inflammation associated with raised blood pressure? Evidence from a National US survey. *J. Hypertens.,* 2010; 28, 2386-2393.

Tugizov, SM; Webster-Cyriaque, JY; Syrianen, S; Chattopadyay, A; Sroussi, H; Zhang, L; Kaushal, A. Mechanisms of viral infections associated with HIV: workshop 2B. *Adv. Dent. Res.,* 2011; 23, 130-136.

Van Essche, M; Quirynen, M; Sliepen, I; Loozen, G; Boon, N; Van Eldere, J; Teughels, W. Killing of anaerobic pathogens by predatory bacteria. *Mol. Oral. Microbiol.,* 2011; 26, 52-61.

Vankeerberghen, A; Nuytten, H; Dierickx, K; Quirynen, M; Cassiman, JJ; Cuppens, H. Differential induction of human beta-defensin expression by periodontal commensals and pathogens in periodontal pocket epithelial cells. *J. Periodontol.,* 2005; 76, 1293-1303.

Zhu, Y; Xiao, L; Shen, D; Hao, Y. Competition between yogurt probiotics and periodontal pathogens *in vitro*. *Acta Odontol. Scand.,* 2010; 68, 261-268.

Zoellner, H. Dental infection and vascular disease. *Semin. Thromb. Hemost.,* 2011; 37, 181-192.

In: Alimentary Microbiome: A PMEU Approach
Editor: Elias Hakalehto
ISBN: 978-1-61942-692-4
© 2012 Nova Science Publishers, Inc.

*Chapter IV*

# Isolations of Gastric Microbes Using the PMEU

***Elias Hakalehto*[1]*, Ilkka Pesola*[2]*,***
***and Kaarlo Jaakkola*[3]**

[1]Department of Biosciences,
University ofEastern Finland, FI, Kuopio, Finland
[2]Clinic of Oral and Maxillofacial Diseases,
Kuopio University Hospital, FI, Kuopio, Finland
[3]Helsinki Antioxidant Clinic,
Kaisaniemenkatu 1 Ba, FI, Helsinki, Finland

## Abstract

The human gastric areas were considered sterile or almost sterile for a long time. Only occasionally incoming bacterial strains were believed to stay alive for extremely short periods of time. With the help of the PMEU research methods, endoscopically collected specimens from the three gastric regions, corpus, antrum and pylorum, were investigated for their contents of lactic acid bacteria (LAB). These samples turned out to contain several strains in case the entire thirteen previously healthy middle-aged individual. It also turned out that only two of these persons were carriers for *Helicobacter* strains.

This study also underlined the huge potential of the PMEU approach to screen microbiologically objects with relatively low contents of cells or cfu as well as of various species.

It was a rather promising finding to discover such a prevalence of beneficial micro-organisms in the stomach, which also gives hope for the further developments in the field of studying and curing diseases of this region. It is also possible to see the gastric condition as a part of a broader picture of our health, and the ways to protect it. In this chapter also some of its relations with oral and intestinal health, as well as the systemic contribution of the microbiota, are considered.

# Introduction

Microbes have enormous metabolic potential. Their capacity to degrade or exploit almost any organic or other substances is often based on cometabolism (Bull and Quayle, 1982). These functions are in many cases operated by several microbial strains together. Then it is possible to carry out biological activities, which otherwise would be impossible for any singular microbe to be achieved alone. Similarly, it seems likely that as communities the micro-organisms are able to penetrate into otherwise not favorable environments. For example, the gastric conditions with a pH around 1-2 were considered for a long time as almost sterile (Danon and Lee, 2001). However, the helicobacteria were proven to colonize the gastric mucosal membranes being capable of inducing diseases to humans. Recently, with the aid of the PMEU it has been revealed that the endoscopic samples from the gastric areas contained numerous lactobacilli and other lactic acid bacteria (LAB) (Hakalehto, *et al.* 2011). This discovery was made from samples of healthy individuals after fasting overnight. Only two out of 13 samples were *Helicobacter* positive. This implies to the means of more numerous eubacteria to be able to survive in the acidic conditions of the stomach. Instead of fimbriae or flagellae easily dissolvable with acids, the lactobacilli often possess protein capsules on their outer surfaces called the S-layers. These structures could help in the attachment onto the gastric surfaces. According to yet unpublished results the stomach of a single person of the test group above could contain as many as 14 different LAB strains. Also yeasts have been detected to extensively colonize the upper gastrointestinal tract.

# Helicobacter Infection and Gastric Ulcer

Helicobacteria (*Helicobacter pylori*, Hp) is a 2.5-4.0 micrometer long rod bacterium whose other end has 2-6 flagella (Schoenhofen, *et al.* 2006). It has only been found from the human gastric wall. It was isolated for the first time in 1982 (Marshall, 2001). Several other helicobacteria strains have been isolated since. Of these about 60% produce toxins. It has also been isolated from tooth plaque and stool. Plaque of teeth has been shown to be a reservoir of *H. pylori* and thus a potential cause for *Helicobacter* reinfections in stomach. On the other hand patients with gastric *Helicobacter* infection have higher incidence for parodontal problems (Esfahanizadeh and Modanlou. 2010). In some studies it has been supposed that *H. pylori* has a role in oral aphthous ulcers (Birek, *et al.* 1999), but there are obviously several etiologic factors behind aphthous lesions.

Helicobacteria live in the mucus layer of the stomach's mucous membrane attached to epithelial cells and inside the mucus on the surface of the epithelium (Kubota, *et al.* 2011). It can inflame the mucous membrane of the stomach. *Helicobacter* can survive in the acidic environment its urease activity (Bury-Moné *et al.*, 2001). After affixing to the surface of the epithelial cells it produces toxic substances. The *H.* pylori LPS has generally a relatively low immunological activity which property may aid the survival of chronically infectious strains (Moran 1995). The immune response of helicobacteria is not able to eliminate the bacteria or infection. The result is chronic gastritis. Potent acute gastritis will most probably lead to the development of atrophic gastritis in which case the glandular structure of the mucous membrane will be damaged and atrophied and lead to the decreased secretion and lack of hydrochloric acid. Vitamin C has been shown to possess inhibitory activity against ureolytic bacteria, notably against *Helicobacter pylori* infections (Krajewska and Brindell, 2011)

The (13)C-urea breath test is an important and fast non-invasive test for identifying the carriers of *Helicobacter pylori* strains (Parente and Bianchi Porro, 2001; Savarino *et al.*, 1999).

Endoscopic methods have been for some time the most important techniques for stomach investigations and for taking specimens from the epithelia thereof. International Maastricht 2 recommendation suggests the endoscopy to be carried out for every patient complaining stomach problems at least at the age of 45-50 years, before starting the therapy for eradicating the helicobacteria (Malfertheiner *et al.*, 2002). The risk for cancers within this region increases rapidly with age. Even experienced practitioners have difficulties in achieving correct diagnosis with only preliminary information, symptoms and clinical investigation. The endoscopic methods reveal also alterations on the mucous membranes, esophageal reflux disease, and damages caused by the pain-killing medicines (the NSAIDs) in the stomach. Helicobacteria can be detected from the specimens taken by the gastroscopic devices. The *Helicobacter* antibodies (IgG) can also be determined from blood. The risks for *Helicobacter* infection and atrofic gastritis as well as gastric cancer and ulcus disease can be estimated by measuring some nutritional factors from the serum. With novel methods it can deduced, if the patient suffers from the gastritis caused by *Helicobacter* infection, or if the gastric epithelium is atrophied. Also the locations of the pathological alterations can be determined.

Finnish researchers have developed a method complementing and in some cases partially replacing the endoscopic research. It has been named as BIOHIT GastroPanel® (Biohit Co., Helsinki, Finland). Its computer program is exploiting the laboratory test values for making a proposition for diagnosis (Lombardo *et al.*, 2010). It also evaluates the risks of the particular patient to get various stomach diseases later in life. Besides, the program is determining the necessity of the eradication of the helicobacteria by antibiotic treatments. In this context, the measurements for folic acid, vitamin B12 and homocysteine are important additional laboratory tests. Also the analysis of celiac disease is often needed.

The growth conditions of helicobacter can be affected by changing the acidity of the stomach mucous membrane surface. Usually the infection weakens and finally dies away when the stomach becomes acid-free because of *e.g.* medication. The acid-freeness of the stomach (achylia) may lead to excessive bacterial growth by other organisms in the small intestines. Achylia can also relate to an increased risk of falling ill with atopy, asthma and rheumatoid arthritis.

In healthy subjects acidic gastric fluid destroys almost all incoming bacteria and other microbes when entering the small intestine and boosts the digestive activity. When the stomach is acid-free there is no efficient barrier preventing the entry of harmful bacteria into the small intestine. This results in the increase of anaerobic bacterial growth. These digest the proteins in the small intestine and use the released amino acids in metabolism (Menozzi and Ossiprandi, 2010). As a result of this exceptional bacterial metabolism, tyrosine, tryptophan and phenylalanine amino acids degradation traces may be detected in urine (Antener, *et al.* 1981). The increase of digestive bacteria may cause bad breath. This unpleasant condition called halitosis relates to bacterial odoriferous substances, usually sulphur compounds (Rosing and Loesche, 2011). Several etiologic factors have been known to cause halitosis (foeter ex ore, bad breath), i.e. chronic parodontitis, tonsillar stones in tonsil crypts

(Tsuneishi, *et al.* 2006) and gastric in patients with dyspepsia or reflux-disease (Canan, *et al.* 2011).

Along with helicobacteria, so-called metaplasia areas may appear on the mucous membranes of the stomach. This is so designated as intestinal metaplasia, in which stomach surface cells have transformed to resemble the surface cells of the small intestine. After successful eradication therapy a part of atrophic gastritis may return to normal. Intestinal metaplasia instead is most probably an irreversible form.

## Gastric Ulcer

The term peptic ulcer disease is used of gastric ulcer. This means both gastric and duodenum ulcers or several ulcers. The Latin term *ulcus* means wound (ulcer). An average of 0.3% of healthy adults, get a helicobacter infection annually (Silva, *et al.* 2010). The infection is related to dense housing and is more common in lower social classes. About 10-15% of helicobacter carriers develop a peptic ulcer disease during their lifetime (Wu, *et al.* 2008).

Helicobacter infection causes acute gastritis which progresses into superficial chronic gastritis ((Andreica-Sandica, *et al.* 2011). Of those infected in about 3 percent it develops further to chronic gastritis and part further to atrophy gastritis, i.e the inflammation and atrophy of the stomach mucous membrane. In about 10-15% of the helicobacter carriers, a gastric ulcer develops as a complication of the infection. Of duodenum ulcer patients about 80% and of gastric ulcer patients about 60% will develop a new ulcer if the helicobacter is not evicted, and 4% of cancers are caused by gastric ulcer. The majority of helicobacter negative gastric ulcers are in people that use NSAIDs. An aged ulcus patient is often both a helicobacter carrier and a user of NSAIDs.

Gastric ulcer is a result from the interaction of hydrochloric acid and pepsin. An ulcer of mucous membrane and its underlying tissues causes often pain that appears in the middle of the upper abdomen. It appears often after meals. In some cases the patient does not complain about pain, instead some kind of discomfort and epigastric disorder. Based on symptoms, it is often difficult to say whether gastritis or gastric ulcer is in question. If the stomach mucous membrane is healthy, the risk of gastric ulcer is small, about 1%. When the mucous membrane becomes septic, the risk of gastric ulcer increases to ten-fold. A helicobacter infection is often in the background of gastric ulcer, but so-called stress ulcus can suddenly appear in connection with a serious illness or crisis (Pietrzak, *et al.* 2011). Its treatment is often a proton pump inhibitor drug for 1 to 3 weeks combined with two microbe drugs for one week. Antioxidants and other supporting drugs are used to promote the repair mucous membrane damage and stress related crisis.

## Gastritis

The infectious condition of the stomach mucous membrane (gastritis) can ultimately be found with endoscopy (Chandrasoma, *et al.* 2011). Chronic gastritis is a change that becomes more common with age and earlier it was not considered to have independent meaning, but nowadays stomach diseases are considered part of all diseases of the gastrointestinal tract

(Salminen, *et al.* 1995). The early diagnosis and right treatment of gastritis is of great importance.

Gastritis caused by helicobacter is probably one of the most common bacterial chronic inflammations. According to some estimations approximately half of the world's population is ill with it. In the industrialized world it is clearly more common than in developing countries. Its infection routes are not well known and the infection can spread from one person to another.

Generous use of salt, vitamin deficiency and smoking are known as factors that lead to the development of atrophic gastritis (Colacci, *et al.* 2011). Atrophic gastritis means the inflammation and atrophy of the stomach's mucous membrane (Deveci and Deveci, 2004). Its pathogenesis is caused by many factors. Acid secretion decreases especially when the mucous membrane of the stomach's structural part degenerates.

Patients with dry mouth or hyposalivation, especially patients of Sjögren syndrome, suffer from mild and sore mucosal surfaces in mouth, nasal cavity and eyes (Yan *et al.*, 2011). This phenomenon is often complicated with candidosis and other microbial exposures causing problems in speech and swallowing (Prelipcean, *et al.* 2007).

During the course of several years, a helicobacter infection leads to atrophic gastritis in half of those who fall ill with it (Bergman, et al. 2001). Also the stomach's secretion functions, such as acid secretion and also hormone secretion and the intrinsic factor and the secretion of vitamin B12, are disturbed (Hou and Schubert, 2006). About 10-15% of the bacteria carriers will develop a gastric ulcer and a small part a stomach cancer. Atrophic gastritis leads to the decrease of hydrochloric acid secretion and lack of acid of the stomach.

Occasionally a so-called autoimmune gastritis develops (Bergman, *et al.* 2001). This means a severe inflammation and atrophy of the stomach's mucous membrane that is limited to the mucous membrane of structural part of the stomach. This often precedes pernicious anemia. The stomach is acid-free when the parietal cells that secrete acid have fully disappeared. This leads to the lack of an intrinsic factor (Kumar, 2007), which in turn causes malabsorption of vitamin B12 and anemia. Predisposition to autoimmune gastritis is often hereditary. A *Helicobacter* infection may trigger its birth as well as mercury loosened from amalgam tooth fillings (Hybenova, *et al.* 2010). Fungus *Candida albicans* can often be found in the duodenum or stomach ulcer of patients that are elderly and in poor health (Ishiguro, *et al.* 2010). Oral candidosis especially in immunosuppressed or elderly patients with dental prosthesis may persist for a long time in oral mucosa and spread with saliva further to oesophagus and stomach (Laurent, *et al.* 2011). Local oral yeast infections are treated with local antimycotic drugs, but when infection spread deeper in pharynx and oesophagus systemic medication is often needed. In mild cases local oral liquid antimycotic drug mixed with saliva is efficient enough both in oral and oesophageal disorders.

The so called peptic ulcer that is related to abundant secretion of hydrochloric acid is a disorder of the gastric or duodenal mucous membrane that penetrates the *muscularis mucosae* layer (Necchi, *et al.* 2007). A key predisposing factor to the disease is chronic gastritis that relates to helicobacter. The symptom is often in the middle of the upper abdomen located deep, blunt, crippling or burning pain sensation that can radiate to he back or shoulders. An e stimated one third of gastric and duodenum ulcers are without symptoms and the healing of the ulcer does not always eliminate symptoms. Empty stomach pain 1 to 3 hours after a meal has been considered a typical symptom of duodenal ulcer. It might ease off with meals or drugs that neutralize acids. This pain may also occur in connection with functional dyspepsia.

# Other Influences of the *Helicobacter* sp.

It has been suggested that infection with HP can be involved in various extra-digestive conditions: respiratory disorders (chronic obstructive pulmonary disease, bronchiectasis, lung cancer, pulmonary tuberculosis, bronchial asthma); vascular disorders (ischaemic heart disease, stroke, primary Raynaud phenomena, primary headache); autoimmune disorders (Sjogren syndrome, Henoch-Schonlein purpura, autoimmune thrombocytopenia, autoimmune thyroiditis, Parkinson's disease, idiopathic chronic urticaria, rosacea, alopecia areata); other disorders (iron deficiency anaemia, growth retardations, liver cirrhosis) (Prelipcean, *et al.* 2007). This multitude of potential links of *Helicobacter* sp. with different disease and malfunctions in our body, is a clear indication about the relevance of the alimentary microbiome, the all parts of it, with our general health.

# The PMEU Study

The PMEU method has been used for isolating the lactic acid bacteria from the gastric mucosa of healthy individuals (Hakalehto, *et al.* 2011). In the harsh conditions during the fasting the bacteria have to exercise several means for survival in the acidic, and hostile, environment. It has been shown that the *Helicobacter pylori* produces colonies which facilitate its survival of the gastric epithelia by the aid of their urealytic activity (Bury-Moné *et al.,* 2001, Lower, *et al.* 2011). Any fimbrial structures or adhesins would be likely to get dissociated by the low pH conditions (Hakalehto, et al. 2000). Therefore it could be understandable that the LAB (lactic acid bacteria) which could attach to the mucosal membranes by eg. protein capsules (S-layers), could be capable of residing in the acidic stomach for relatively long periods as indicated by our preliminary results (Hakalehto, 2011). Still it requires an explanation, how they protect themselves against the low pH. On the other hand, any bacterial colonies could be powerful in keeping up the suitable conditions for survival even amongst unfavourable surroundings. This kinds of functions have been recently documented in the case of Gram negative microbes surviving the osmotic stress by regulating the volume of the periplasmic space (Nester, 1982).

The experiments with the endoscopic samples were carried out from 13 healthy individuals (Hakalehto, *et al.* 2011). They did not have any chronic disease, and gastric ulcers or malignances where not found. Only two out of thirteen patients were the carriers of a *Helicobacter* strain. In this study it was revealed that in all three regions of the stomach, namely cardia, antrum and pylorus, several lactic acid strains were found by the PMEU. In Finland commercially available species of *Lactobacillus reuteri* was detected in epithelial samples from 0, 3 and 4 individuals, respectively. This bacterium seemed to have been surviving in the lower gastric areas. This finding underlined the possibility of such health promoting strains surviving for longer periods in the gastric mucosa. This observation seemed to suggest that the probiotic strains have the capacity to maintain their cellular integrity and vitality within the acidic gastric regions. This is an encouraging discovery and implied to the possibility of such health promoting strains surviving for longer periods of time on the gastric mucosa. Some species, such as *Streptococcus salivarius*, were likely to originate from the mouth. A few enterococci were also found. with the PMEU cultivator all patients were proved to possess gastric LAB's. Only one patient out of 13 did have the antrum germ-free in this experiment, and two patients did not give indication of bacterial growth in the pylorus.

# Host Body Functions and the Incoming Bacteria

Microbial surface structures facilitate their attachment onto the mucosal membranes (Hakalehto, 2000). The various protrudements, such as fimbriae or flagellins, as well as protein or polysaccharide capsules integrate them closely onto the epithelia. The extracllular polymeric substances (EPS) together with various metabolic products and other material form a matrix around the bacterial cells (Donlan, 2002). These biofilms are interspecies communities of micro-organisms on intestinal walls, and in the cavities therein.

The association of gut microbiological functions with the symphatic and parasymphatic nervous systems deserves a delicate attention (Nijhuis *et al.*, 2010). Any physiological conditions, such as malnutrition, stress or disease are having their neurological influences. These, in turn, provoke various effects on the gut flora. For example, continuous stress is stimulating acid production in the stomach, which cause a continuous strain for the duodenal microflora. Also counter-effects from the microbial populations on the host body could be important in describing the overall balance. In this case of excessive acidity, the acid and gas production by bacterial or yeast overgrowth could easily create overpressure, which is discouraging the apetite. Consequently, the eating habits could be somewhat influenced by the microflora. Intestinal ethanol production could cause neurological effects and thus increase eating.

The rise in the gastric pH increased both gastric and pulmonary bacterial number in young rats (Lee, 1995). It has been documented that there is a link between inflammations in the lugn and the alimentary tract. For example, tubular addition of 2,3 –butanediol into the mice intestines was preventing pulmonary infections. This substance is produced also by some other pathogens, such as *Serratia marcescens* and *Vibrio cholera* (Rao *et al.*, 2011). For a long time it was believed that the gastric areas were not containing bacterial communities. Firstly, it was established that *Helicobacter pylori* was a significant pathogen residing in the acidic human or animal stomach protecting itself by basic substances producing ureolytic enzymes. In squirrel monkeys gastric biopsy specimens contained also other ureolytic bacteria in 2 animals out of 12 (Khanolkar-Gaitonde, *et al.* 2000). More recently, the gastric mucosal membranes has been shown to contain several other bacterial phyla (44) besides the *H. pylori* in the case of South American patients, for instance (Maldonado-Contreras, *et al.* 2011). In animal studies it was documented much earlier that there have been found other permanent bacteria on the gastric mucosa of various animals, besides the *H. pylori* (Lee and O'Rourke, 1993).

# The Continuum of the Alimentary Processing of Food and Human Protection

Already in 1927 it was stated: "The duodenal tract of normal individual is an important part of the alimentary canal, not only because it receives secretions from neighboring organs, but also because it is the place of initial development of that vast column of bacteria numbering some thirty trillions, which is excreted daily in the fecal mass." (Kendall, 1927).

In the duodenum high microbial bile acid tolerance is almost a necessity for survival (Hakalehto, *et al.* 2010). In the bile samples around 1/3 of the microbial isolates were facultative coliforms, a half of which were mixed-acid fermenters and the other half more neutral substances producing strains, about 1/3 were enterococci. In the case of 14-day-old chicks the duodenal flora resembled that of human origin consisting of the facultative anaerobic strains of *Streptococcus*, *Staphylococcus*, *E. coli* and *Lactobacillus* (Salanitro, *et al.* 1978). Besides, these facultatives 9-39 % of obligate species were found in the ileum belonging to such genera as *Eubacterium*, *Clostridium*, *Propionibacterium and Fusobacterium.* In cecum of the young chicks, these strict anaerobes dominate. It is suprising to see the similarities between the microflora content of the chicken and human individuals. The former were fed by corn based diet, whereas the newborn children get their nutrition of the early days mostly from the milk. For some reason, the human intestines seem to get occupied with microbial strains in a much lower pace with periods of several months and even years needed for the maturation of the gut microflora (Sandin, *et al.* 2009).

Whether this prolonged duration of the settling down the intestinal microflora is due to the slower maturation of the immune system or to the infant special milk diet, or to both these factors, remains to be further elucidated. It could be hypothesized that the numerous substances of the breast milk are naturally protecting the gastrointestinal and respiratory tracts of the babies against infections. For example, lactoferrin is a natural host defense protein found in milk, having numerous physiological functions, such as antimicrobial, antitumor, antioxidant, and immunodeletory effects (Valenti and Antonini. 2005). The stomach pepsin enzymes are partially degrading the lactoferrin, producing even more potent antimicrobials, lactoferricin peptides. The activities of lactoferrin derived inhibition of various bacteria is based on their binding on the bacterial surfaces by electrochemical interactions, and subsequent disruption of the membranes of *E. coli*, *S. aureus* and *C. albicans* (Chen, *et al.* 2006). These activities have been shown to be functional within the alimentary tract (Chen, *et al.* 2010).

It has also been documented that more than 25 % of human milk sample contained risin – producing *Lactococcus lactis* (Beasley and Saris. 2004). This observation exemplified a case of reciprocally beneficial cooperation between man and his or her microbes. This risin-producing symbiotic strain could protect the mother and infant against harmful microbes, too. During the early months of life, the normal microflora has the opportunity to slowly develop, and to adapt to nutritional shifts later on. In case of chicken in eggs the newborn ones have been protected by the antibodies produced into the egg yolk, and surrounded by the avidin rich egg white blocking the microbial growth by binding all biotin necessary for microbial development [Hakalehto, 1998]. Then the IgY antibodies of the bird egg yolk could be added to the human intestines for its protection (Kovacs-Nolan and Mine. 2005). Similar functions have the antibodies produced into colostrums milk (Parreno *et al.*, 2010). Then the neonatal alimentary tract could also be protected with maternal flow of breast milk with probiotics, increasing the production of anti-infective molecules (Prescott, *et al.* 2008). In any case the host body produces also such effective antibiotic substances as the human defenses (Schröder and Harder, 1999). The antimicrobial activities of the HBD-2 (humanβ-defensin2) have recently been studied in the PMEU (Hakalehto, 2012).

# Conclusion

In order to fully understand the starting point for the microbiome development, we need to consider the material arriving and going out of the stomach. The gastric low pH is a blockade to many outside intrusions, because it destroys a majority of microbial cells within the food. However, in studies from endoscopic samples using the PMEU (Portable Microbe Enrichment Unit) we have discovered that many micro-organisms are able to survive on the gastric epithelia. These species include the harmful *Helicobacter* sp. but also beneficial lactic acid bacteria (LAB). There is a good reason to suppose that even more strains are to be found from this "deserted" region. There also the food substances meet body mechanisms for nutrient refinement for later uptake, and almost miraculously, regain the nearly neutral pH when arriving into duodenum after leaving the gastric area. Many microbes participate in the digestion thereof. There are also several immunological protection mechanism available for the safe food consumption starting from the many antimicrobials and antibodies in the breast milk.

# References

Andreica-Sandica, B; Panaete, A; Pascanu, R; Sarban, C; Andreica, V. The association between *Helicobacter pylori* chronic gastritis, psychological trauma and somatization disorder. A case report. *J. Gastrointestin. Liver Dis.,* 2011; 20, 311-313.

Antener, I; Tonney, G; Verwilghen, AM; Mauron, J. Biochemical study of malnutrition. Part IV. Determination of amino acids in the serum, erythrocytes, urine and stool ultrafiltrates. *Int. J. Vitam. Nutr. Res.,* 1981; 51, 64-78.

Beasley, SS; Saris, PE. Nisin-producing *Lactococcus lactis* strains isolated from human milk. *Appl. Environ. Microbiol.,* 2004; 70, 5051-5053.

Bergman, MP; Faller, G; D'Elios, MM; Del Prete, G; Vandenbroucke-Grauls, CMJE; Appelmelk, BJ. Gastric Autoimmunity. In: Mobley HLT, Mendz GL, Hazell SL, editors. *Helicobacter pylori: Physiology and Genetics.* Washington (DC): ASM Press; 2001.

Birek, C; Grandhi, R; McNeill, K; Singer, D; Ficarra, G; Bowden, G. Detection of *Helicobacter pylori* in oral aphthous ulcers. *J. Oral Pathol. Med.,* 1999; 28, 197-203.

Bull, AT; Quayle, JR. New dimensions in microbiology: an introduction. *Philos. Trans. R. Soc. Lond. B. Biol. Sci.,* 1982; 297, 447-457.Bury-Moné, S; Skouloubris, S; Labigne, A; De Reuse, H. The *Helicobacter pylori* UreI protein: role in adaptation to acidity and identification of residues essential for its activity and for acid activation. *Mol Microbiol.,* 2001; 42, 1021-34.

Canan, O; Ozcay, F; Ozbay-Hosnut, F; Yazici, C; Bilezikci, B. Value of the Likert dyspepsia scale in differentiation of functional and organic dyspepsia in children. *J. Pediatr. Gastroenterol. Nutr.,* 2011; 52, 392-398.

Chandrasoma, P; Wijetunge, S; Ma, Y; Demeester, S; Hagen, J; Demeester, T. The dilated distal esophagus: a new entity that is the pathologic basis of early gastroesophageal reflux disease. *Am. J. Surg. Pathol.,* 2011.

Chen, HL; Lai, YW; Chen, CS; Chu, TW; Lin, W; Yen, CC; Lin, MF; Tu, MY; Chen, CM. Probiotic *Lactobacillus casei* expressing human lactoferrin elevates antibacterial activity in the gastrointestinal tract. *Biometals,* 2010; 23, 543-554.

Chen, HL; Yen, CC; Lu, CY; Yu, CH; Chen, CM. Synthetic porcine lactoferricin with a 20-residue peptide exhibits antimicrobial activity against *Escherichia coli, Staphylococcus aureus*, and *Candida albicans. J. Agric. Food. Chem.,* 2006; 54, 3277-3282.

Colacci, E; Pasquali, A; Severi, C. Exocrine gastric secretion and gastritis: pathophysiological and clinical relationships. *Clin. Ter.,* 2011; 162, e19-25.

Danon, SJ; Lee, A. Other Gastric helicobacters and spiral organisms. In: Mobley HLT, Mendz GL, Hazell SL, editors. *Helicobacter pylori: Physiology and Genetics.* Washington (DC): ASM Press; 2001.

Deveci, MS; Deveci, G. Altered distribution of metaplastic Paneth, gastrin and pancreatic acinar cells in atrophic gastritic mucosa with endocrine cell lesions. *Tohoku J. Exp. Med.,* 2004; 202, 13-22.

Donlan, RM. Biofilms: microbial life on surfaces. *Emerg. Infect. Dis.,* 2002; 8, 881-890.

Esfahanizadeh, N; Modanlou, R. Correlation between oral hygiene and *Helicobacter pylori* infection. *Acta Med. Iran,* 2010; 48, 42-46.

Hakalehto, E, inventor. Syringe comprising an adhering substrate for microbes. US Patent No. 5,846,209. 1998 .

Hakalehto, E. Characterization of *Pectinatus cerevisiiphilus* and *P. frisingiensis* surface components. Use of synthetic peptides in the detection of some Gram-negative bacteria. Kuopio, Finland: Kuopio University Publications C. Natural and Environmental Sciences 112; 2000. Doctoral dissertation.

Hakalehto, E. Simulation of enhanced growth and metabolism of intestinal *Escherichia coli* in the Portable Microbe Enrichment Unit (PMEU). In: Rogers MC, Peterson ND, editors. *E. coli infections: causes, treatment and prevention.* New York, USA: Nova Publishers; 2011.

Hakalehto. E. Antibiotic resistant traits of facultative *Enterobacter cloacae* strain studied with the PMEU (Portable Microbe Enrichment Unit). In: Méndez-Vilas, A., editor. Science against microbial pathogens: communicating current research and technological advances. *Microbiology book series* Nr. 3. Formatex Research Center. Badajoz, Spain. In Press. 2012.

Hakalehto, E; Hujakka, H; Airaksinen, S; Ratilainen, J; Närvänen, A. Growth-phase limited expression and rapid detection of *Salmonella* type 1 fimbriae. In: Hakalehto E. *Characterization of Pectinatus cerevisiiphilus and P. frisingiensis surface components. Use of synthetic peptides in the detection fo some Gram-negative bacteria.* Kuopio, Finland: Kuopio University Publications C. Natural and Environmental Sciences 112; 2000. Doctoral dissertation.Hakalehto, E; Hell, M; Bernhofer, C; Heitto, A; Pesola, J; Humppi, T; Paakkanen, H. Growth and gaseous emissions of pure and mixed small intestinal bacterial cultures: Effects of bile and vancomycin. *Pathophysiology,* 2010; 17, 45-53.

Hakalehto, E; Vilpponen-Salmela, T; Kinnunen, K; von Wright, A. Lactic acid bacteria enriched from human gastric biopsies. *ISRN Gastroenterol,* 2011; 2011, 109183.

Hou, W; Schubert, ML. Gastric secretion. *Curr. Opin. Gastroenterol.,* 2006; 22, 593-598.

Hybenova, M; Hrda, P; Prochazkova, J; Stejskal, V; Sterzl, I. The role of environmental factors in autoimmune thyroiditis. *Neuro. Endocrinol. Lett.,* 2010; 31, 283-289.

Ishiguro, T; Takayanagi, N; Ikeya, T; Yoshioka, H; Yanagisawa, T; Hoshi, E; Hoshi, T; Sugita, Y; Kawabata, Y. Isolation of *Candida* species is an important clue for suspecting gastrointestinal tract perforation as a cause of empyema. *Intern. Med.,* 2010; 49, 1957-1964.

Kendall, J. The abuse of water. *Science,* 1927; 66, 610-611.

Khanolkar-Gaitonde, SS; Reubish, GK,Jr; Lee, CK; Stadtlander, CT. Isolation of bacteria other than *Helicobacter pylori* from stomachs of squirrel monkeys (*Saimiri* spp.) with gastritis. *Dig. Dis. Sci.,* 2000; 45, 272-280.

Kovacs-Nolan, J; Mine, Y. Microencapsulation for the gastric passage and controlled intestinal release of immunoglobulin Y. *J. Immunol. Methods,* 2005; 296, 199-209.

Krajewska, B; Brindell, M. Urease activity and L-ascorbic acid. *J. Enzyme Inhib. Med. Chem.*, 2011; 26, 309-18

Kubota, S; Yamauchi, K; Sugano, M; Kawasaki, K; Sugiyama, A; Matsuzawa, K; Akamatsu, T; Ohmoto, Y; Ota, H. Pathophysiological investigation of the gastric surface mucous gel layer of patients with *Helicobacter pylori* infection by using immunoassays for trefoil factor family 2 and gastric gland mucous cell-type mucin in gastric juice. *Dig. Dis. Sci.,* 2011.

Kumar, V. Pernicious anemia. *MLO Med. Lab. Obs.,* 2007; 39, 28, 30-1.

Laurent, M; Gogly, B; Tahmasebi, F; Paillaud, E. Oropharyngeal candidiasis in elderly patients. *Geriatr. Psychol. Neuropsychiatr. Vieil.,* 2011; 9, 21-28.

Lee, A. Animal models and vaccine development. *Baillieres Clin. Gastroenterol.,* 1995; 9, 615-632.

Lee, A; O'Rourke, J. Gastric bacteria other than *Helicobacter pylori. Gastroenterol. Clin. North Am.,* 1993; 22, 21-42.Lombardo, L; Leto, R; Molinaro, G; Migliardi, M; Ravarino, N; Rocca, R; Torchio, B. Prevalence of atrophic gastritis in dyspeptic patients in Piedmont. A survey using the GastroPanel test. *Clin. Chem. Lab. Med.,* 2010; 48, 1327-1332.

Lower, M; Geppert, T; Schneider, P; Hoy, B; Wessler, S; Schneider, G. Inhibitors of *Helicobacter pylori* protease HtrA found by 'virtual ligand' screening combat bacterial invasion of epithelia. *PLoS One,* 2011; 6, e17986.

Maldonado-Contreras, A; Goldfarb, KC; Godoy-Vitorino, F; Karaoz, U; Contreras, M; Blaser, MJ; Brodie, EL; Dominguez-Bello, MG. Structure of the human gastric bacterial community in relation to *Helicobacter pylori* status. *ISME J.,* 2011; 5, 574-579.Malfertheiner, P; Mégraud, F; O'Morain, C; Hungin, AP; Jones, R; Axon, A; Graham, DY; Tytgat, G. European *Helicobacter Pylori* Study Group (EHPSG). Current concepts in the management of *Helicobacter pylori* infection--the Maastricht 2-2000 ConsensusReport.*Aliment. Pharmacol. Ther.*, 2002; 16, 167-180. Marshall, BJ. One hundred years of discovery and rediscovery of *Helicobacter pylori* and its association with peptic ulcer disease. In: Mobley HLT, Mendz GL, Hazell SL, editors. *Helicobacter pylori: Physiology and Genetics.* Washington (DC). ASM Press. 2001; Chapter 3.

Menozzi, A; Ossiprandi, MC. Assessment of enteral bacteria. *Curr. Protoc. Toxicol.,* 2010; Chapter 21, Unit 21.3.Moran, AP. Cell surface characteristics of *Helicobacter pylori. FEMS Immunol. Med. Microbiol.* 1995; 10, 271-80.

Necchi, V; Candusso, ME; Tava, F; Luinetti, O; Ventura, U; Fiocca, R; Ricci, V; Solcia, E. Intracellular, intercellular, and stromal invasion of gastric mucosa, preneoplastic lesions, and cancer by *Helicobacter pylori. Gastroenterology,* 2007; 132, 1009-1023.

Nester EW, Pearsall NN, Roberts JB, Roberts CE. *The Microbial Perspective.* Philadelphia, PA, USA. CBS College Publishing. 1982.

Nijhuis, LE; Olivier, BJ; de Jonge, WJ. Neurogenic regulation of dendritic cells in the intestine. *Biochem. Pharmacol.,* 2010; 80, 2002-2008.Parente, F; Bianchi Porro, G. The (13)C-urea breath test for non-invasive diagnosis of *Helicobacter pylori* infection: which procedure and which measuring equipment? *Eur. J. Gastroenterol. Hepatol.*, 2001; 13, 803-806.

Parreno, V; Marcoppido, G; Vega, C; Garaicoechea, L; Rodriguez, D; Saif, L; Fernandez, F. Milk supplemented with immune colostrum: protection against rotavirus diarrhea and modulatory effect on the systemic and mucosal antibody responses in calves experimentally challenged with bovine rotavirus. *Vet. Immunol. Immunopathol.,* 2010; 136, 12-27.

Pietrzak, RH; Goldstein, RB; Southwick, SM; Grant, BF. Medical comorbidity of full and partial posttraumatic stress disorder in US adults: results from wave 2 of the National Epidemiologic Survey on Alcohol and Related Conditions. *Psychosom. Med.,* 2011; 73, 697-707.

Prelipcean, CC; Mihai, C; Gogalniceanu, P; Mitrica, D; Drug, VL; Stanciu, C. Extragastric manifestations of *Helicobacter pylori* infection. *Rev. Med. Chir. Soc. Med. Nat. Iasi,* 2007; 111, 575-583.

Prescott, SL; Wickens, K; Westcott, L; Jung, W; Currie, H; Black, PN; Stanley, TV; Mitchell, EA; Fitzharris, P; Siebers, R; Wu, L; Crane, J; Probiotic Study Group. Supplementation with *Lactobacillus rhamnosus* or *Bifidobacterium lactis* probiotics in pregnancy increases cord blood interferon-gamma and breast milk transforming growth factor-beta and immunoglobin A detection. *Clin. Exp. Allergy,* 2008; 38, 1606-1614.

Rao, B; Zhang, LY; Sun, J; Su, G; Wei, D; Chu, J; Zhu, J; Shen, Y. Characterization and regulation of the 2,3-butanediol pathway in *Serratia marcescens. Appl. Microbiol. Biotechnol.,* 2011.

Rosing, CK; Loesche, W. Halitosis: an overview of epidemiology, etiology and clinical management. *Braz. Oral. Res.,* 2011; 25, 466-471.

Salanitro, JP; Blake, IG; Muirehead, PA; Maglio, M; Goodman, JR. Bacteria isolated from the duodenum, ileum, and cecum of young chicks. *Appl. Environ. Microbiol.,* 1978; 35, 782-790.

Salminen, S; Isolauri, E; Onnela, T. Gut flora in normal and disordered states. *Chemotherapy,* 1995; 41 Suppl 1, 5-15.

Sandin, A; Bråback, L; Norin, E; Björkstén, B. Faecal short chain fatty acid pattern and allergy in early childhood. *Acta Paediatr.,* 2009; 98, 823-827.Savarino, V; Vigneri, S; Celle, G. The 13C urea breath test in the diagnosis of *Helicobacter pylori* infection. *Gut,* 1999;45, I18-122.

Schoenhofen, IC; Lunin, VV; Julien, JP; Li, Y; Ajamian, E; Matte, A; Cygler, M; Brisson, JR; Aubry, A; Logan, SM; Bhatia, S; Wakarchuk, WW; Young, NM. Structural and functional characterization of PseC, an aminotransferase involved in the biosynthesis of pseudaminic acid, an essential flagellar modification in *Helicobacter pylori*. *J. Biol. Chem.,* 2006; 281, 8907-8916.

Schröder, JM; Harder, J. Human beta-defensin-2. *Int. J. Biochem. Cell Biol.,* 1999; 31, 645-651.

Silva, FM; Navarro-Rodriguez, T; Barbuti, RC; Mattar, R; Hashimoto, CL; Eisig, JN. *Helicobacter pylori* reinfection in Brazilian patients with peptic ulcer disease: a 5-year follow-up. *Helicobacter,* 2010; 15, 46-52.

Tsuneishi, M; Yamamoto, T; Kokeguchi, S; Tamaki, N; Fukui, K; Watanabe, T. Composition of the bacterial flora in tonsilloliths. *Microbes Infect.,* 2006; 8, 2384-2389.

Valenti, P; Antonini, G. Lactoferrin: an important host defence against microbial and viral attack. *Cell Mol. Life Sci.,* 2005; 62, 2576-2587.

Wu, MS; Chow, LP; Lin, JT; Chiou, SH. Proteomic identification of biomarkers related to *Helicobacter pylori*-associated gastroduodenal disease: challenges and opportunities. *J. Gastroenterol. Hepatol.,* 2008; 23, 1657-1661.

Yan, Z; Young, AL; Hua, H; Xu, Y. Multiple oral *Candida* infections in patients with Sjögren's syndrome -- prevalence and clinical and drug susceptibility profiles. *J. Rheumatol.,* 2011; 38, 2428-2431.

In: Alimentary Microbiome: A PMEU Approach
Editor: Elias Hakalehto
ISBN: 978-1-61942-692-4
© 2012 Nova Science Publishers, Inc.

*Chapter V*

# Bacterial Interactions in the Small Intestine and Their Contribution to Host Nutrient Uptake

***Elias Hakalehto[1], Jouni Pesola[2], Eva M. del Amo[3], and Osmo Hänninen[4]***
[1]Department of Biosciences, University of Eastern Finland, FI, Kuopio, Finland
[2]Clinic of Children and Adolescents, Kuopio University Hospital, Kuopio, Finland
[3]Center for Drug Research, University of Helsinki, FIN, Helsinki, Finland
[4]Department of Pathology, University of Eastern Finland, FI, Kuopio, Finland

## Abstract

Bacteria and other microbes in the upper gastrointestinal tract, stomach and duodenum, have often been considered as almost indifferent, or even non-existent. However, the microbiology of these areas is the clue for understanding the entire microbial ecosystems of the alimentary tract. The low numbers of the microbes especially in the duodenum does not imply that the microflora and its composition there would be less important or insignificant. As a matter of fact, the situation is quite the contrary. In this chapter the essential importance of the duodenal microflora for the formation of all intestinal microbial community is elucidated. In these small intestinal environments also the highest amounts of nutrients are taken up by the body. Already at 1927 Kendall and coworkers made a statement: "The duodenal tract of normal individual is an important part of the alimentary tract, not only it receives secretions from the neighboring glandular organs, but also because it is the place of the initial development of that vast column of bacteria, numbering some thirty millions, which is excreted daily in the fecal mass." These huge masses of microscopic organisms essentially contribute on the nutrient absorption in the small intestines.

# Introduction

It is now widely recognized that the gut microflora participates in the nutrient uptake, and in their circulation in the alimentary ecosystems. In the PMEU studies we have simulated the interactions of various duodenal bacteria, such as the facultative anaerobes (Hakalehto *et al.*, 2008). These closely related strains, however, use variable pathways to ferment sugars: the members of the *Klebsiella/Enterobacter* group and other species commit the 2,3-butanediol (butylene glycol) fermentation balancing the more acidic products of the mixed acid fermentation carried out *e.g. Escherichia coli* into neutral substances such as butanediol and ethanol [Cheng, 2010],. They also produce remarkable amounts of hydrogen, which is removed. On the other hand, the carbon dioxide emission of this reaction could be lowered in specific duodenal conditions [Hakalehto *et al.*,, 2012]. By these metabolic operations these organisms carrying out 2,3-butanediol fermentation help to maintain the pH balance beneficial for all parties of this symbiosis, and also for the host. Simultaneously they decrease the osmotic pressure in the microenvironments. This metabolic cooperation could be the foundation of the BIB (Bacterial Intestinal Balance) [Hakalehto, 2011]. The members in that succession are all bile resistant, and they regulate the levels of other bacteria occurring in the mixed cultures [Hakalehto *et al.*, 2010]. It could well be postulated that in the BIB the common strains are not interfering with each other, but the diverse microflora quarantees the option for all strains for persistence. Several microbes are able to manipulate their microenvironment in order to facilitate their survival. By doing so they also maintain the status quo in the intestines which helps the host in the nutrient uptake.

The digestion of nutrients and much of the drug metabolism is carried out by the GI tract. Within the walls of the digestive tract there are numerous small glands secreting mucus and digestive fluids. Large glandular organs secreting digestive enzymes are large salivary glands, liver and pancreas. Also intestinal microbiota has multiple enzymes that are involved in the digestion of nutrients and also in the uptake of drugs. Each part of the alimentary tract is needed for the uptake to happen in a proper way. The specific characteristics of different parts – *e.g.* the acidity of the ventricle [Bonfils, 1979], secretion of biliary and pancreatic juices in duodenum, rapid passage through the small intestines with extensive absorption capacity and finally the rich microbiota of the colon – have been an important area of pharmaceutical research [Thelen, 2011]. There are major challenges but new kind of formulas and ways of drug actions have been found [Szabo, 2011].

When the partially digested food arrives as an acidic chyme into the duodenum, the bicarbonates secreted by pancreas, liver and gut neutralize the acidity [Shyr, 2002]. The bile acid compounds together with other defenses and effective hydrolysing enzymes restrict the microbial growth in this area [Johansson and Hansson, 2011]. Several antimicrobial proteins play an important role in keeping the bacteria at a distance. In the PMEU, the human defensin hBD2 was proven about 100 times more influential on bacterial cultures than the clinically used antibiotics [Hakalehto, 2011]. Then the normal flora of the mucosa walls keeps the pH around 6 by the interactions of the facultative coliformic bacteria [Hakalehto *et al.*, 2008]. These microbes together with the enterococci can withstand the bile acids in that area [Hakalehto *et al.*, 2010]. Any intruding bacteria in normal conditions are outpowered by the strong normal population prevailing in the gut. For example, as mentioned above such strains as the *Staphylococcus aureus* or *Bacillus cereus* strains could find out a niche beneath the symbiotic *E. coli* and *Klebsiella mobilis* strains in the PMEU. In these conditions the different bacteria establish an ecosystem, which maintains the balance between the members, and serves as protection against the challenges caused by environmental conditions [Hakalehto, 2011]. Together these strains form a more persistent community and ecosystem. The capabilities of different strains add defensive capability of the mixed population. Most likely the strains have been selected by the developing immune system in the early childhood [Pesola and Hakalehto, 2011]. The normal population is usually re-established after *e.g.* antibiotic medications or nutritional alterations according to the balance principle between different bacteria. As a general rule it could be stated: the more diverse the microbial community is, the more sustainable it is by maintaining the BIB. Thus it also resists the various environmental stresses in the gut. For the host the balanced intestinal flora is extremely important in maintaining the health.

## Bacterial Intestinal Balance (BIB) Is Essential for Our Welfare

When an individual bacterium is moving on the mucosa or in the intestinal fluids, it meets lots of influences that continuously are caused by other microbes, by the nutritional condition of the host, or by the health status of the host. Infections, for example, could be considered as the net effects of these numerous parameters. The bacterium has to be prepared to sudden changes on its way. The same is also true, when the same strain attaches onto the membranes, as the intestinal contents are in a constant movement. In fact, cells require for their orientation two cellular organelles, or appendages, flagellae and fimbriae [Curry *et al.*, 2006]. They provide tools for finding a suitable niche. The fibrils of the microbes are a part of the sensing system of the cell; this needs to be elucidated in detail for better understanding the behavior of an individual strain as a member of microbial community [McNab and Lamont, 2003]. After provisional attachment the microbes rapidly develop microcultures in layers, and eventually they form structured biofilms [Gomez-Aguado *et al.*, 2011].

Cell surfaces are the interphases, which the bacteria use also in communication. Some messages are addressed to the same species of cells, but it seems obvious that various bacteria could interpret also the molecular information originated from other species or groups, as well as the host's molecular communication [Lazar, 2011]. The alimentary microbiome forms an information network, which directs the entire system resembling a separate organism or organ [Lederberg, 2000]. Therefore, balance is desirable for the participating strains in the intestinal community. For our body, it is of utmost importance to build up a microbiological ecosystem into the alimentary tract in order to protect, or buffer, against diseases. The bacteria and other microbes participating in the establishment of this delicate network are crucial for our health. In fact, they seem to form the basis of it. There is an old Indian saying that constipation is the mother of all diseases. This could be broadened into a motto: "Any disturbance in the BIB is a threat for our well-being, and causes reactions on the entire body system level". Therefore, the bacterial interactions are important to be understood anywhere in the body. On the microbial level they constitute the interspecies metabolic machinery that is continuously in action in our alimentary tract. The bacteria do not recognize day or night, which makes them behave differently from our own cells. However, it is most likely, and also documented that our nervous systems can extend their regulation onto the microbial level, and *vice versa*. Similarly, the microbes are not outsides for our hormonal system, but send and receive signals in integrated fashion.

In order to be involved in the BIB, on the prevailing balance, the microorganisms, instead of being hostile, seem to most often strive for interdependence. This mode of action is constructive in nature. If the principal balance is disturbed, it is possible that the gut microbes, for example, form a "pathological balance". This is counterproductive for the host, and is threatening his or her health. For example, the yeasts are universal members of the human microbiota, but if acquiring a position out of control of the other normal flora, they will become causative agents for the yeast syndrome and dysbiosis [Caramia, 2009]. Then they become aggressors also against the entire ecosystem of the host.

## Neutralization of Gastric Hydrochloric Acid

Background is provided by the oral health. The alimentary tract is a column of different ecosystems [Kendall *et al.*, 1927]. The normally acidic stomach is a barrier in that chain. The chance has to be abrupt for the delicate joint defense between the microbiome and the host immunological and other mechanisms to have a "clean table". If this is not the case, a serious overgrowth of some strains could be the result.

Acidic gastric juice is effectively restricting the invasion of microbes into the lower alimentary tract. A mild hydrochloric acid (0.05M) has been used for detaching and dissociating the flagellae, and to isolate the flagellin protein molecules [Hakalehto *et al.*, 1984]. In these experimental conditions the flagellin proteins seemed to be firstly associated to the outer membrane vesicles. From these structures they were then liberated by boiling in the sodium dodecyl sulphate solution. This indicates the nonpolar nature of the unit proteins. In fact, this observation *in vitro* could indicate more robust structural integrity of the bacteria against acidic conditions than has been generally believed. Nevertheless, these influences on microbial cells and on their functionality degrade the abilities of infectious particles. Simultaneously, ingestion of food requires higher pH in the duodenum. Therefore, the pancreatic secretion of bicarbonate is rapidly elevating the pH, which is supported by the maintenance of the dualistic pH balance by the facultative microflora [Hakalehto *et al.*, 2008]. This cooperation between the body system and its microbes illustrates the great principle in the alimentary tract: decent health precludes good management of the microbiome.

Already in stomach enzymes start proteolysis in acidic conditions [Martos, 2010]. In duodenum other enzymes continue the process together with bacterial enzymes. For them it is of crucial importance that the environment permits the microbial cocultures in which delicate interactions formulate the metabolic pattern. The microbial community does not excessively strain the host tissues but live in harmony with it. Therefore, in a healthy gastrointestinal tract pathogens do not successfully grow. Intestines are the starting point for a joint struggle between the host and the micro-organisms to maintain the balance. The main human nutrients are polysaccharides, which also serve as the bacterial energy and carbon source. Both partners of the host-microbe interaction contribute to their hydrolysis, as human enzymes cannot degrade *e.g.* cellulose. Pancreatic lipases have a decisive role in fat hydrolysis [de la Garza, 2011].

# Host Defense Mechanisms Select the Intestinal Microflora

There are several mechanisms which the host uses to keep the epithelial surfaces clean enough for nutrient uptake. In practice this means first of all prevention of biofilm formation in those areas. Most likely the host is selecting immunologically the bacteria that are favored by it [Blandino, 2008]. This selection of "approved" organisms of the normal flora are "inherited", as the mother delivers many of the strains to the newborn baby during the birth [Jara, 2011]. During the early months mother and the infantexchange microbes effectively [Luoto, 2010]. The more diverse the normal flora is, the healthier the neonate usually is. During that period the individual immune system is developing, and the "original" strains get covered with some kind of protective "umbrella" being thus accepted as the constitutive members of the microbiome. In a recent study, the interactions of the neurological and immunosystems behind the overall protection are considered [Trakhtenberg and Goldberg, 2011]. The noradrenaline secreted from the peripheral nerve terminals in the spleen and liver regulate the cytokine secretion by stimulating specialized T cells. All these mechanisms form a network in a body system.

In the gut microbiology studies, the less dense population of the small intestine has often been neglected. Some reasons for that misleading lack of information are the difficulties in establishing reliable methods for studying *e.g.* the duodenal microbes. However, their influences on our nutrition and health may exceed their small numbers. Namely, during the nutrient uptake the food mass treated with hydrolytic enzymes passes the intestinal surfaces covered by villi and microvilli [Brouns, 1993]. For an illustration of the small bowel surface, see the Figures 1 and 2. The bacteria and other microbes in these areas and on the surfaces multiply rapidly, and also penetrate the food in the intestinal fluid. They are processing the nutrients during the transport of the chyme and the particles in it. In that point, it is of utmost importance for our health that the balance of the microecosystem is maintained.

In studies with the PMEU, where the conditions of the small intestines have been simulated, it was found that the various enterobacterial strains participate in maintaining the pH balance inside the gut [Hakalehto *et al.*, 2008]. These strains are bile resistant, and one important function of the bile seems to be the controlling the microbial flora. After the harsh restriction by the gastric hydrochloric acid secretion, the presence of bile substances is maintaining the duodenal microflora structure. In the PMEU simulations it has also been detected that any incoming strain is controlled by the dominating flora as well, not necessarily by destroying the newcomer, but by the balancing action [Hakalehto *et al.*, 2010]. This is made possible in the conditions of excessive and rapidly flowing nutrients, which is simulated in the PMEU [Hakalehto *et al.*, 2012]. In the studies with neonates, the influences of the diet were clearly observable in a PMEU survey of the family *Enterobacteriaceae* in the fecal microflora [Pesola *et al.*, 2009].

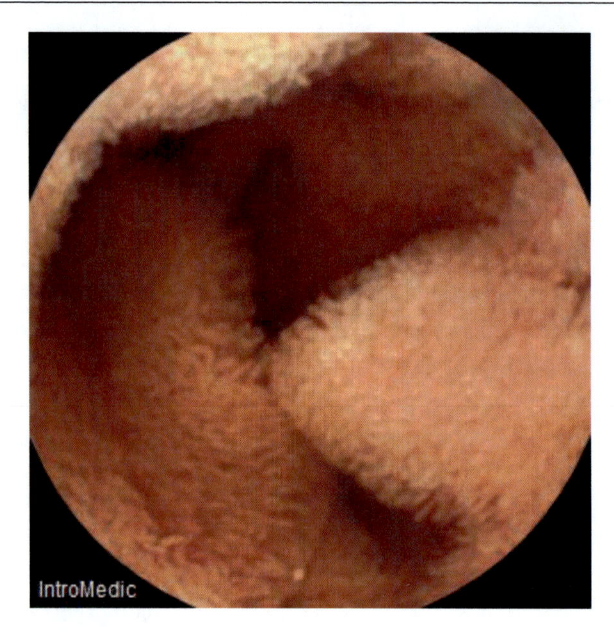


Figure 1. The photo taken by a capsule camera system of the folded small intestines covered with villi. The enormous intestinal surface area is facilitating the adsorption of vital nutrients, and is also provoking mutual symbiotic relationships between the host and the microbiome.

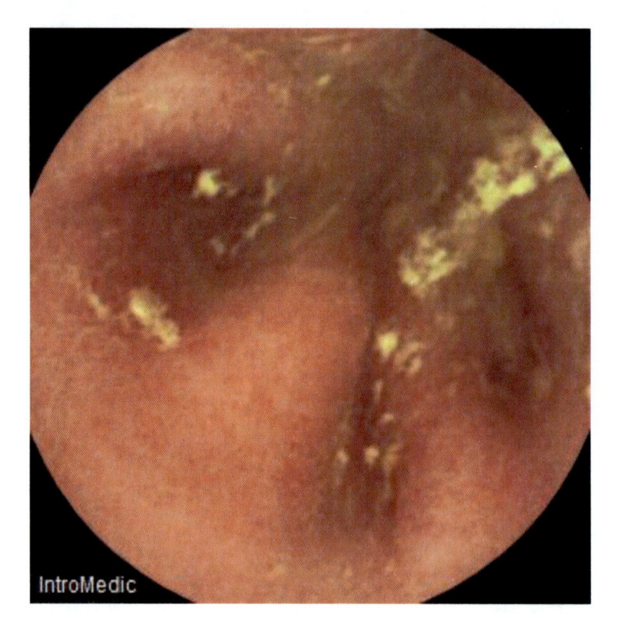


Figure 2. The small intestinal chime. Photo taken by capsule camera system. The chime is traversing through and onto the surfaces where the nutrient adsorption (and also microbiological action) are taking place with ultimate speed.

Besides the immunological selection, bile influence and intestinal movements several other host factors are controlling the microbial numbers and the microflora composition [Abeysuriya, 2008; Hakalehto, 2011; Johansson and Hansson, 2011].

Defensins are effective antimicrobial peptides produced by different human organs, including the intestines [Masuda, 2011]. They have been shown to be functional in lower concentrations than the antibiotics [Hakalehto, 2011]. The host body is secreting slime, which is preventing microbes physically, and by its various molecular components [Johansson and Hansson, 2011].

# Host Enzymes Contribute to Microbial Community

Enzymes dramatically lower the amount of energy needed for reactions to occur. Without them, many reactions simply would not take place in our body. Also the gastrointestinal tract is secreting numerous enzymes to enhance nutrient absorption [Peuhkuri, 2011]. The enzymes of the human digestive tract hydrolyze the macromolecules, modify or transfer the nutrients for uptake, protect the body against harmful substances [Andrievsky, 2009]. Also several enzymes of microbial origin take part in the food processing in the alimentary tract [Thomas, 2011]. Enzymes need other substances to accomplish their tasks, such as vitamins and certain minerals. These substances often serve as coenzymes and enzyme activators. For the enzymatic action the physicochemical conditions in the gut should be favorable for the reactions.

## Digestion of Nutrients

*Carbohydrates:* The most important part of digestible nutritional carbohydrates are starch and glycogen. Salivary and pancreatic amylases mostly release maltose that is digested to two glucose molecules by maltase of the microvilli of the mucosal cells of the upper part of the small intestine [Mifflin, 1985]. Bacterial enzymes also contribute to this reaction. Cellulose is not hydrolysed by human enzymes but intestinal microbes digest it in some extent. Also some other plant fibres such as hemicellulose and lignin, are digested by the intestinal microbes. Saccharose is digested to glucose and fructose by saccharase of the microvilli of the small intestine [Bouhnik, 1996]. Lactose is digested to glucose and galactose [Englyst, 1985].

*Proteins:* Proteolytic enzymes are 1) pepsin in the gastric juice, and 2) trypsin and chymotrypsin in the pancreatic fluid. Besides food substance, also microorganisms are subjected to gut proteasis [Trěek, 2011]. These enzymes digest the proteins to peptides containing 2-6 amino acids. These are digested further by peptidases of the microvilli of the small intestine. Again bacterial enzymes also contribute to this.

*Lipids*: The lipase of the pancreatic fluid is able to digest the triglyserides emulsified by bile salts to monoglyserides and free fatty acids [Musso, 2011].

The daily volumes of intestinal juices secreted to the alimentary tract are shown in Table 1. Bacteria in gut lumen meet and mix with these secretions. The most important digestive enzymes are collected to Table 2.

Microbes carry out an important task in the intestines by adjusting the conditions in the gastrointestinal tract. Similarly, in the case of malfunctioning microflora, the disadvantageous effects are directed also toward the enzyme systems of the host. Therefore, the balanced interaction between humans and our microbes is extremely important for the overall health, and for our nutrition.

**Table 1. Daily secretions of intestinal juices [Guyton, 1986]**

| Intestinal juice | Daily volume (ml) | pH |
|---|---|---|
| Saliva | 1000 | 6.0-7.0 |
| Gastric secretion | 1500 | 1.0-3.5 |
| Pancreatic secretion | 1000 | 8.0-8.3 |
| Bile | 1000 | 7.8 |
| Small intestinal secretion | 1800 | 7.5-8.0 |
| Brunner's gland secretion | 200 | 8.0-8.9 |
| Large intestinal secretion | 200 | 7.5-8.0 |
| Total | 6700 | |

**Table 2. The most important digestive enzymes of the alimentary tract [Nienstedt *et al.*, 1989]. The digestive enzymes of minor importance are shown in brackets. The microbial enzymes complement these**

| | Saliva | Ventricle | Bile | Pancreatic fluid | Intestinal microvilli and intestinal fluid |
|---|---|---|---|---|---|
| **Carbohydrates** | Amylase | | | Amylase | Disaccharidases: maltase, lactase, saccharase etc., (intestinal amylase) |
| **Proteins** | | Pepsin | | Trypsin etc. | Peptidases |
| **Lipids** | (lipase of lingual glands) | (lipase of ventricle) | Bile salts | Lipases Phospholipases | |

## Secretion of the Saliva

Saliva is secreted from small and large salivary glands. The small ones are within the mucosa of lips, cheeks, palatinum and tongue. The big ones are parotic glands, submandibular glands and sublingual glands. The salivary glands contain different cells secreting either mucosal or serosal saliva. The serosal saliva contains also amylase that is capable to digest starch. Mouth fluid is rich in bacteria.

During eating the salivary secretion is biphasic. In the beginning the saliva consists of mucus, amylase and a lot of sodium and bicarbonate ions. If the secretion is strong, the final saliva is quite unchanged. However, if the flow of the saliva is slow, some sodium, bicarbonate and chloride ions are replaced by potassium ions within the ducts of the salivary organs [Holmgren and Ohlsson, 2011].

Because food is swallowed so quickly, the salivary amylase normally works mostly in ventricles. Before the low pH of the ventricle inactivates the enzyme it remains active inside the food bulk being able to digest even 40% of dietary starch to disaccharides [Gentilcore, 2011].

## Tonsils of the Pharynx

There are four tonsils around the pharynx: a pair of palatinal tonsils, lingual tonsil and pharyngeal tonsil. Their mucosal surface is covered by stratified epithelium and they consist mostly of lymphoid tissue with a lot of lymphoid cells also within the epithelium. There are many crypts or holes in the epithelial surface of tonsils. These structures are important in the recognition and presentation of microbes entering the body via oral or nasal route for the lymphatic system [Yamanaka, 2011]. They may also be like hiding places for some microbes.

## Secretion of Gastric Fluid

The daily amount of secreted gastric fluid is 1.5-3 liters. There are two different types of tubular glands within the gastric wall. The oxyntic glands, which mean acid-forming glands, secrete hydrochloric acid, pepsinogen, intrinsic factor and mucus to crypts of the ventricular wall of corpus and fundus of the ventricle. The pyloric glands located in the antral and pyloric portion of the stomach secrete mainly mucus for protection of the pyloric mucosa but also some pepsinogen and hormone gastrin [Ikeda, 2011].

The oxyntic (or parietal) cells of the oxyntic glands secrete hydrochloric acid with pH 0.9. They also secrete intrinsic factor that binds $B_{12}$-vitamin and is essential for the absorption of this vitamin from the ileum. The peptic (or chief) cells and some other cells secrete pepsinogen that transforms to active proteolytic enzyme, pepsin, in the presence of hydrochloric acid. Pepsin remains active only in acidic environment. Mucous neck cells of the oxyntic glands secrete mucus [Abduinour-Nakhoul, 2011].

The acidity of the gastric juice is an important infection defense mechanism. Some microbes may however pass through the ventricle and enter to the duodenum. This may happen when some contents of the ventricle remain less acidic because of large meal that can neutralize and dilute the gastric juice. The passage is even easier for microbes when effective antacidic drugs, like $H_2$ antagonists or proton-pump inhibitors are used.

Some microbes, like mycobacteria, can stand the acidity of the ventricle better than others. *Helicobacter pylori* is frequently detected in the samples taken from gastric mucosa. It can resist the gastric acidity by several mechanisms [Sachs *et al.*, 2003].

## Liver and Bile Secretion

Liver is involved in the digestion by secreting bile that is needed for the emulcification of nutritional fats. If no gall enters the gut the absorption of both the dietary fats and also fat-soluble vitamins is deficient.Some waste materials, like bilirubin diglucuronide, which is a metabolite of hemoglobin and myoglobin is secreted in the bile [Gravante, 2010]. Bile enters the duodenum through the biliary tract at the anatomical structure called papilla Vateri. 0,5 – 1 liters of bile is secreted from the liver per day. It is stored in the gall bladder and concentrated. After meal the gall bladder is emptied. The contraction of the muscles of the wall of the gall bladder is mediated by an intestinal hormone called cholecystokinin. Fats and proteins of food trigger the mucous membranes of the intestinal wall to secrete this hormone.

The gall contains only minor amounts of enzymes. But the bile salts prepared from cholesterol by the liver cells are essential for the absorption of the dietary fats. The lipid molecules join the bile salts and other substances and form small micelles that mix with the watery fluid of the small intestines. As a result an emulsion is formed that is important for the action of pancreatic lipase that is capable of digesting fats only in the fat-water borderline. After digestion the fats are absorbed to the cells of the mucous membrane of upper part of small intestines. Bile substances promote the fat uptake. Entero-hepatic circulation is important with respect to the bile salts [Johnston, 2011]. Over 90 % of bile salts are absorbed in the lower part of the small intestines. Intestinal bacteria transform bilirubin diglucuronide to urobilinogen that is partly absorbed back to the gall [Kataoka, 2011].

The venous blood flow from mesenterial area is directed to the liver by portal vein. A big proportion of the cells of the liver belong to the reticuloendothelial cells. They are important defense mechanisms against the pathogens entering from intestines and also in the removal of old and damaged blood cells.

### Pancreatic Fluid

Pancreas has both endocrine and exocrine functions [Chen, 2011]. The endocrine cells form only 2 % of the pancreas that secrete insulin, glucagone and somatostatin. 0,5 – 2 liters of pancreatic fluid is produced daily and secreted via pancreatic duct to duodenum. There are two different types of exocrine cells in the pancreas, duct cells and acinar cells. The secretion of the flat duct cells is rich in bicarbonates and thus alkaline that is important in the neutralization of the acidic digesta entering the duodenum from the ventricle. The secrete of acinar cells contains different enzymes capable of digesting all the most important dietary compounds: trypsin and chymotrypsin are proteolytic, pancreatic amylase digests carbohydrates as well as the salivary amylase and lipase digests the lipids emulsified by the bile salts. The pancreatic fluid contain also nucleic acid digesting enzymes

The proteolytic enzymes of the pancreas are secreted through the pancreatic duct in an inactive form and activated in the duodenum by enterokinase secreted from the intestinal wall. Pancreatic enzymes are very potent having also strong antimicrobial properties.

## Structure and Function of Small Intestines

The length of the small intestines is some 3 meters. The first 20-30 cm belong to the duodenum, next 2/5 of the small intestines to the jejunum and the rest of the small intestines to ileum. In fact, the most important sections of the small bowel with respect to nutrient uptake are the two first parts. Especially it should be emphasized that the digesta needs to be as completely homogenized as possible. If the digesta entering the ileum is containing undigested food, this becomes somewhat poisonous for the organism.

The upper part of the duodenum is acidic because of the acidic digesta entering from the ventricle. The lower part of the duodenum is already neutral because of the addition of biliary and pancreatic fluids to the digesta [Olsen *et al.*, 1985]. Duodenal facultative bacteria are also maintaining the correct pH in the upper intestines [Hakalehto *et al.*, 2008].

90% of digestion and absorption takes place in the small intestine. Most of the absorption takes place in duodenum and the upper part of the jejunum. As a sign of active absorption a lot of ring-like gatherings of the mucosa are found in the jejunal wall.

In all parts of the small intestines the mucous membranes have also Peyer's patches containing lymphoid tissue. However, the lymphoid tissue is especially prominent within the wall of ileum [Storey, 2009]. These traits showing immunological activity of the ileum reflect to large microbial population in the area.

In addition to the glandular organs of the alimentary tract like gall bladder and pancreas, also the cells of the intestinal wall secrete digestive enzymes [Okawa, 2009]. When these cells are dropped off from the mucosal epithelium the enzymes of the cells remain active for some time. The enzymes secreted by these cells are peptidases, that digest polypeptides to tripeptides, dipeptides and aminoacids, and disaccharidases, that digest disaccharides (maltose, lactose, saccharose) to monosaccharides.

## Structure and Function of Colon

The length of colon is a bit more than a meter. The principal parts of it: cecum, colon ascendens, colon tranversum, colon descendens, sigmoid colon and rectum. From the lower part of the cecum starts appendix that has a narrow lumen and walls consisting plenty of lymphoid tissue that resembles that of the tonsils of the pharynx. In appendicitis the lumen of the appendix may get swollen when the inflammatory secretions cannot get out. If not operated early enough this may lead to a perforation of appendix resulting peritonitis. During an observation period of 21 years (1987–2007), as many as 186,558 appendectomies were performed in Finland, of which 137,528 (74%) cases were reported as acute appendicitis [Ilves *et al.*, 2011].

The goblet cells of the colon secrete mucus that makes the feces slippery and protects the mucous membranes [Lesuffleur, 1991]. Practically only sodium chloride and water are absorbed from the colon.

The colon contains a large amount of fecal mass with the richest and most numerous microbiota of intestines. However, it reflects the flora of the more proximal intestines. That is why the fecal samples can be used for an indirect estimate of the microbiota of the intestines. However, monitoring of the microbiome of the small intestines is difficult by fecal analysis only.

## Hormones of the Alimentary Tract

Tens of hormones of alimentary tract are known [Sam, 2011]. Duodenal mucosa secretes secretin produced when acidic digesta enters duodenum. Secretin induces the secretion of ductal cell pancreatic fluid that contains a lot of sodiumbicarbonate but only scarcely enzymes. Fats and partly proteolysed proteins induce the production of pancreozymin from the wall of the duodenum. This hormone induces secretion of acinar cell pancreatic fluid that is rich in the enzymes. Also parasympathetic activity stimulates the acinar pancreatic fluid secretion.

The different enzymes functioning in the GI tract include enzymes of both host and microbe origin. Amylases and other starch degrading enzymes start their operations already in the mouth, where these hydrolyzing enzymes are secreted by the salivary glands [Beltzer,

2010]. The gastric protein digesting enzymes are working well in the low pH [Schubert, 2011]. Lipases of the pancreas are secreted for breaking up the fats [Smith, 2005]. Different enzymes are processing various disaccharides, such as sucrose, lactose and maltose, which are present in different foods [van Wijck, 2011]. Cellulases of microbial origin are breaking down some of the cellulose fibers, even though most of these structures are rather slowly degrading [Mittal, 2011]. The multitude of enzymatic action in the gut is serving the rapid exploitation and uptake of nutrients by the host body. Deficiencies in their activities could lead to various symptoms, such as constipation, bloating, flatulence and acid reflux [Celik, 2006]. These effects, in turn, tend to interfere with the balances between micro-organisms. This could cause yeast and bacterial overgrowth in the intestines. If the microbes grow actively, and without limits, the volatiles and gases produced by them could cause overpressures, which prevents eating in the worst cases. Experiencing glucose satiety is operating by "pressure sensors" in the stomach with tissue hormons involved [Rayner, 2000]. Extensive production of acids could also be a consequence of unbalanced microflora composition and its missdirected action, which have negative influences on the host health [van Vliet, 2001].

## Establishment of Bioprocess Technologies on the Basis of Intestinal Microbial Metabolism

Such concepts as bioprocess or biochemical engineering as well as industrial microbiology or microbial biotechnology are closely related with the organisms familiar from the intestinal microbiome. Their physiology and metabolism is therefore highly interesting also from the point of developing technologies for biorefinery applications [Hakalehto *et al.* 2012]. In the industrial processes, as like in the alimentary tract, the microbial activities are boosted with the digestive enzymes. Research on the interactions in the intestines could thus produce useful applications for sustainable industrial processes.

## Small Intestinal/Bowel Bacterial Overgrowth (SIBO, SBBO)

Regardless of the many host defense mechanisms, the human small intestines offer space for some microflora, which could have many specific and essential functions for the host well-being [Hakalehto *et al.*, 2008; Hakalehto, 2011; Hakalehto, 2012]. It is rich of nutrients, and there prevail permissive conditions for rapid bacterial growth when the food is available. Therefore, the microbial strains present in this area are able to speed up their growth and metabolism in a manner resembling that of the microbial ecosystems in the tropical rainforest soil. The small intestines have a surface area equalling with one football field [le Roux, 2010]. In the normal conditions, this part of the intestines harbour low bacterial contents (less than 1000 cells / ml), which is believed to be crucial for nutritional uptake and epithelial barrier integrity [Croft, 1973]. However, in the case of food delivery, this bacterial population mainly consisting of bile-resistant strains is believed to be in a rapidly growing and metabolizing state [Hakalehto *et al.*, 2008; Hakalehto *et al.*, 2010; Hakalehto, 2011]. In normal conditions, this occasional overgrowth soon disappears when the nutrients are not any more available, and the bacterial flora is flushed away downward in the alimentary tract. The defense systems of the body, such as antibodies and defensins and bile digesting enzymes then clear the surfaces from excessive growth [Kim, 2011]. Only a minor reservoir of organisms remains attached for the next overflow of the food substances during the subsequent meal portion. Despite the defenses, in some cases Small Intestinal Bacterial Overgrowth (SIBO) can result by abnormally high presence of diverse species of bacteria in the small intestine. SIBO is defined as a bacterial count in excess of $10^5$ CFU/ml of the jejunal fluid [Toskes, 1993]. The common reasons for SIBO have been believed to be anatomic blind loops and reduced peristaltic movement of the small intestine [Wang, 2005]. The disturbed balance of intestinal microflora could well be added to these causes of illness. In fact, it is disputable whether the bacterial or other microbial overgrowth is ultimately leading to inflammation, or *vice versa*. In any case, the complexity of the phenomena in the microbiome can well explain the multiple views arising around this matter.

Terminal ileum or proximal area of ileoceacal valve shows a state of transition and represents bacterial flora of colonic origin like *Bacteroides* sp. or members of *Enterobacteriaceae* and strict anaerobes in range of $10^5$-$10^7$ cells/ml of ileal juice [Rambaud, 1992]. The composition of this flora is also a result of the flow from the upper gastrointestinal areas, and thus a result of earlier metabolic events and interactions (Hakalehto *et al.*, 2010). Conventional microbiological methods are not free from problems in applying for the identification and isolation of the small intestinal flora. Much of the intestinal flora is difficult to collect and even non-cultivable. More accurate culturing methods could introduce new options for diagnosis and understanding the pathophysiology of SIBO.

We have used the PMEU technologies for simulating the conditions in the intestines, and to monitor microbial interactions there [Hakalehto *et al.*, 2010; Hakalehto, 2011]. These methods could also be used for pre-enrichment prior to the genetic analysis [Pitkänen *et al.*. 2009]. The flora so obtained could also be genotyped using techniques like Pyrosequencing/Denaturing Gradient Gel electrophoresis (DGGE) or Pulsed-field gel electrophoresis (PFGE) [Kim, 2011]. In fact, by these techniques it is possible to get enriched as many as 14 LAB (Lactic Acid Bacteria) strains from such hostile samples as the gastric biopsies of one individual (unpublished results).

Even though the etiology of SIBO is not known, it is able to cause very severe consequences. This disease affects a large global population and leads to malnutrition, morbidity and Inflammatory Bowel Disease (IBD) [Gerasimidis, 2011]. According to some estimates 1.80 million deaths occur worldwide due to diarrheal diseases [Morens *et al.,* 2004]. For the diagnosis of SIBO, non-invasive hydrogen breath tests have been developed [Walters and Vanner, 2005]. In the PMEU technology rapid verification of volatiles and exhaust gases released by the microflora is accomplished by the use of the PMEU Scentrion® device [Hakalehto *et al.*, 2009]. The effects of various other pathological conditions of the gastrointestinal tract are presented in Chapter 8 of this book. An overall principle is that many such conditions could be consequences of the bacterial or other microbial strains growing fiercely in the intestines, but without attempting to penetrate into the host tissues. In this case these microbes remain somewhat unnoticed by the host immune system and other defences. However, such fermentation process inside the body could cause many of the symptoms, which are often called as "functional".

# Bile Acids and Role of Microbes in Bile Circulation

In the intestines there are several interfaces which all have utmost importance for the survival and activity of all micro-organisms. They inhabit the fat-water interfaces having hydrophobic surfaces [Ellwood *et al.,* 1982]. These strains are important in the degradation and assimilation of lipids in the gastrointestinal tract [Armand, 1999]. For example, there is a sensitive balance between the supply and catabolism of the cholesterol [Dietschy *et al.,* 1993]. Excessive cholesterol in the body is mainly converted to bile acids in the liver [Russell and Setchell, 1992]. After meals, gallbladder secretes bile acids, conjugated with taurine or glycine in liver, to duodenum through papilla Vateri. They are converted to secondary and tertiary bile acids by intestinal bacteria. The bile acids emulsify the lipids in the intestines forming the above-mentioned interphases [Hylemon *et al.,* 2009]. They also act as signal molecules for the lipid uptake.

Many bacteria have the capacity to deconjugate the bile acids [Brashears, 1998]. The primary bile acids, cholic and chenodeoxycholic acid are synthesized *de novo* in the liver from cholesterol. The hydrophobic steroid nucleus is conjugated as an N-acyl amidate with either glycine (glycoconjugated) or taurine (tauroconjugated) prior to secretion [Mihalik *et al.*, 2002]. Conjugation makes bile acids amphipathic and solubilizes lipids to form mixed micelles that are then absorbed in the intestines [Kohan, 2010].

Some bacteria in the colon flora that possess bile salt hydrolase (BSH) enzymes have capacity to deconjugate these bile salts [Masco *et al.*, 2007]. The mechanism involves the hydrolysis of peptide linkage of conjugated bile acids between bile acid carboxyl and amino group of taurine or glycine, thereby liberating the amino acid moiety. This is also called as the "gateway" reaction since it affects a much wider pathway of host physiology by the gut flora [Rugutt and Rugutt, 2011]. Any malfunctioning of the bile acids has thus significant impacts on the host besides the nutrient uptake. Therefore, it could be supposed that the body system to some extent selects the microbial species in the gut. The underlying mechanisms are most likely related to the development of our immunosystem, and they need to be further elucidated. Therefore, it is important to study further the flora in case of dysregulation in the bile axis. Many pathogens, but also commensal or probiotic strains are able to deconjugate the bile salts. One important aspect could be that the deconjugation should be restricted to the areas where the bile substances are secreted from the papilla Vateri into the duodenum for securing a properly functioning bile circulation mechanism. That is promoted by the delicate bacterial balance [Hakalehto *et al.,* 2010]. It has also been shown that excessive small bowel bacterial overgrowth is causing excessive bile salt deconjugation and dehydroxylation, which prevent effectively the bile acid reuse, and cause also potentially adverse effects in the colon [Morotomi *et al.*, 1996]. It is noteworthy that the coliforms participating in the duodenal pH regulation are not deconjugating the bile [Shindo and Fukushima, 1976].

The facultative anaerobic coliforms have been shown to contribute to the maintenance of a dualistic balance in the duodenal tract, and keep the pH around 6 [Hakalehto *et al.,* 2008]. It can also be deduced that any discrepancies in the dualistic balance may also hamper the fat uptake and other digestive body functions, for instance by preventing adequate formation of fat micelles [Bouchoux, 2011]. These imbalances may indicate the SIBO. Consequently, this kind of change or disturbance has an impact on the microflora as a whole, which might accelerate the vicious circle toward imbalance, both from the host point of view and in the bacterial perspective. However, bile acids function as signalling molecules regulating their own biosynthesis, and they both control cholesterol homeostasis and exhibit potent innate immune functions, thereby affecting body functions much wider than on the digestion only [De Fabiani, 2010].

Deoxycholic acid is a secondary microbial product of the primary biliary cholic acid via 7-alpha-dehydroxylation in the intestines. It has been observed that gallstone patients have higher levels of 7-alpha-dehydroxylating fecal bacteria and appear to harbor only members of the genus *Clostridium* with this activity [Wells *et al.,* 2000]. It has also been shown recently in a PMEU simulation that carbon dioxide produced by the host normal flora members, such as lactobacilli, could prevent the excessive colonization of clostridia in the gut by boosting them into active growth phase [Hakalehto and Hänninen, 2012]. This is an example of the microbial interaction in the intestines favouring the host, and maintaining his or her health.

The bile substances are circulated, the ileum being their main reabsorption site [Van Deest *et al.,* 1968]. Some results indicate that one of the predominant bacteria in human feces, *Bacteroides* sp., comprise one of the main bacterial groups responsible for the deconjugation of bile acids [Narushima *et al.,* 2006]. Since deoxycholic acid, for instance, is dissolved into ethanol and acetate, these metabolites of the facultative anaerobic bacteria may thus contribute to the fate of the lipids in the intestinal environment in an intriguing manner. In one study fat absorption was studied in 10 patients recovering from an episode of acute infectious gastroenteritis who failed to gain weight despite adequate caloric intake [Jonas *et al.,* 1979]. The administration of a test meal demonstrated a marked deficiency of duodenal bile acid concentration and of fat incorporation into the micellar phase in patients. Ileal dysfunction and associated bile acid loss are possible causes of disturbed fat assimilation following acute intestinal infection in children.

As surfactants, bile acids are potentially toxic to the cells. Therefore their levels need to be regulated. It has been postulated that excess bile acids in large intestine may contribute to the colon cancer [Kasbo *et al.,* 2002]. Their reabsorption and the collection of the majority of the bile acids in the ileum are important, both for more effective reuse and for avoiding these adverse effects in the colon area. In fact, the bile acid pool was diminished by 88% in the small intestine of a dog in 6 hours, which portion was obviously recollected from the gut for reuse [Scott *et al.,* 1983]. Patients with colorectal cancer have the highest concentrations of neutral animal sterols, the lowest degree of esterification of neutral sterols, the lowest relative amount of saponifiable bile acids, and the highest concentrations of unconjugated primary bile acids [Korpela *et al.,* 1988].

Besides many *Bacteroides* species, also some lactobacilli have been shown to carry out this function of bile acid deconjugation, for which the optimal pH was 6.0 [Gilliland and Speck, 1977]. It is close to the final pH values achieved in our experiments in anaerobis with mixed cultures of *E. coli* and *K. mobilis* after the active growth phases in the PMEU [Hakalehto *et al.,* 2008]. This makes it lucrative to believe that the action of facultative bacteria by producing such solvents as ethanol and 2,3-butanediol could assist in bile regulation by both adjusting the pH and by making the liquid contents in the small bowel more fluidic for the fats and bile substances, which consequently would assist their assimilation in the micellar form. On the other hand, it has been shown that excessive small bowel bacterial overgrowth is causing bile salt deconjugation and dehydroxylation. It can also be deduced that discrepancies in the dualistic pH balance may also hamper this body function, thus preventing adequate bile reabsorption and excretion [Hakalehto *et al.*, 2008].

Consequently, this kind of negative change has an impact on the microflora as a whole, which might accelerate the vicious circle toward imbalance, both from the host point of view and from the bacterial perspective.

The nutritional factors also influence strongly on the development of the microflora. For example the African children in Burkina Faso have almost no proteins or animal fats in their diets, and concequently their intestines are colonised with *Prevotella* sp. bacterium and related species [De Filippo *et al.*, 2010]. Some microbes cannot stand the lipases, phospholipases, lysolecithines, free fatty acids and proteinase or peptidases. Therefore, they are not necessarily present or abundant in our western diets. Our intestines are correspondingly often colonised with high *Bacteroides* levels. These above-mentioned factors then partially explain the scarcity of microbes in the upper intestines – together with bile secretion. The various microbiome types are introduced by Gophna [2011] in a recent article reviewing the work [Arumugam *et al.*, 2011]. Gnotobiotic, germ-free animals, SPF (Specified Pathogen Free) animals, and specified microflora associated animals could be used as tools in the studies of microbial colonization and pathogenesis [Hänninen *et al.*, 1979; Hänninen *et al.*, 1987; Pelkonen and Ylitalo, 1989]. With these systems it could also be possible to observe the formation of antibodies secreted into the gut [Tsuda *et al.*, 2010]. The complicated relationship between the tolerance and allergies has been reviewed recently [Brandtzang *et al.*, 2005). Also the effects of such influences as heavy sucrose and other sugar loads, as well as salty foods, onto the duodenal and other intestinal microflora deserves special attention [Hakalehto *et al.*, 2012]. We have experimented with the instant effects of consuming sugarous drinks, such as soft drinks, on the alcohol production of the GI tract microbes [Hakalehto and Hänninen, 2012]. This could be one factor behind the increased obesity in the industrialised countries. Also the assimilation of fructose in the liver only, but not in the other tissues, could be provoking overweight [Stanhope, 2011]. The microflora is an essential mediator in all these influences.

The fiber content determines the gut motility. Slow motility increases the time the harmful metabolites stay in the gut. The low fiber content is one of the common problems of the present diets and reason of the constipation. The higher the fiber intake, the lower the colon malignancy rates.

## Dualistic Balance, Obesity and Bacterial Energy Metabolism

In the microbiome ecological succession is continuously in operation. Often this succession is trying to balance the changes in order to preserve optimal conditions for the entire population in long term. The host´s own mechanisms are also taking part in the regulation. In small bowel, where the bile acid reabsorption takes place, the regulation is interconnected to the dualistic microbe balance between the facultatives [Hakalehto *et al.*, 2008]. This mechanism offers also a link to the tendencies for obesity in some individuals. Bile acids contribute to the fat metabolism, and therefore any bacterial contribution to these phenomena has influences on the metabolic status of the host. Moreover, it seems likely that as butanediol fermentation is producing larger amounts of easily assimilated ethanol than the mixed acid fermentation. It is thus contributing to an improved and more efficient food uptake also by that way. It has also been postulated that levels of 2,3-butanediol in alcoholics are generally higher than in normal population, also demonstrating the assimilation of bacterial fermentation products by the host [Montgomery *et al.,* 1993].

Acetoin is a precursor of 2,3-butanediol in its biosynthesis. The cleavage of acetoin is a thiolytic reaction forming first a thiohemiacetal, which is subsequently oxidized to acetyl CoA [Goedeke, 2011]. Anaerobic *Pelobacter carbinolicus* cells grown on acetoin produced acetyl-CoA and acetone only in the presence of CoASH from methylacetoin [Oppermann *et al.,* 1989]. It has also been shown that the acetoin dehydrogenase systems of *Clostridium magnum* and *Alcaligenes eutrophus* resemble that of the *Pelobacter* sp. [Lorenzl *et al.,* 1993; Krüger *et al.,* 1994]. However, it has been shown that the pH by the *Klebsiella* sp. in the PMEU studies could rise up even two units in a couple of hours [Hakalehto *et al.,* 2008]. This kind of major activity is not likely to be carried out by a anabolic pathway inside the cells. Therefore, we have suggested on the basis of some NMR (Nuclear Magnetic Resonance) evidence that the production of 2,3-butanediol could take place very rapidly also in the periplasmic space of some Gram –negative bacteria using another metabolic route of overflow type of metabolism [Hakalehto *et al.,* 2012]. There is significant evidence from the PMEU studies that in elevated osmotic pressure the 2,3- butanediol is produced without population growth and with less gas generation. In the intestinal environment lowered carbon dioxide production could prevent pH drop. On the other hand, hydrogen molecules emitted in connection with the butanediol fermentation remove protons from the gut microenvironment, and also cause compensation for the acidification caused by ordinary acid fermentations. Carbon dioxide is quickly removed as it passes easily different membranes, but it is still important as a signal molecule.

Turnbaugh and Ley and their coworkers have presented that the intestinal microflora vary with respect to the obesity tendencies of different individuals [Turnbaugh *et al.*, 2006; Ley *et al.*, 2006]. The statistical correlation between butyrate bacteria in the cecum with obesity tendency was revealed. This was shown to lead to increased ratio of the Firmicutes bacterial phylum compared with the *Bacteroidetes*. On the other hand, the cecum contains only the last remaining 20% of the nutrients of the human diet. The elevated level of butyrate producers possibly reflects the earlier events in the upper parts of the gastrointestinal system where the facultatively anaerobic strains maintain the balance between acidic and more neutral conditions. In fact, in the forage production a too weak lactic acid fermentation delays pH drop, which gives room for the butyric acid bacteria. Correspondingly, the pH neutralising effect of the butanediol fermentation in the small bowel could give a better start for the butyrate bacteria in the obese person′s cecum. Similar phenomenon has been observed also in corn silage where butyric acid bacteria were concentrated into anaerobic niches with an increased pH just below the surface [Vissiers *et al.*, 2007].

Because acetate activates many steps in the butanediol production [Johansen *et al.*,1975], this mechanism is an important regulatory factor as it prevents an excessive pH drop in the small intestine. In our experiments the lowering pH seemed to induce the butanediol fermentation. This finding is in line with earlier observations [Lorenzl *et al.,* 1993; Krüger *et al.,* 1994]. If pH drop would fall out of control, it could hamper the effective food adsorption, and create conditions where the relatively slowly exploitable acid fermentations like acetate fermentation, and in particular lactate fermentation, would take too big share of the total energy load out of the reach of the host (and other bacteria). It would eventually also turn down the bacterial activities of the gut by excessively lowering the pH. The dualistic balance of facultatively anaerobic enteric bacteria in the small intestine based on the butanediol fermentation is also one possible mechanism behind the increased amount of Firmicutes in obese individuals, due to the pH regulation of 2,3-butanediol fermentation. This could link the earlier observation of Turnbaugh and co-workers relating Firmicutes with obesity tendencies [Turnbaugh *et al.,* 2006]. In the human small intestine and colon many different microbes compete for the carbon and energy sources. It seems evident that the neutral end products in butanediol fermentation provide both human digestion and gut microflora somewhat higher energetic level containing sources than the different acid fermentations on average. This fermentation type is also likely to favour the butyric acid bacteria, which belong to the Firmicute division, by delaying the pH drop in the cecum. The colon receives substrate remnants. It houses a vast number of different bacteria that process the waste further for their own and host usage. The disturbances in this system may mainly be due to the imbalanced or disturbed small bowel host-microbe interactions.

Short chain fatty acids are produced by microflora in colon, and they are important energy source for mucosal cells. Some of them have also antimicrobial properties.

The production of neutral end products, with the rise in rapidly exploitable ethanol with towards the colon traversing 2,3-butanediol production also protects the bacteria which perform mixed acid fermentation from the feedback regulation of excess acid formation ]Hakalehto *et al.,* 2008; Hakalehto *et al.,* 2012]. Hence, the butanediol fermentation has a vital role for the host body as well as the entire gut microflora. Interestingly, the emerged worldwide pandemic causing *Vibrio cholerae* El Tor biotype has been defeating and overcoming the classical biotypes by being able to perform the 2,3-butanediol fermentation [Yoon and Mekalanos, 2006]. Therefore, the El Tor type strains are not as easily ruled out by their own acid fermentation like the classical *Vibrio cholerae* type in carbohydrate containing environment. Consequently it also survives better if carbohydrate foods, such as rice, are present in their environment. The better survival in water of the El Tor biotype is based on biofilm formation on chitin polymers in the environment, in turn based on the metabolic capabilities of the El Tor biotype. Production of 2,3-butanediol by *Klebsiella oxytoca* is influenced by oxygen limitation [Beronio and Tsao, 1993]. During batch culture studies, two phases of growth were observed: energy-coupled growth, during which cell growth and oxygen supply are coupled; and, energy-uncoupled growth, which arises when the degree of oxygen limitation reaches a critical value. Optimal 2,3-butanediol productivity occurs during the energy-coupled growth phase. In that article, a control system, which maintains the batch culture at a constant level of oxygen limitation in the energy-coupled growth regime was designed. Control, which involves feedback control on the oxygen transfer coefficient, is achieved by continually increasing the partial pressure of oxygen in the feed gas, which in turn continually increases the oxygen transfer rate. Control has resulted in a balanced state of growth, a repression of ethanol formation, and an increase in 2,3-butanediol productivity of 18%. Thus, the limitation of ethanol formation in the butanediol fermentation obviously enhanced the butanediol production. These mechanisms indicate that various modes of butanediol fermentation may exist, and that in the oxygen limited conditions in the gut the pattern includes also a substantial ethanol formation. In another set of fermenter studies it was shown that a higher oxygen supply favours the formation of cell mass at the expense of butanediol [Jansen *et al.,* 1984]. Decreasing the oxygen supply increases the butanediol yield, but decreases the overall conversion rate due to a lower cell concentration.

# Bacterial Cell Attachment and Movement in Colonization

There are several interfaces which all have utmost importance the survival and activity of all intestinal micro-organisms. They inhabit the fat-water interfaces having hydrophobic surfaces [Ellwood *et al.*, 1982]. These strains are important in the degradation and assimilation of lipids in the gastrointestinal tract [Zhao and Yang, 2005]. For example there is a sensitive balance between the supply and catabolism of the cholesterol [Dietschy *et al.*, 1993]. Excess cholesterol in the body is mainly converted to bile acids in the liver [Russell and Setchell, 1992]. After the meals, gallbladder secretes bile acids to duodenum through papilla Vateri. They are conjugated with taurine or glycine. In the duodenum they are converted to secondary and tertiary bile acids by intestinal bacteria. The bile acids emulsify the lipids in the intestines forming the above-mentioned interphases [Hylemon *et al.*, 2009]. They also act as signal molecules for the uptake of fat substances. Also the pathogens have sophisticated methods for facilitating their entry into the body system. For example, the MMTV virus uses lipopolysaccharides (LPS) from gut bacteria to avoid the human immune system [Pennis, 2011].

# Antibiotics and Intestinal Microbiota

In the use of antibiotics many unpleasant health and economical consequences may result [Hakalehto, 2012]. Recently, it has been urged that the U.S. should follow Denmark (and other Scandinavian countries) and stop delivering antibiotics to fatten up farm animals (Scientific American, April 2011, Science Agenda by the Editors). Many U.S. bodies, such as the American Medical Association, and the Infections Diseases Society of America have given advices to that direction. In our research with water-borne infective agents we have got alarmed with the waste-water outlets from hospitals to the water treatment plants where these strains may get effectively enriched. The control of the antibiotic strains should thus be implemented as a general practice inside the hospital systems, where the PMEU units could facilitate the microbiological control [Hakalehto *et al.*, 2009; Hakalehto, 2010, Hakalehto *et al.*, 2010, Hakalehto, 2012]. The detection of emissions of volatile compounds from enriched samples could reveal even the low numbers of the resistant bacteria in few hours. The same methods are applicable for *e.g.* screening single *Salmonella* cells in the flow of the tap water [Hakalehto *et al.*, 2011].

The most tolerable bacterial groups, such as coliforms and enterococci seem to colonize most often also the bilial tract [Hakalehto *et al.*, 2010]. Some intestinal bacteria carry out the deconjugation of bile acids. For example, all the enterococci perform this function, whereas none of the *E. coli* or *Klebsiella mobilis* (*Aerobacter aerogenes* by their former name) can accomplish it [Shindo and Fukushima, 1976].

The excessive growth of the bacteria in the intestinal areas cause the production of toxic compounds into the gut [Delzenne, 2011]. They also transfer useful compounds into poisonous forms. In the intestinal sludge the conditions can allow rapid spread of genetic elements in a mixed populations causing *e.g.* distribution of antibiotic resistance plasmids [McKenna, 2011].

This is increasing the risks of emerging new infections of previously known pathogens uncontrollable by the traditional antibiotic cure approach [Nordmann *et al.*, 2011]. That epidemiological development could worsen the capabilities of the healthcare system for controlling the communicable diseases, and consequently produce severe consequences and increase carriers of the antibiotic strains in the hospitals [Hakalehto, 2006]. New effective tools and measures need to be invented and adopted for the fight against the spreading infective strains.

The medium-chain fatty acids have been shown to have a bacteriostatic effect [Hassinen *et al.*, 1951]. They occur in milk protecting the neonates against infections together with immunoglobulins, lactoferrin, lactoperoxiddase, lysozyme and some other proteins. Also some other lipids and oligosaccharides restrict the microbes in the milk, the latter inhibiting the bacterial attachment to the sublimentary tract surfaces [Kunz *et al.*, 2000]. On the other hand, it has also been shown that some bacterial toxins could be active even against other strains of their producing species, such as the cerulide of the *Bacillus cereus* dairy strain (Shaheen, 2009).

## Protective Microbiome and Its Manipulative Functions

Many biological regulators contribute and modulate the metabolism of micro-organisms, and their behavior. For example, molecules produced by other microbes may encourage or make pace of their growth. Many microbes produce antibiotic substances, which could slow down, hinder or prevent the growth of the others. This formation of different organic acids, if excessive, could influence remarkably also on the metabolism of the acid-producing microbes themselves. For example, *E. coli* belongs to the mixed-acid fermenting bacteria. It secretes big amounts of *e.g.* formic and acetic acids. Simultaneously, a significant drop in the pH results from its own activities and causes a self-limitation. In the duodenum there has been revealed to exist an intriguing mechanism regulating the pH levels [Hakalehto *et al.*, 2008]. Besides *E. coli* and other mixed-acid fermenters in these areas occur strains of the *Klebsiella / Enterobacter* group, which perform 2,3 –butanediol fermentation. In this metabolic pattern, 2,3 –butanediol, ethanol, $CO_2$ and hydrogen ($H_2$) are formed besides some acidic compounds. The propositions of various solvents in the chime, in turn influence on the epithelial permeability for endotoxins [Mitzscherling, 2009]. The neutral substances, 2,3 –butanediol and ethanol slow down the pH drop by binding the carbon skeletons into these compounds instead of acids. This activity of the biochemical pathways is fast, and is counteracting the acid production of some other species. Meanwhile, the gaseous $H_2$ "leaks away" the protons, which would otherwise by causing the decrease of the pH. In the experiments with *Klebsiella mobilis* we have detected that the $CO_2$ is in some cases produced less, if the production of the neutral substances is increasing [Hakalehto, 2011; Hakalehto *et al.*, 2012]. This observation implies to the existence of another metabolic route for the production of 2,3 –butanediol besides the sideline of the anabolic production of the amino acid valine (Blomqvist *et al.*, 1993). The idea is supported *e.g.* by the finding that the 2,3 –butanediol is produced in large amounts without concomitant cell growth, which could be explained by a form of previously unknown overflow metabolism [Hakalehto *et al.*, 2011]. It could be deduced, in anyway, that the $H_2$ production is keeping up the gradient of protons across the cell membranes, which is crucial for ALL bacteria for the energy metabolism. The power house of the cells, the proton pump, is functioning in connection with the membranes, and is also most likely the foundation of most electrochemical phenomena within the cells [Smirnov *et al.*, 2011]. In order to preserve this option the klebsiellas and their kind of facultative bacteria are carrying out an important task not only for themselves, but for the entire microbial community.

In the particular symbiotic relation between the mixed-acid fermenters, such as *E. coli*, and the 2,3 –butanediol producing species, such as *Klebsiella mobilis* (*Enterobacter aerogenes* or *Aerobacter aerogenes* as the older designation ) grow in a mixed culture equally well than in the separate cultures [Hakalehto *et al.*, 2008]. This finding has been repeatedly documented, and the symbiotic relationship between the relative facultative bacteria is:

1. promoting the energy metabolism of the entire microbial population
2. regulating the duodenal pH (at around 6)
3. keeping the mixed-acid fermenters and the 2,3 –butanediol producers in a dualistic balance, which well could be the foundation of the BIB (Bacterial Intestinal Balance).

In further studies with the PMEU (Portable Microbe Enrichment Unit) we have documented that the newcoming strains of *Staphylococcus aureus* and/or *Bacillus cereus* do not interrupt the balance between *E. coli* and *K. mobilis* [Hakalehto *et al.*, 2010]. The facultative bacteria were also relatively indifferent to the effects of oxgall (a bile substance) or vancomycin. Interestingly, these coliformic bacteria, together with enterococci, are the most commonly used bacterial hygienic indicators [Hakalehto *et al.*, 2011]. These species are also the most resistant ones toward the antimicrobial action of the bile [Hakalehto *et al.*, 2010]. Therefore, they survive the best in the duodenum, and get their place as the most abundant, or common, bacterial group in the fecal samples, and the (gastro)intestinal areas. The production of the 2,3 –butanediol by the klebsiellas has shown to be activated by the acetic acid (Hakalehto, 2008). Therefore, the symbiosis with *E. coli* is strengthened also by this signal effect. The more microbial activities do occur, the more activity of the *Klebsiella* / *Enterobacter* group is likely to take place, especially in nutritional situations where excessive sugar or starch is suddenly appearing. In these circumstances, the 2,3 –butanediol producers have been shown to increase dramatically their related activities without simultaneous populations growth [Hakalehto et al., 2008; Hakalehto *et al.*, 2012]. This function is protecting the bacterial population against the pH drop and destructive osmotic pressures. The proposed mechanism of the Gram-negative bacteria in question includes some periplasmic enzymes. The ethanol formed together with the 2,3 –butanediol is instantaneously consumed by the human body, and its rigorous uptake prevents its accumulation inside the gut, and the ethanol inhibition which could restrict the normal flora. This kind of selection could then lead to hazardous situations, such as candidosis, which occasionally leads to dysbiosis [Bolkov, 2005].

Also the gaseous compounds may transmit information among bacterial populations. *E. coli* cultures have been shown to get stimulated with a shortened lag phase, or the earlier onset of growth, by the production of the $CO_2$ of the microbes themselves [Hakalehto, 2011].

This stimulative or signaling effect has been documented also in the case of interspecies association between lactobacilli and butyric acid bacteria [Hakalehto and Hänninen, 2012]. This bacterial coculture is found also in the silage, where getting the optimal result requires that the butyric acid levels remain low enough. In the intestines the stimulation and activation of the clostridia could serve as a preventive factor for their spore formation, or attachment, and thus indirectly prevent the colonization of the intestinal walls by these health threatening, putrefacient bacteria. The anaerobic clostridia and especially the butyric acid producing ones, nevertheless, grow extensively in the cecum, the proximal part of the human large intestines. There they have been found more numerous in obese individuals [Vrieze *et al.*, 2010]. This could imply to a mechanism which is taking into use the 2,3 –butanediol, which otherwise is relatively slowly consumed by the bacterial communities. In soil, for example, the 2,3 –butanediol is degraded into simple organic compounds, such as acetic acid and ethanol [Han, 2011]. The site for its assimilation in the gut areas could well be the colon, where clostridia are carrying out this task. However, for the benefit of the human body and its entire microbiota, it is more beneficial that the clostridia do not conquer too much space. This kind of undesired situation is occurring in the case of *Clostridium difficile* colonizing the intestinal tract after heavy antibiotic medications [Coakley, 2011]. Such infections could be life-threatening, and it is of utmost importance to monitor *C. difficile* most carefully [Hell *et al.*, 2010].

The cells energize their membranes by the electron flow across the lipid bilayer. Therefore, the membrane potential plays an important role in bacterial energy metabolism. The electric flow across the cell surface is recharging the phosphorylated energy reserves as the ATP (adenoside triphosphate) [Malakooti *et al.*, 2011]. In broader sense, these electron transport chains through the bacterial cell walls are the power houses of the cells. Their functioning is entirely dependent on the gradients formed across the cell membranes. Therefore, the diffusion of nutrients and waste substances close to the cells has an enormous impact on the cell metabolism. It has been generally believed that the anaerobic bacterial metabolism would be much less effective than the aerobic one [Hentges, 1996]. This has been explained by the mobilization of the citric acid cycle (Krebs cycle) enzymes in the latter one. However, in the experiments with the PMEU cultivator it has been documented that the anaerobic metabolic activity of *e.g.* some facultative bacteria is at the same level as it is in the aerobic growth mode [Hakalehto *et al.*, 2007; Hakalehto, 2011].

We have documented that without any population growth the facultative klebsiellas can carry out rapid pH shifts [Hakalehto *et al.,* 2008]. It is not likely that these extensive reactions where remarkable amount of 2,3-butanediol is produced could occur as a branch of a biosynthetic pathway as generally suggested [Rao *et al.*, 2011]. In the experiments with PMEU we also proved out that these reactions, which rapidly consume the sugar from the medium, did not combine with either population growth or increase in gas production [Hakalehto *et al.,* 2012]. In fact, the volatiles emission was diminished regardless of the increased production of butanediol. Therefore we suggest that with Gram-negative facultatively anaerobic bacteria there could exist a previously unknown route for the overflow metabolism. This rapid mechanism would protect the bacterial cells from high osmotic pressures caused by elevated sugar concentrations, and also from the lowering of the pH in their microenvironment.

## Drug Delivery by the Action of Host/Bacterial Enzymes

Different oral drug delivery systems have been investigated with the aim to ensure the therapeutic action of the drug in their target following different approaches such as (1) masking drug properties that limit bioavailability, (2) overcoming chemical/enzymatical instability of the drug or (3) subdue the lack of site specificity [Fasinu *et al.*, 2011]. One possible strategy to activate the delivery of the drug from the system is utilizing intestinal enzymes. The present section is focusing in the enzyme-activated drug delivery system by the approach of classifying them according to the origin of the enzyme involved, either from the host or from the microflora of the gastrointestinal tract (GIT).

Throughout the GIT, host enzymes are secreted from the mouth to the distal end of the ileum (end part of the small intestine). The epithelial cells in the large intestine contain almost no these enzymes [Guyton, 1986]. This region, on the other hand, is rich in bacterial flora, having much higher concentration of microflora than the upper regions. The main bacterial population present is that of oxygen-intolerant anaerobic bacteria of various types [Sinha *et al.*, 2003]. Bacterial fermentation in the large intestine, specifically in colon is a symbiotic relationship where the host derives many benefits involved in the digestion of carbohydrates and proteins, and as well as exogenous compounds such as drugs [Sinha *et al.*, 2003]. Generally the small intestine is considered as the primary site for drug absorption but colon is believed to be a suitable alternative to direct drugs as well.

## Host Intestinal Enzyme-Activated Drug Delivery Systems

Esterases, phosphatases, peptidases are some of the many enzymes present in the upper tract of the GIT of the human. These are the host enzymes that are targeted for prodrug design. Prodrugs are bioreversible derivatives of drug molecules that undergo an enzymatic and/or chemical transformation *in vivo* to release the active parent drug [Narang and Mahato, 2010]. This approach is intended to improve the oral bioavailability of the parent drug, in other words to ensure that the maximum fraction of unchanged drug is absorbed and reaches the systemic circulation. For this purpose, the prodrug design will aim to increase the solubility of such drugs that are poorly soluble and improve the lipophilicity of molecules that sparsely permeates the intestinal wall. Esters are the most common prodrug used generally to enhance the lipophilicity of water soluble drugs by masking charged groups of the parent drug [Rautio *et al.*, 2008]. The prodrugs are converted back to the active parent drug via the ubiquitous esterases presenting in GIT [Fasinu, 2010]. In other cases, the prodrugs aim to improve aqueous solubility by adding a phosphate group (ionizable group) to the parent compound. They are rapidly hydrolyzed to the parent drugs by endogenous alkaline phosphatases of the host at the intestinal cell surface during absorption [Rautio *et al.*, 2008]. However, the targeting activation of prodrugs in specific regions of the GIT is not possible due to the ubiquitous distribution of these host enzymes all along the tract.

# Bacterial Enzyme-Activated Drug Delivery Systems

As commented above, the human colon is rich in bacteria with over 400 distinct species as resident flora [Philip and Philip, 2010]. This flora produces a vast number of reductive and hydrolytic enzymes (see Table 1) among them azoreductases, glycosidases, and glucuronidases are being exploited for formulation of colon-specific drug delivery systems [Narang *et al.*, 2010]. The fact that these enzymes are specifically present in the colon at higher concentration than any other region of the GIT makes possible the development of colonic-targeted drug delivery systems. These include prodrugs and systems based upon biodegradable polymers which are specifically degraded by these enzymes aiming to local or systemic therapeutic effect. Some of the approaches of site-specific drug delivery in colon systems may be:

- to treat local illness in that region (ulcerative colitis, Crohn's disease, and colorectal cancer). So, colon-specific delivery systems can treat efficiently these illnesses, with fewer side-effects, lower required dose and improved patient compliance.
- to improve the oral bioavailability of drugs, avoiding degradation and/or lack of absorption in the upper gastrointestinal tract and deliver the drug for site-specifically absorption in the proximal colon into the systemic circulation.
- to prevent side effect of gastric irritation induce by some drugs such as non-steroidal anti-inflammatory agents (NSAIDs) [Philip *et al.,* 2010).
- to achieve a delay in drug absorption required from a therapeutic point of view *e.g.* in case of nocturnal asthma, angina etc. [Sinha *et al.,* 2003].

**Table 1. Description of the microflora in the colon, types of bacterial enzymes secreted and substrates of the metabolic reactions**

| Colonic microflora | Bacterial enzymes | Substrates |
|---|---|---|
| Predominant species:<br>*Bacteroides*, *Bifidobacterium*, *Eubacterium.*<br>Also:<br>Cocci, Clostridia, Enterococci, *Enterobacteriaceae*, *Diphtheroides*, Coliforms, *Staphylococci*, *Lactobacillus*, *Spirochetes*, yeasts, *Proteus*, *Pseudomonas*, *Bacillus subtilis*, *Actinomyces*, *Borrelia*, *Fusobacterium* and *Clostridium.* | Reductive enzymes:<br>nitroreductase,<br>azoreductase,<br>deaminase,<br>urea dehydroxylase<br><br>Hydrolytic enzymes:<br>β-glucuronidase,<br>β-xylosidase,<br>α-arabinosidase,<br>β-galactosidase. | Undigested carbohydrates (polysaccharides such as cellulose, xylan, and pectin as well as starch) and proteins.<br><br>Xenobiotics and drugs such as sulphasalazine, isonicotinuric and salicyluric acids, L-dopa, digoxin, cyclamates, lactulose. |

## Prodrugs

A prodrug is a pharmacologically inactive derivative of parent drug molecule that requires spontaneous or enzymatic transformation *in vivo* to release the active drug. For colonic delivery, the hydrophilic prodrug is designed to undergo minimal hydrolysis in the upper tracts of GIT and be susceptible to cleavage by the bacterial enzymes, (so azo-, glucuronide-, glycoside- bond derivatives or cyclodextrin-derivatives) into a drug (more lipophilic than the prodrug) that will be absorbed in the colon [Stella *et al.*, 2007]. Metabolism of azo- compounds by intestinal bacteria is one of the most extensively studied bacterial metabolic process. The best known colon-targeting prodrugs are azo-bond derivatives of 5-aminosalicylic acid, an anti-inflammatory drug used for the treatment of inflammatory bowel diseases [Narang *et al.*, 2010]. A detail list of prodrugs evaluated for colon specific drug delivery is reviewed elsewhere [Sinha *et al.*, 2003; Philip *et al.*, 2010].

## Systems Based upon Biodegradable Polymers

Colon-targeting biodegradable polymer systems contain polymers that are specifically degraded in the colon region of the GIT and additionally, may protect the drug from the environments of stomach and small intestine. Once these polymers reach the colon, they undergo assimilation by micro-organism, or degradation by enzyme or break down of the polymer back bone leading to a subsequent reduction in their molecular weight and thereby loss of mechanical strength. Consequently, the drug is delivered from the system. The colon-targeted biodegradable polymer systems activated by the metabolism of the bacterial enzymes can be classify into

- azo-polymeric prodrugs and azo-polymeric coating. This approach consists in using polymers that are bound to the drug moiety by azo-linkage. The azo bonds of the polymer are reduced by the colonic bacterial azo-reductases and the drug is specifically delivered in this region [Sinha *et al.*, 2003].
- polysaccharide based delivery systems. The colonic microflora fulfills its energy needs by fermenting various types of substrates that have been left undigested in the small intestine (Table 1). The strategy in this drug delivery system is to use the same type of polysaccharides substrates to carry the drug. Once in the colon, the polysaccharides are broken down by bacterial hydrolytic enzymes and the drug is released [Sinha *et al.*, 2003].
- a novel colon targeted delivery system CODESTM. CODESTM technology combines the use of polysaccharides (namely lactulose) and materials such as methacrylic acid copolymers (Eudragits) in the elaboration of the tablet system. Lactulose acts as a trigger for site specific drug release in the colon while the different methacrylic acid copolymers protect the drug from the pH degradation in stomach and small intestine [Philip *et al.*, 2010].

# Conclusion

The small intestinal mucosa forms a large surface that is essential for the absorption of nutrients. It is also important in the drug metabolism that continues in the colon. Although, the number of bacteria is lower in the small intestines than in the colon, the microbes of the small intestines have a major impact on the human welfare. Different species of the microbiota participate in the maintenance of the intestinal immunity against foreign pathogens entering the intestines and in the metabolism of nutrients and drugs. They also interact with each other maintaining the metabolic equilibrium suitable for all the members of the microbiota. This metabolic co-operation called Bacterial Intestinal Balance is a cornerstone in the resistance of the microbiota against disturbances caused by environmental and microbiological changes in the small intestinal ecosystem. Understanding more about the small intestinal microbiome could open new windows for the treatment of intestinal problems such as dyspepsia, malabsorption, bacterial overgrowth, irritable bowel syndrome, inflammatory diseases and infections.

# References

Abdulnour-Nakhoul, S; Nakhoul, HN; Kalliny, MI; Gyftopoulos, A; Rabon, E; Doetjes, R; Brown, K; Nakhoul, NL. Ion transport mechanisms linked to bicarbonate secretion in the esophageal submucosal glands. *Am. J. Physiol. Regul. Integr. Comp. Physiol.*, 2011; 301, R83-96.

Abeysuriya, V; Deen, KI; Wijesuriya, T; Salgado, SS. Microbiology of gallbladder bile in uncomplicated symptomatic cholelithiasis. *Hepatobiliary Pancreat. Dis. Int.*, 2008; 7, 633-637.

Andrievsky, GV; Bruskov, VI; Tykhomyrov, AA; Gudkov, SV. Peculiarities of the antioxidant and radioprotective effects of hydrated C60 fullerene nanostuctures *in vitro* and *in vivo*. *Free Radic. Biol. Med.*, 2009; 47, 786-793.

Armand, M; Pasquier, B; André, M; Borel, P; Senft, M; Peyrot, J; Salducci, J; Portugal, H; Jaussan, V; Lairon, D. Digestion and absorption of 2 fat emulsions with different droplet sizes in the human digestive tract. *Am. J. Clin. Nutr.*, 1999; 70, 1096-1106.

Arumugam, M; Raes, J; Pelletier, E; Le Paslier, D; Yamada, T; Mende, DR; Fernandes, GR; Tap, J; Bruls, T; Batto, JM; Bertalan, M; Borruel, N; Casellas, F; Fernandez, L; Gautier, L; Hansen, T; Hattori, M; Hayashi, T; Kleerebezem, M; Kurokawa, K; Leclerc, M; Levenez, F; Manichanh, C; Nielsen, HB; Nielsen, T; Pons, N; Poulain, J; Qin, J; Sicheritz-Ponten, T; Tims, S; Torrents, D; Ugarte, E; Zoetendal, EG; Wang, J; Guarner, F; Pedersen, O; de Vos, WM; Brunak, S; Dore, J; MetaHIT Consortium; Antolin, M; Artiguenave, F; Blottiere, HM; Almeida, M; Brechot, C; Cara, C; Chervaux, C; Cultrone, A; Delorme, C; Denariaz, G; Dervyn, R; Foerstner, KU; Friss, C; van de Guchte, M; Guedon, E; Haimet, F; Huber, W; van Hylckama-Vlieg, J; Jamet, A; Juste, C; Kaci, G; Knol, J; Lakhdari, O; Layec, S; Le Roux, K; Maguin, E; Merieux, A; Melo Minardi, R; M'rini, C; Muller, J; Oozeer, R; Parkhill, J; Renault, P; Rescigno, M; Sanchez, N; Sunagawa, S; Torrejon, A; Turner, K; Vandemeulebrouck, G; Varela, E; Winogradsky,

Y; Zeller, G; Weissenbach, J; Ehrlich, SD; Bork, P. Enterotypes of the human gut microbiome. *Nature,* 2011; 473, 174-180.

Beltzer, EK; Fortunato, CK; Guaderrama, MM; Peckins, MK; Garramone, BM; Granger, DA. Salivary flow and alpha-amylase: collection technique, duration, and oral fluid type. *Physiol. Behav.*, 2010;101, 289-296.

Beronio, PB Jr; Tsao, GT. Optimization of 2,3-butanediol production by *Klebsiella oxytoca* through oxygen transfer rate control. *Biotechnol. Bioeng.*, 1993;42, 1263-1269.

Blandino, G; Fazio, D; Di Marco, R. Probiotics: overview of microbiological and immunological characteristics. Expert Rev. *Anti Infect Ther.*, 2008; 6, 497-508.

Blomqvist, K; Nikkola, M; Lehtovaara, P; Suihko, ML; Airaksinen, U; Stråby, KB; Knowles, JK, Penttilä, ME. Characterization of the genes of the 2,3-butanediol operons from *Klebsiella terrigena* and *Enterobacter aerogenes*. *J. Bacteriol.*, 1993; 175, 1392-1404.

Boĭkov, SS; Moroz, AF; Babaeva, EE. Association of Candida albicans fungi with some opportunistic microorganisms in intestinal dysbiosis in patients of different age groups. *Zh. Mikrobiol. Epidemiol. Immunobiol.*, 2005; 2, 65-69.

Bonfils, S; Mignon, M; Rozé, C. Vagal control of gastric secretion. *Int Rev Physiol.*, 1979; 19, 59-106.

Bouchoux, J; Beilstein, F; Pauquai, T; Guerrera, IC; Chateau, D; Ly, N; Alqub, M; Klein, C; Chambaz, J; Rousset, M; Lacorte, JM; Morel, E; Demignot, S. The proteome of cytosolic lipid droplets isolated from differentiated Caco-2/TC7 enterocytes reveals cell-specific characteristics. *Biol. Cell.*, 2011;103, 499-517.

Bouhnik, Y; Flourié, B; Riottot, M; Bisetti, N; Gailing, MF; Guibert, A; Bornet, F; Rambaud, JC. Effects of fructo-oligosaccharides ingestion on fecal bifidobacteria and selected metabolic indexes of colon carcinogenesis in healthy humans. *Nutr. Cancer.*, 1996; 26, 21-29.

Brandtzang, P; Isolauri, E; Prescott, SL, editors. *Microbial-host interaction: Tolerance versus allergy*. Nestle Nutrition Institute Workshop Series Pediatric Program, 2009; 64. Karger, Basel, Switzerland.

Brashears, MM; Gilliland, SE; Buck, LM. Bile salt deconjugation and cholesterol removal from media by Lactobacillus casei. *J. Dairy Sci.*, 1998; 81, 2103-2110.

Brouns, F; Beckers, E. Is the gut an athletic organ? Digestion, absorption and exercise. *Sports Med.*, 1993; 15, 242-257.

Caramia, G. Gastroenteric pathology and probiotics: from myth to scientific evidence. Current aspects. *Minerva Gastroenterol. Dietol.*, 2009; 55, 237-272.

Celik, M; Ercan, I. Diagnosis and management of laryngopharyngeal reflux disease. Curr. Opin. Otolaryngol. *Head Neck Surg.*, 2006; 14, 150-155.

Chen, N; Unnikrishnan, IR; Anjana, RM; Mohan, V; Pitchumoni, CS. The complex exocrine-endocrine relationship and secondary diabetes in exocrine pancreatic disorders. *J. Clin. Gastroenterol.*, 2011; 45, 850-861.

Cheng, J., editor. Renewable energy processes. 2010. Boca Raton, FL, USA. CRC Press, Taylor & Fransis Group.

Coakley, BA; Sussman, ES; Wolfson, TS; Bhagavath, AS; Choi, JJ; Ranasinghe, NE; Lynn, ET; Divino, CM. Postoperative antibiotics correlate with worse outcomes after appendectomy for nonperforated appendicitis. J. Am. Coll. Surg., 2011 Dec;213(6), 778-783.

Croft, DN; Cotton, PB. Gastro-intestinal cell loss in man. Its measurement and significance.

*Digestion*, 1973;8, 144-160.

Curry, A; Appleton, H; Dowsett, B. Application of transmission electron microscopy to the clinical study of viral and bacterial infections: present and future. *Micron*, 2006; 37, 91-106.

De Fabiani, E; Mitro, N; Gilardi, F; Crestani, M. Sterol-protein interactions in cholesterol and bile acid synthesis. *Subcell Biochem.* 2010;51:109-135.

De Filippo, C; Cavalieri, D; Di Paola, M; Ramazzotti, M; Poullet, JB; Massart, S; Collini, S; Pieraccini, G; Lionetti, P. Impact of diet in shaping gut microbiota revealed by a comparative study in children from Europe and rural Africa. *Proc. Natl. Acad. Sci. USA,* 2010; 107, 14691-14696.

De la Garza, AL; Milagro, FI; Boque, N; Campión, J; Martínez, JA. Natural inhibitors of pancreatic lipase as new players in obesity treatment. *Planta Med.*, 2011; 77, 773-785.

Delzenne, NM; Neyrinck, AM; Cani, PD. Modulation of the gut microbiota by nutrients with prebiotic properties: consequences for host health in the context of obesity and metabolic syndrome. *Microb Cell Fact.* 2011;10 Suppl 1,10.

Dietschy, JM; Turley, SD; Spady, DK. Role of liver in the maintenance of cholesterol and low density lipoprotein homeostasis in different animal species, including humans. *J. Lipid. Res.*, 1993; 34, 1637-1659.

Ellwood, DC; Keevil, CW; Marsh, PD; Brown, CM; Wardell, JN. Surface-associated growth. *Philos. Trans. R Soc. Lond. B Biol. Sci,.* 1982;297, 517-532.

Englyst, HN; Cummings, JH. Digestion of the polysaccharides of some cereal foods in the human small intestine. *Am. J. Clin. Nutr.*, 1985; 42, 778-787.

Fasinu, P; Pillay, V; Ndesendo, VM; du Toit, LC; Choonara, YE. Diverse approaches for the enhancement of oral drug bioavailability. *Biopharm. Drug Dispos. 2011;* 32, 185–209.

Gentilcore, D; Vanis, L; Teng, JC; Wishart, JM; Buckley, JD; Rayner, CK; Horowitz, M; Jones, KL. The oligosaccharide α-cyclodextrin has modest effects to slow gastric emptying and modify the glycaemic response to sucrose in healthy older adults. *Br. J. Nutr.*, 2011; 106, 583-587.

Gerasimidis, K; McGrogan, P; Edwards, CA. The aetiology and impact of malnutrition in paediatric inflammatory bowel disease. *J. Hum. Nutr. Diet.*, 2011; 24, 313-326.

Gilliland, SE; Speck, ML. Deconjugation of bile acids by intestinal lactobacilli. *Appl. Environ. Microbiol.,* 1977;33, 15-18.

Goedeke, L; Fernández-Hernando, C. Regulation of cholesterol homeostasis. *Cell. Mol. Life. Sci.*, 2011 Oct 19. [Epub ahead of print].

Gómez-Aguado, F; Alou, L; Corcuera, MT; Sevillano, D; Alonso, MJ; Gómez-Lus, ML; Prieto, J. Evolving architectural patterns in microbial colonies development. *Microsc. Res. Tech.*, 2011; 74, 925-930.

Gophna, U. Microbiology. The guts of dietary habits. *Science*, 2011; 334, 45-46.

Gravante, G; Knowles, T; Ong, SL; Al-Taan, O; Metcalfe, M; Dennison, AR; Lloyd, DM. Bile changes after liver surgery: experimental and clinical lessons for future applications. *Dig. Surg.*, 2010; 27, 450-460.

Guyton, A. Text book of medical physiology. Saunders 7th edition. 1986.

Hakalehto, E. Hygiene monitoring with the Portable Microbe Enrichment Unit (PMEU). 2010, VTT (State Research Centre of Finland), Espoo, Finland.

Hakalehto, E. Antibiotic resistant traits of facultative *Enterobacter cloacae* strain studied with the PMEU (Portable Microbe Enrichment Unit). In Press. In: Méndez-Vilas A, editor.

*Science against microbial pathogens: communicating current research and technological advances. Microbiology book series Nr. 3.* Badajoz, Spain: Formatex Research Center; 2012.

Hakalehto, E. Semmelweis' present day follow-up: Updating bacterial sampling and enrichment in clinical hygiene. *Pathophysiology,* 2006;13, 257-67.

Hakalehto, E. Simulation of enhanced growth and metabolism of intestinal *Escherichia coli* in the Portable Microbe Enrichment Unit (PMEU). In: Rogers MC, Peterson ND, editors. *E. coli infections: causes, treatment and prevention.* New York, USA: Nova Publishers; 2011.

Hakalehto, E; Haikara, A; Enari, TM; Lounatmaa, K. Hydrochloric acid extractable protein patterns of *Pectinatus cerevisiophilus* strains. *Food Microbiol.,* 1984; 1, 209-216.

Hakalehto, E; Hänninen, O. Lactobacillic $CO_2$ signal initiate growth of butyric acid bacteria in mixed PMEU cultures. Manuscript in preparation. 2012.

Hakalehto, E; Heitto, L; Heitto, A; Humppi, T; Rissanen, K; Jääskeläinen, A; Paakkanen, H; Hänninen, O. Fast monitoring of water distribution system with portable enrichment unit – Measurement of volatile compounds of coliforms and *Salmonella* sp. in tap water. *JTEHS,* 2011; 3, 223-233.

Hakalehto, E; Hell, M; Bernhofer, C; Heitto, A; Pesola, J; Humppi, T; Paakkanen, H. Growth and gaseous emissions of pure and mixed small intestinal bacterial cultures: Effects of bile and vancomycin. *Pathophysiology*, 2010; 17, 45-53.

Hakalehto, E; Humppi, T; Paakkanen, H. Dualistic acidic and neutral glucose fermentation balance in small intestine: Simulation *in vitro*. *Pathophysiology,* 2008; 15, 211-220.

Hakalehto, E; Pesola, J; Heitto, A; Deo, BB; Rissanen, K; Sankilampi, U; Humppi, T; Paakkanen, H. Fast detection of bacterial growth by using Portable Microbe Enrichment Unit (PMEU) and ChemPro100i((R)) gas sensor. *Pathophysiology*, 2009; 16, 57-62.

Hakalehto, E; Pesola, J; Heitto, L; Närvänen, A; Heitto, A. Aerobic and anaerobic growth modes and expression of type 1 fimbriae in *Salmonella*. *Pathophysiology*, 2007;14, 61-69.

Hakalehto, E; Tiainen, M; Laatikainen, R; Paakkanen, H; Humppi, T; Hänninen, O. Rapid bacterial metabolic activity by the production of ethanol and 2,3 -butanediol without population growth protects intestinal flora against osmotic stress. *Manuscript in preparation,* 2012.

Han, W; Wang, Z; Chen, H; Yao, X; Li, Y. Simultaneous biohydrogen and bioethanol production from anaerobic fermentation with immobilized sludge. *J. Biomed. Biotechnol.*, 2011;2011:343791. Epub 2011 Jul 3.

Hänninen, O; Aitio, A; Hietanen, E. Physiological defence against xenobiotics at their portals of entry to the organism. *Med. Biol.,* 1979; 57, 251-255.

Hänninen, O; Lindstrom-Seppä, P; Pelkonen, K. Role of gut in xenobiotic metabolism. *Arch. Toxicol.,* 1987; 60, 34-36.

Hassinen, JB; Durbin, GT; Bernhardt, FW. The bacteriostatic effect of saturated fatty acids. *Arch. Biochem. Biophys.*, 1951; 31, 183-189.

Hell, M; Bernhofer, C; Huhulescu, S; Indra, A; Allerberger, F; Maass, M; Hakalehto, E. How safe is colonoscope-reprocessing regarding *Clostridium difficile* spores? *J. Hosp.Inf.*, 2010; 76, 21-22.

Hentges, DJ. Anaerobes: General Characteristics. Medical Microbiology. 4th edition. Galveston (TX): University of Texas Medical Branch at Galveston; 1996. Chapter 17.

Holmgren, S; Olsson, C. Autonomic control of glands and secretion: a comparative view. *Auton. Neurosci.*, 2011; 165, 102-112.

Hylemon, PB; Zhou, H; Pandak, WM; Ren, S; Gil, G; Dent, P. Bile acids as regulatory molecules. *J. Lipid. Res.*, 2009; 50, 1509-1520.

Ikeda, K; Chiba, T; Sugai, T; Kangawa, K; Hosoda, H; Suzuki, K. Correlation between plasma or mucosal ghrelin levels and chronic gastritis. *Hepatogastroenterology*. 2011; 58, 1622-1627.

Ilves, I; Paajanen, HE; Herzig, KH; Fagerström, A; Miettinen, PJ. Changing incidence of acute appendicitis and nonspecific abdominal pain between 1987 and 2007 in Finland. *World J. Surg.*, 2011; 35, 731-738.

Jansen, NB; Flickinger, MC; Tsao, GT. Production of 2,3-butanediol from D-xylose by *Klebsiella oxytoca* ATCC 8724. *Biotechnol. Bioeng.*, 1984; 26, 362-369.

Jara, S; Sánchez, M; Vera, R; Cofré, J; Castro, E. The inhibitory activity of Lactobacillus spp. isolated from breast milk on gastrointestinal pathogenic bacteria of nosocomial origin. *Anaerobe*, 2011; 17, 474-477.

Johansen, L; Bryn, K; Stormer, FC. Physiological and biochemical role of the butanediol pathway in *Aerobacter* (*Enterobacter*) *aerogenes*. *J. Bacteriol.*, 1975 ;123, 1124-1130.

Johansson, ME; Hansson, GC. Keeping Bacteria at a Distance. *Microbiology, perspectives,* 2011; 334, 182-183.

Jonas, A; Weiser, S; Segal, P; Katznelson, D. Oral fat loading test: a reliable procedure for the study of fat malabsorption in children. *Arch. Dis. Child.*, 1979; 54, 770-772.

Kasbo, J; Saleem, M; Perwaiz, S; Mignault, D; Lamireau, T; Tuchweber, B; Yousef, I. Biliary, fecal and plasma deoxycholic acid in rabbit, hamster, guinea pig, and rat: comparative study and implication in colon cancer. *Biol. Pharm. Bull.*, 2002; 25, 1381-1384.

Kataoka, R; Kimata, A; Yamamoto, K; Hirosawa, N; Ueyama, J; Kondo, T; Okada, R; Kawai, S; Hishida, A; Naito, M; Morita, E; Wakai, K; Hamajima, N. Association of UGT1A1 Gly71Arg with urine urobilinogen. *Nagoya J. Med. Sci.*, 2011; 73, 33-40.

Kendall, AI; Alexander, AD; Walker, AW; Haner, BC. The Bacteriology and Chemistry of adult duodenal contents. Studies in Bacterial Metabolism LXXXII. *J. Infect. Dis.*, 1927; 40, 677-688.

Kim, BS; Kim, JN; Cerniglia, CE. *In vitro* culture conditions for maintaining a complex population of human gastrointestinal tract microbiota. *J. Biomed. Biotechnol.*, 2011; 2011:838040 Epub 2011 Jul 24.

Kohan, A; Yoder, S; Tso, P. Lymphatics in intestinal transport of nutrients and gastrointestinal hormones. *Ann. N Y Acad. Sci.*, 2010;1207 Suppl 1:E44-51.

Korpela, JT; Adlercreutz, H; Turunen, MJ. Fecal free and conjugated bile acids and neutral sterols in vegetarians, omnivores, and patients with colorectal cancer. *Scand. J. Gastroenterol.*, 1988; 23, 277-283.

Krüger, N; Oppermann, FB; Lorenzl, H; Steinbüchel, A. Biochemical and molecular characterization of the Clostridium magnum acetoin dehydrogenase enzyme system. *J. Bacteriol.*, 1994;176, 3614-3630.

Kunz, C; Rudloff, S; Baier, W; Klein, N; Strobel, S. Oligosaccharides in human milk: structural, functional, and metabolic aspects. *Annu. Rev. Nutr.*, 2000;20, 699-722.

Lazar, V. Quorum sensing in biofilms - How to destroy the bacterial citadels or their cohesion/power? *Anaerobe*, 2011; 17, 280-285.

le Roux, CW; Borg, C; Wallis, K; Vincent, RP; Bueter, M; Goodlad, R; Ghatei, MA; Patel, A; Bloom, SR; Aylwin, SJ. Gut hypertrophy after gastric bypass is associated with increased glucagon-like peptide 2 and intestinal crypt cell proliferation. *Ann. Surg.*, 2010; 252, 50-56.

Lederberg, J. Infectious history. *Science*, 2000; 14, 288, 287-293.

Lesuffleur, T; Kornowski, A; Luccioni, C; Muleris, M; Barbat, A; Beaumatin, J; Dussaulx, E; Dutrillaux, B; Zweibaum, A. Adaptation to 5-fluorouracil of the heterogeneous human colon tumor cell line HT-29 results in the selection of cells committed to differentiation. *Int. J. Cancer.* 1991;49, 721-730.

Ley, RE; Turnbaugh, PJ; Klein, S; Gordon, JI. Microbial ecology: human gut microbes associated with obesity. *Nature*, 2006;444, 1022-1023.

Li, ZJ; Jian, J; Wei, XX; Shen, XW; Chen, GQ. Volley, G; Rettger, LF. The influence of carbon dioxide on bacteria. *Appl. Microbiol. Biotechnol.* 2010, 87, 2001-2009.

Lorenzl, H; Oppermann, FB; Schmidt, B.; Steinbüchel, A. Purification and characterization of the El component of the *Clostridium magnum* acetoin dehydrogenase enzyme system. *Antonie van Leeuwenhoek,* 1993; 63, 219-225.

Luoto, R; Laitinen, K; Nermes, M; Isolauri, E. Impact of maternal probiotic-supplemented dietary counselling on pregnancy outcome and prenatal and postnatal growth: a double-blind, placebo-controlled study. *Br. J. Nutr.*, 2010; 103, 1792-1799.

Malakooti, J; Saksena, S; Gill, RK; Dudeja, PK. Transcriptional regulation of the intestinal luminal Na and Cl transporters. *Biochem.J.*, 2011; 435, 313-325.

Martos, G; Contreras, P; Molina, E; López-Fandiño, R. Egg white ovalbumin digestion mimicking physiological conditions. *J. Agric. Food Chem.*, 2010; 58(9), 5640-5648.

Masco, L; Crockaert, C; Van Hoorde, K; Swings, J; Huys, G. *In vitro* assessment of the gastrointestinal transit tolerance of taxonomic reference strains from human origin and probiotic product isolates of *Bifidobacterium. J.Dairy Sci.*, 2007; 90, 3572-3578.

Masuda, K; Nakamura, K; Yoshioka, S; Fukaya, R; Sakai, N; Ayabe, T. Regulation of microbiota by antimicrobial peptides in the gut. *Adv. Otorhinolaryngol.*, 2011; 72, 97-99.

McKenna, M. The enemy within. *Sci.Am.*, 2011; 304, 46-53.

McNab, R; Lamont, RJ. Microbial dinner-party conversations: the role of LuxS in interspecies communication. *J. Med. Microbiol.,* 2003; 52, 541-545.

Mifflin, TE; Benjamin, DC; Bruns, DE. Rapid quantitative, specific measurement of pancreatic amylase in serum with use of a monoclonal antibody. *Clin. Chem.*, 1985 ; 31, 1283-1288.

Mihalik, SJ; Steinberg, SJ; Pei, Z; Park, J; Kim, DG; Heinzer, AK; Dacremont, G; Wanders, RJ; Cuebas, DA; Smith, KD; Watkins, PA. Participation of two members of the very long-chain acyl-CoA synthetase family in bile acid synthesis and recycling. *J.Biol.Chem.*, 2002; 277, 24771-24779.

Mittal, A; Katahira, R; Himmel, ME; Johnson, DK. Effects of alkaline or liquid-ammonia treatment on crystalline cellulose: changes in crystalline structure and effects on enzymatic digestibility. *Biotechnol. Biofuels.*, 2011; 4, 41.

Mitzscherling, K; Volynets, V; Parlesak, A. Phosphatidylcholine reverses ethanol-induced increase in transepithelial endotoxin permeability and abolishes transepithelial leukocyte activation. *Alcohol.Clin.Exp.Res.,* 2009; 33, 557-562.

Montgomery, JA; David, F; Garneau, M; Brunengraberl, H. Metabolism of 2,3-butanediol stereoisomers in the perfused rat liver. *J. Biol. Chem.*, 1993; 268, 20185-20190.

Morens, DM; Folkers, GK; Fauci AS. The challenge of emerging and re-emerging infectious diseases. *Nature*, 2004; 430, 242-249.

Morotomi, M; Sakaitani, Y; Satou, M; Takahashi, T; Makino, T. Production of free bile acids by bacterial deconjugation from conjugated bile acids derived from peptone in a medium. *Nihon Saikingaku Zasshi.*, 1996; 51, 1043-1047.

Musso, G; Gambino, R; Cassader, M. Interactions between gut microbiota and host metabolism predisposing to obesity and diabetes. *Annu. Rev. Med.*, 2011; 62, 361-380.

Narang, AS; Mahato, RI, editors. *Targeted delivery of small and macromolecular drugs.* Chapter 13, 2010. CRC Press. Inc, FL.

Narushima, S; Itoh, K; Miyamoto, Y; Park, SH; Nagata, K; Kuruma, K; Uchida, K. Deoxycholic acid formation in gnotobiotic mice associated with human intestinal bacteria. *Lipids*, 2006; 41:835-843.

Nienstedt, W; Hänninen, H; Arstila, A; Björkqvist, SE. *Ihmisen fysiologia ja anatomia* (in Finnish), 1989; Werner Söderström Osakeyhtiö. Porvoo, Finland.

Nordmann, P; Naas,T; Poirel, L. Global spread of carbapenemase-producing *Enterobacteriaceae*. *Emerg.Infect.Dis*, 2011; 17, 1791-1798.

Okawa, M; Fujii, K; Ohbuchi, K; Okumoto, M; Aragane, K; Sato, H; Tamai, Y; Seo, T; Itoh, Y; Yoshimoto, R. Role of MGAT2 and DGAT1 in the release of gut peptides after triglyceride ingestion. *Biochem. Biophys. Res. Commun.*, 2009; 390, 377-381.

Olsen, PS; Poulsen, SS; Kirkegaard, P. Adrenergic effects on secretion of epidermal growth factor from Brunner's glands. *Gut*, 1985; 26, 920-927.

Oppermann, FB; Steinbüchel, A; Schlegel, HG. Evidence for oxidative thiolytic cleavage of acetoin in *Pelobacter carbinolicus* analogous to aerobic oxidative decarboxylation of pyruvate. *FEMS Micribiology Letters* , 1989; 60, 113-118.

Pelkonen, K; Ylitalo, P. Absorption of oral enalapril in germ-free and microbially-associated rats. *Eur. J. Drug Metab. Pharmacokinet.,* 1989; 14, 101-105.

Pennis, E. Gut Bacteria Lend a Molecular Hand to Viruses. *Science, Microbiology,* 2011; 334, 168.

Pesola, J; Hakalehto, E. Enterobacterial microflora in infancy - a case study with enhanced enrichment. *Indian J. Pediatr.*, 2011; 78, 562-8.

Pesola, J; Vaarala, O; Heitto, A; Hakalehto, E. Use of portable enrichment unit in rapid characterization of intantile intestinal enterobacterial microbiota. *Microb. Ecol. Health Dis.*, 2009; 21, 203-210.

Peuhkuri, K; Sihvola, N; Korpela, R. Dietary proteins and food-related reward signals. *Food Nutr. Res.*, 2011; 55. doi: 10.3402/fnr.v55i0.5955.

Philip, AK; Philip, B. Colon Targeted Drug Delivery Systems: A Review on Primary and Novel Approaches. *OMJ.* 2010; 25, 70-78.

Pitkänen, T; Bräcker, J; Miettinen, IT; Heitto, A; Pesola, J; Hakalehto, E. Enhanced enrichment and detection of thermotolerant *Campylobacter* species from water using the Portable Microbe Enrichment Unit and real-time PCR. *Can. J. Microbiol.*, 2009; 55, 849-858.

*pylori. Annu. Rev. Physiol.* 2003; 65, 349-369.

Rambaud, JC. Bacterial ecology of the digestive tract and defense of the body. *Ann. Gastroenterol. Hepatol. (Paris).*, 1992; 28, 263-266.

Rao, B; Zhang, LY; Sun, J; Su, G; Wei, D; Chu, J; Zhu, J; Shen, Y. Characterization and regulation of the 2,3-butanediol pathway in *Serratia marcescens*. *Appl. Microbiol. Biotechnol.,* 2011; Oct 9, [Epub ahead of print].

Rautio, J; Kumpulainen, H; Heimbach, T; Oliyai, R; Oh, D; Järvinen, T; Savolainen J. Prodrugs: design and clinical applications. *Nat. Rev. Drug Discov.*, 2008; 7, 255-70.

Rayner, CK; Park, HS; Doran, SM; Chapman, IM; Horowitz, M. Effects of cholecystokinin on appetite and pyloric motility during physiological hyperglycemia. *Am. J. Physiol. Gastrointest. Liver Physiol.*, 2000; 278, G98-G104.

Rugutt, JK; Rugutt, KJ. Antimycobacterial activity of steroids, long-chain alcohols and lytic peptides. *Nat.Prod.Res.,* 2011; [Epub ahead of print]

Russell, DW; Setchell, KD. Bile acid biosynthesis. *Biochemistry*, 1992; 31, 4737-4749.

Sachs, G; Weeks, DL; Melchers, K; Scott, DR. The gastric biology of *Helicobacter*

Sam, AH; Troke, RC; Tan, TM; Bewick, GA. The role of the gut/brain axis in modulating food intake. *Neuropharmacology*. 2011 Oct 21. [Epub ahead of print].

Schubert ML. Gastric secretion. *Curr. Opin. Gastroenterol.*, 2011; 27, 536-542.

Scott, RB; Strasberg, SM; El-Sharkawy, TY; Diamant, NE. Regulation of the fasting enterohepatic circulation of bile acids by the migrating myoelectric complex in dogs. *J. Clin. Invest.*, 1983; 71, 644-654.

Shaheen, R. *Bacillus cereus* spores and cereulide in food-borne illness. Thesis (PhD). University of Helsinki, Faculty of Agriculture and Forestry, 2009; 34. Helsinki, Finland.

Shindo, K; Fukushima, K. Deconjugation of bile acids by human intestinal bacteria. *Gastroenterol. Jpn.*, 1976; 11, 167-174.

Shyr, YM; Su, CH; Wu, CW; Lui, WY. Gastric pH and amylase and safety for non-stented pancreaticogastrostomy. *Hepatogastroenterology*, 2002; 49, 1747-1750.

Sinha, VR et al. Microbially triggered drug delivery to the colon. *European Journal of Pharmaceutical Sciences*, *2003;* 18, 3–18.

Smirnov, AY; Mourokh, LG; Nori, F. Electrostatic models of electron-driven proton transfer across a lipid membrane. *J. Phys. Condens Matter.*, 2011; 23, 234101.

Smith, RC; Southwell-Keely, J; Chesher, D. Should serum pancreatic lipase replace serum amylase as a biomarker of acute pancreatitis? *ANZ J. Surg.*, 2005;75, 399-404.

Stanhope, KL. Role of Fructose-Containing Sugars in the Epidemics of Obesity and Metabolic Syndrome. *Annu. Rev. Med.,* 2011; [Epub ahead of print].

Stella, V; Borchardt, R; Hageman, M; Oliyai, R; Maag, H; Tilley, J., editors. *Prodrugs: challenges and rewards. Part 1.*, 2007. Springer, New York.

Storey, R; Gatt, M; Bradford, I. Mucosa associated lymphoid tissue lymphoma presenting within a solitary anti-mesenteric dilated segment of ileum: a case report. *J. Med. Case Reports*, 2009; 3, 6.

Szabo, S; Deng, X; Tolstanova, G; Khomenko, T; Paunovic, B; Chen, L; Jadus, M; Sandor, Z. Angiogenic and anti-angiogenic therapy for gastrointestinal ulcers: new challenges for rational therapeutic predictions and drug design. *Curr. Pharm. Des.*, 2011; 17, 1633-1642.

Thelen, K; Coboeken, K; Willmann, S; Burghaus, R; Dressman, JB; Lippert, J. Evolution of a detailed physiological model to simulate the gastrointestinal transit and absorption process in humans, part 1: oral solutions. *J. Pharm. Sci.* 2011; 100, 5324-5345.

Thomas, F; Hehemann, JH; Rebuffet, E; Czjzek, M; Michel, G. Environmental and gut bacteroidetes: the food connection. *Front. Microbiol.*, 2011; 2, 93.

Toskes, PP. Bacterial overgrowth of the gastrointestinal tract. *Adv. Intern. Med.*, 1993; 38, 387-407.

Trakhtenberg, EF; Goldberg, JL. Neuroimmune Communication. *Science, Immunology,* 2011; 334, 47-48.

Trček, J; Oellerich, MF; Niedung, K; Ebel, F; Freund, S; Trülzsch, K. Gut proteases target Yersinia invasin *in vivo*. *BMC Res. Notes.* , 2011; 4, 129.

Tsuda, M; Hosono, A; Yanagibashi, T; Kihara-Fujioka, M; Hachimura, S; Itoh, K; Hirayama, K; Takahashi, K; Kaminogawa, S. Intestinal commensal bacteria promote T cell hyporesponsiveness and down-regulate the serum antibody responses induced by dietary antigen. *Immunol. Lett.,* 2010; 132, 45-52.

Turnbaugh, PJ; Ley, RE; Mahowald, MA; Magrini, V; Mardis, ER; Gordon, JI. An obesity-associated gut microbiome with increased capacity for energy harvest. *Nature*, 2006; 444, 1027-1031.

van Vliet, AHM; Bereswill, S; Kusters, JG. Ion Metabolism and Transport. In: Mobley HLT; Mendz GL; Hazell SL, editors. *Helicobacter pylori*: *Physiology and Genetics.* Washington (DC): ASM Press; 2001. Chapter 17.

van Wijck, K; van Eijk, HM; Buurman, WA; Dejong, CH; Lenaerts, K. Novel analytical approach to a multi-sugar whole gut permeability assay. *J. Chromatogr. B. Analyt. Technol. Biomed. Life Sci.*, 2011; 879, 2794-2801.

Vissers, MM; Driehuis, F; Te Giffel, MC; De Jong, P; Lankveld, JM. Minimizing the level of butyric acid bacteria spores in farm tank milk. *J.Dairy Sci.*, 2007; 90, 3278-3285.

Vrieze, A; Holleman, F; Zoetendal, EG; de Vos, WM; Hoekstra, JB; Nieuwdorp, M. The environment within: how gut microbiota may influence metabolism and body composition. *Diabetologia.*, 2010; 53; 606-613.

Walters, B;Vanner, SJ. Detection of bacterial overgrowth in IBS using the lactulose H2 breath test: comparison with 14C-D-xylose and healthy controls. *Am.J.Gastroenterol.*, 2005; 100, 1566-1570.

Wang, XY; Lammers, WJ; Bercik, P; Huizinga, JD. Lack of pyloric interstitial cells of Cajal explains distinct peristaltic motor patterns in stomach and small intestine. *Am. J. Physiol. Gastrointest. Liver Physiol.*, 2005; 289, 539-549.

Wells, JE; Berr, F; Thomas, LA; Dowling, RH; Hylemon, PB. Isolation and characterization of cholic acid 7α-dehydroxylating fecal bacteria from cholesterol gallstone patients. *J. Hepatol.*, 2000; 32, 4-10.

Yamanaka, N. Moving towards a new era in the research of tonsils and mucosal barriers. *Adv. Otorhinolaryngol.*, 2011; 72, 16-19.

Yoon, SS; Mekalanos, JJ. 2,3-butanediol synthesis and the emergence of the Vibrio cholerae El Tor biotype. *Infect.Immun.*, 2006; 74, 6547-6556.

Zhao, JR; Yang, H. Progress in the effect of probiotics on cholesterol and its mechanism. *Wei Sheng Wu Xue Bao*, 2005; 45;315-319.

In: Alimentary Microbiome: A PMEU Approach
Editor: Elias Hakalehto ISBN: 978-1-61942-692-4
© 2012 Nova Science Publishers, Inc.

*Chapter VI*

# Development of Microbial Ecosystems

***Elias Hakalehto***
Department of Biosciences, University of Eastern Finland,
FI, Kuopio, Finland

## Abstract

The ecology of micro-organisms is differing from the ordinary zoological common concepts of population ecology. In the latter one individual animals are often divided into predators and prey or herbivores. Essentially, the interactions between animals are in any case taking place only occasionally. The microbes, on the contrary, live in the "chemical universe" where their own cellular metabolism is interconnected and dependent on many other cells or other unit structures. They continuously exercise molecular signaling, thus resembling together a multi-cellular organism. Therefore, Prof. Joshua Lederberg has launched the term "microbiome" to emphasize the true nature of the alimentary micro-organisms in our body, which constitute merely an organ of the body, rather than a separate entity.

The molecular communication between the microbes and man, or the host, is the fundamental criterium for our health. The microbial strains are forming the closest contacts with our tissues. We live surrounded by them, and they cover up our mucosal membranes and skin. Actually, our individual health could be claimed to be dependent on our relation with our own microbes. This is the starting point and the basis of this book, which also is trying to look for the features of our coexistence with these little but influential creatures from their perspective. Although they do not have thoughts or consciousness by themselves, we could open up our own vision by imaginatively shrinking our own dimensions by escaping the common place proportions into the fascinating ecosystem of our microbial partners.

In nature different ecosystems have entirely different basis for building up the balance within themselves. Therefore, it is of utmost importance not to confuse the general ideas of ecology with the exceptional essence of microbial ecology, and especially that of the alimentary tract. For developing our understanding with respect to these issues, we ultimately need novel research tools. Therefore, the PMEU (Portable Microbe Enrichment Unit) is often referred to an instrument for simulating and simplifying the microbial ecosystem. Any change has a direction. Consequently the parameters for a microbial strain make a sense if they are cut into single influences at one time, and then deduced from a behavior of a multitude of cells in a homogenous state.

The generally accepted ideas of microbial coexistence in the alimentary tract include the assumption that the bacteria and other microbes would be in a continuous battle for space and nutrients with each other and with the host. However, on the basis of numerous cultivations with mixed flora in the PMEU (Portable Microbe Enrichment Unit) we have learnt that this is not always the case. Various microbial strains in permissive conditions tend to metabolize actively, and often benefiting from the activities of the other members of the culture. This cooperation takes naturally place within the limits of the environment, but the microbes often also strive for improving the conditions by themselves. This activity is not always producing any population growth, but is directed clearly toward the betterment of the surrounding niches for the microbe in question. Moreover, the action as such is many times beneficial for the entire microbial ecosystem. These ecosystems can be roughly classified according to their stage of competition and by other parameters. This division into various types is explaining the relation of the microbial strains to each other, for example in the use of antibiotics as molecular weapons. In the alimentary milieu the microbes aim at balanced cooperation instead of direct competition.

## Introduction

According to the simulation studies using the PMEU (Portable Microbe Enrichment Unit) cultivation system with different gas compositions and partial pressures the formation of microbial communities take place in a balanced fashion. The intestinal bacteria tend to control their environments by using molecular communication and symbiotic associations. The members of the normal flora could protect each other from pH alterations, osmotic pressure, antibiosis, or other outside pressures.

They attach to surfaces forming colonies and biofilms. Inside these structures the cells jointly exploit the nutritional resources. Their action has an influence on the host body system, which, in turn attempts to regulate the ecosystems inside the body in order to preserve health and beneficial conditions for *e.g.* nutritional uptake. The methods for the control function include bile substances, immunological means, intestinal movements, molecular mechanisms such as defensins, regulating body temperature, and many other ways of adapting to the normal microflora.

This adaptation commences immediately after the birth with respect to the adjustment to the bacteria and other microbes invading the alimentary tract. Even earlier, it is possible that the mother is contributing with immunomolecules to the formation of the defense systems of the neonate. In the breast milk many antibodies and other protective molecules are transmitted to the newborn baby. The PMEU system has proven out to be a valuable instrument in monitoring the normal flora as well as in researching its interactions. This chapter is describing the different studies on this matter using the PMEU. The contributions of these studies on the host health are discussed.

As the baseline of intestinal microbial interactions (and the BIB, Bacterial Intestinal Balance) is laid in the duodenal growth permissive conditions, we have simulated the interactions of facultative anaerobes there using the PMEU technology. Three functions related to carbon metabolism seem to set the frames for microbial co-operation (Figure 1).

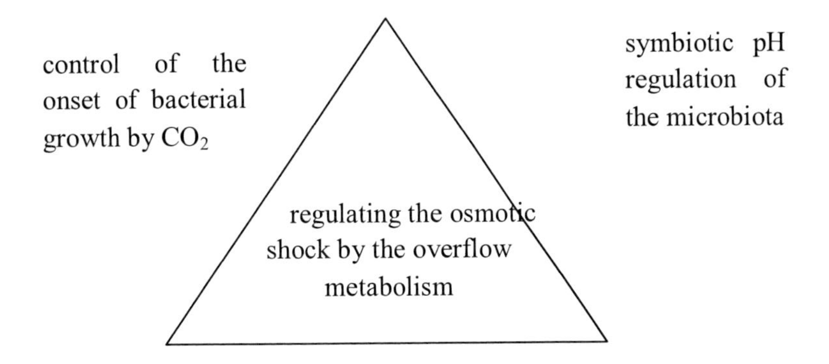


Figure 1. Three functions of the microbial ecosystems related to the carbon metabolism according to the PMEU studies.

The intestinal ecosystems are based on some general principles, which could be illustrated by this triangle. These parameters are functioning both in submerged cultures in the alimentary fluids as well as on the surfaces and in the biofilms. The microbial inoculants to the lower regions of the GI tract derive from the upper parts of the digestive tract, and their actions are in a healthy individual coordinated with the different digestive and metabolic functions of the host body. It is also typical for the microflora in the alimentary tract that its members are in a resting state, which is awakening to sudden growth or other activities. This onset of growth could be triggered with various signals, such as carbon dioxide. This gaseous component was shown to promote the growth of *E.coli* in the PMEU [Hakalehto and Hänninen, 2012]. It has been also found that at suboptimal $CO_2$ concentration the *E. coli* growth rate is controlled by the $CO_2$ concentration [Repuske and Clayton, 1968]. In fact almost a century ago, after studies on many bacterial species, Volley and Rettger (1927) stated:"The removal of carbon dioxide from an environment, which is otherwise favorable, results in complete cessation of bacterial growth." Very seldom the bacteria in normal conditions aim for overgrowth within the intestines. For example, the $CO_2$ produced by the normal flora is supposed to keep the clostridia active in the colon thus possibly preventing them from excessive colonization of the surfaces [Hakalehto and Hänninen, 2012].

The intestinal ecosystems are formed when the microbial strains benefit from each other, and when the host body system is living in a mutual symbiotic relationship with the microflora. In order to keep the balance and to maintain the beneficial conditions, all members of the ecosystems are signaling to each other. They also share the nutrients in a fashion that reflects the "common good" of the biome. It is not the ultimate benefit of any micro-organism to totally outgrow the others. If so were the things, the fittest would obviously have been ruled out the others. Consequently, in case of a microbial mixed culture in the alimentary tract, the strive is not toward a monoculture but into the direction of increasing biodiversity. This principle is safeguarding the health of the host human being, whose regulatory systems are favoring the multitude of microbe strains within the gastrointestinal tract. The more monotonous the microbial mixed populations are the more vulnerable they are to changes. Therefore, any new-coming species are integrated into the existing community [Hakalehto *et al.*, 2010]. This principle greatly supports the balanced expansion of the ecosystems into novel niches within the body. Any disintegrated parts inside the system form remarkable risks, such as diverticulosis, of developing into diseased condition. On the other hand, there are natural dividers in the gastrointestinal tract such as the ileocecal valve, which make up zones or sections into the structure. For example, the pylorus is serving as a borderline between the acidic stomach and duodenum, where nutrients are taken up more readily. The conditions are favorable also for many bacteria, but due to body's defense between mixed acid fermenting and 2,3-butanediol producing facultative strains maintain and support the pH balance [Hakalehto *et al.*, 2010]. In case of reflux disease, the bilial material could burst into the wrong direction causing heartburn together with the highly acidic gastric juice. Also the hydrochloric acid originating from the gastric membranes could overflow into the duodenum causing detrimental conditions both for the normal flora and for the host.

Also the division between terminal ileum and cecum is somewhat sharp micro-biologically, although some microbes are supposed to leak also upstream from the large intestines into the small bowel there. Between the two ends of the small intestines there is an ingenious loop of bile substances, which are secreted in to the duodenum through *papilla Vateri*, and recollected in the terminal ileum [Johnston *et al.*, 2011]. Meanwhile, they participate in the uptake of fat substances. On their journey along the small intestines (duodenum, jejunum and ileum) micro-organisms take part in this important metabolic passageway. The situation in the cecum resembles to some extent that of the silage production where also the microbiological balance between the lactic acid bacteria (LAB) and the butyric acid bacteria (BAB) plays an important role. For example, aerobic conditions during the silage fermentation may cause the formation of BAB spores, which then spoil the cheese quality in food processing [Vissers *et al.*, 2007].

## Common Goals of an Microbial Population in the Intestinal and Other "Biospheres"

According to Floch [2011] the intestinal ecosystem consist of four components (see Table 1).

**Table 1. Definition of intestinal microbial ecosystem according to [Floch, 2011]**

| | |
|---|---|
| 1 | Luminal gastrointestinal tract and its epithelium |
| 2 | Secretions of the tract |
| 3 | Nutrients and foods entering the tract |
| 4 | Microflora |

**Table 2. Different modes or categories of microbial metabolism**

| mode of metabolism | major objective |
|---|---|
| Catabolism | maintenance of life functions |
| Anabolism | propagation -> population growth |
| Over-flow | improvement of living conditions |
| Sparing of resources | spores and other specific forms for survival |

Based on our studies with the PMEU system, the outlines of the microflora existence could be classified to some extent. It seems obvious that the microbes carry out at least four modes of metabolism with different major objectives (Table 2).

When observing a living system, such as host matrix, or a host-microbial ecosystem it seems obvious that the microbial strains carry out their functions in cooperation with other forms of life. This includes the host organism as well as the other microbes. Any attempt to overgrow the other microbes totally in this environment will eventually lead to diminishing options for the "aggressor" strain itself. The overall diversity is playing an important role for the survival of all strains. Variable conditions need variable solutions and metabolic potential from the entire microbial community. If not so, the microbial successions and coexistence would be replaced with monocultures ultimately incapable of dealing with altering environmental conditions. Therefore, the two leading principles in the life of microbes in the intestines are:

1. strive for cooperation, and
2. manipulation of the surroundings into more favorable direction.

In a spoiled bowl of a dessert ("kiisseli" in Finnish) the sweet food is both having high osmotic pressure due to sugar addition, as well as low pH as a result of the acidic berries used for preparing it. This food is, however, getting spoiled by the microbes, such as moulds (Figure 2). Somewhat similar conditions occur in the duodenum when food is arriving the tract. It is acidic after the gastric region, and also often containing high hydroscopic pressures. In any case, the bacteria and other microbes need to deal with it [Hakalehto *et al.*, 2008]. They also use some type of overflow metabolism in order to convert the substances into less stressing ethanol (absorbed by the host), and 2,3-butanediol, which is neither assimilated fast by the microbes or the host nor degrade by them. [Hakalehto *et al.*, 2012]

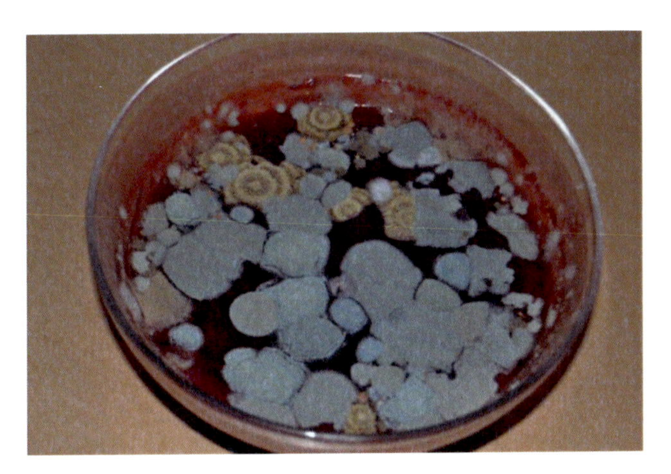

Figure 2. Moulds in a spoiled bow of a dessert.

These leading motifs of microbial life in the host-microbe biosphere are direct consequences of the nature of this environment, where individual cells and the strains need to maintain the balance in the specific conditions of the microbiome. The fundamental characteristics of this ecosystem are determined by

- metabolic cooperation
- signals between cells
- gaseous impacts and interchange
- removal of waste materials
- preconditioning and refinement of the environment
- surface interphases
- energy generation
- participation in host functions
- circulation of matter within the communities
- molecular communication with the host

In real-life ecosystems all these activities are directed toward the accomplishment of the above-mentioned two main principles or aims of microbes in the alimentary tract surroundings in general. These conditions essentially differ from the circumstances in the ordinary laboratory cultures. In the PMEU environment the simple basic parameters resemble the true conditions around the microbial cell. It is of secondary importance whether this cell is belonging to a biofilm or is in a submerged state. In any case, the fundamentals of microbial life are depending mostly on the ten above-mentioned characteristics. Any microbial cell is responding to its proximal surroundings. Its accomplishments in specific conditions could be studied in the PMEU. There all cells are in equal metabolic status. Accordingly, any conditions resulting in attachment could be recorded from the culture samples [Hakalehto *et al.*, 2007].

## Allegory of Microbial Ecosystem with Rainforest Vegetation

If we consider the energy-generating machinery of a bacterial cell, it is possible to study it as a reflection of surface-associated phenomena of the cells. The electric potential across the cell walls and membranes constitute a power house of the cell, which could be disrupted by such antimicrobial peptides as magainins [Westerhoff *et al.*, 1989]. Therefore, for an individual cell it is of utmost important what is the condition (or chemical sphere, or phase) around the cell. The cells communicate with each other taking into account this condition. It is subjected to the so called diffusion limitation known in the fermentation science [Guedon *et al.*, 1999]. In practice this means that around active cells the amount of nutrients tend to sink down while the waste substances accumulate. This "limitation" leads to natural deprivation of growth potential. Overcoming this results in growth enhancement. This unique set up is achieved on the duodenal epithelia where the host body is speedily taking up and absorbing the nutrients, while the bacteria participate [Hakalehto *et al.*, 2008]. The gaseous substances and volatile compounds of the latter ones are resulting from their metabolic activities. This multitude of reactions is highly dependent on the interphases between solid surfaces and liquid solutions on the other hand, and the partial pressures of the gases on the other. The microbes extremely quickly liberate the huge chemical energies stored in the food substances mainly for the host usage. This situation is somewhat parallel to the jungle soil [Sanchez *et al.*, 1982]. Also there in the tropics, the material is converted into the fuels and building blocks of the huge variety of different organisms and living forms through nutritional chains [Kutsch *et al.*, 2011]. All this variability in life and its forms and in their interactions is built upon the basis of microbial life in soil, where *the nutrients are at least adequately or even excessively available.*

## Counter-Effect of Competition and the Ground It Rises from (in the Alimentary Tract)

If we look soil as such, and in normal condition, it is a place poor in nutrients. This is true especially if we compare the real metabolic outcome with *the enormous potentials of various microbes and their communities to process the materials.* These resources are bound to remain unrealized in the circumstances, where the scarcity of nutrients is a prevailing condition, and where these microbes are subjected to a continuous stress. The organic compounds in soil constitute an inadequate supply for the microbes, and they become an object for fierce competition. This is not the case in the duodenum after the meals, for instance [Hakalehto *et al.*, 2008]. Therefore, the antibiotic action and the production of antibiotics is not belonging to the latter ecosystem as such a self-evident prerequisite for survival as it is in the soil ecosystem. In order to understand this difference and its contribution to the structure and function of the microbiome, and consequently to our health we need some further consideration. *The intestinal ecosystem is not benefiting from the diseased gut, only individual pathogenic strains do so temporarily.*

A big proposition of the microbes in soil produces antimicrobial substances in order to struggle for space and to compete for the inadequate nutritional resources. This competition is a normally prevailing condition in soil with having such exceptions as co-metabolism [Dalton and Stirling, 1982; Wolfaardt *et al.*, 2004], where two or more species together deal with recalcitrant or other tediously consumable chemical substances. In these cases, the common principle or strive for co-operation in the microbial world is demonstrated. Otherwise, the struggle will continue. Nevertheless, the soil usually retains its capacity for degrading various substances by microbial strains remaining latent in a scattered or even resting state ecosystem (see also Table 3). In duodenal enhanced ecosystems, the nutrients are available much more frequently, usually in a cycled fashion (Hakalehto *et al.* 2008).

In human healthcare, one major emerging risk is the formation of antibiotic resistant strains, which influence the entire healthcare system [Hakalehto, 2006]. The microbes are omnivalent entities and if they are not within the control of the body system and/or its normal microflora, this brings along a real problem of infectious diseases, or at least harmful overgrowth of microbes. The antibiotic resistance traits have been increasingly abundant within microbial populations, and they are spreading from harmless strains to tedious pathogens [Hakalehto, 2012].

The bacteria in the small intestines are not usually involved in such a huge competition as like the organisms in the soil have to take part in. Actinomycetes produce more than 90% of the medically exploited antibiotics, and they belong to the latter type of environment, soil microbial ecosystem [Davelos *et al.*, 2004]. Even though there are many examples on the degradation of recalcitrants by collaboration of various strains using co-metabolism, direct competition is much more abundant in soil than in the case of human microbiome. In the duodenum relatively scarce bacterial cells reside between the meals in the mucus and on the surfaces. The *Enterobacteriaceae* family is well represented on the intestinal walls [Peach *et al.*, 1978; Marks *et al.*, 1979].

As indicated by our studies with mixed cultures in the PMEU it seems likely that any given microbial strains in the duodenal conditions are actively seeking for a balance [Hakalehto *et al.*, 2010; Hakalehto *et al.*, 2012]. Any hegemony by single strains would be leading to monocultures and eventually become counter-effective and disastrous also for the dominant strain itself. In fact, the choice of cultivation methods in microbiology has produced some misleading concepts, such as the application of the term, “survival of the fittest”. Also the view of particular microbes having a niche in the gut is determined by their favorable nutrient [Chang *et al.*, 2004] should be re-examined on the basis of our recent findings [Hakalehto *et al.*, 2012]. Namely, in the PMEU in favorable conditions resembling the duodenal chime or membranes, the growth of all components of the microflora is taking place at the optimal rate. This allows enhanced exchange of gases and exploitation of nutrients, as well removal of metabolic wastes. As a consequence, growth speeds in the populations, as well as metabolic interactions take place in an accelerated pace. However, this enhanced metabolism is highly interactive as can be seen in the case of the symbiotic interactions of *E. coli* and *Klebsiella mobilis* introduced also elsewhere in this book [Hakalehto *et al.*, 2008]. This co-operation is not disturbed by the incoming *S. aureus* or *B. cereus* when the latter ones were inoculated either separately or together into the co-culture in the above mentioned two facultatives [Hakalehto *et al.*, 2010]. As a consequence of this intrusion of novel strains the cell concentrations (CFU’s) of the two coliforms were lowered onto one fifth of the level in their dualistic culture. However, they remained as major strains and maintained their mutual collaboration and the newcoming strains settled on a level of more minor presence. This kind of mutual or common control of microbial ecosystems is likely to be a common phenomenon in the intestines. Consequently, it could explain the high degree of order in the healthy microbiome. In the above mentioned examples klebsiellas and colibacteria are able to together establish a functionable symbiotic relationship including enhanced metabolic potential for both bacteria. The members of this subcommunity are not directly competing against any other species in the gut. Instead, they are able to control the fundamental conditions in the duodenal tract, and thus determine the structure of the entire microbial community and the ecosystem.

The antibiotic substances are not used in the small intestines as like they are facilitating competition in the soil. This is a consequence of the universal strive of gut microbes for:

- co-operation between strains
- controlling their environment

The antibiotics, such as colicin (*E. coli*) or lantibiotics (*Streptococcus sp.*) are produced for maintaining the balance, not necessarily for dominating the other microbes [Bures *et al.*, 1979; Khalil, 2009]. Such distinction makes a big difference in the microbial ecology considerations. Same general principles are as applicable to all situations. In the gut the microbes could function at extreme speed (especially in the duodenum where high nutrient concentrations are occasionally met for short periods of time). These enhanced growth simulations are effectively being carried out in the PMEU [Hakalehto *et al.*, 2010].

Population maintenance and cooperation with the host are not dependable on the growth of different bacterial populations, but on the balance between them, and between the microbiome and the host body system.

As a result of the above-mentioned general principles in the small intestinal ecosystem, the strains in these areas are not striving for eliminating the others, but merely for structural cooperation. This mode of behavior is favored by and sustained with the host body functions, such as immunological or neurological activities [Lyte *et al.*, 2006]. In order to understand more about the alimentary microbiome and its contributions to our health, we need to investigate the cooperation between active strains.

It is remarkable that the antibiotic resistant markers are spreading particularly in the biofilm conditions, or in the conditions where remarkable organic ingredients form a sludge or equivalent with considerable diffusion limitation. There the emphasis on competition is ambivalent. In fact, microbial communities are possible to be categorized into four groups (see Table 3).

The complete energetic efficiency of a microbiome is based on liberating chemical energies bound in the food or feed substances.

Our views on microbial degradation processes are often restricted to the two first types. Therefore, the full capacity of the microbial metabolic activities could remain experienced as distant or unknown to us. However, if thinking about a ruminant such as cow, we could get an idea of the efficiencies of the combined host-microbiome ecosystem in exploiting the reserves in such feed as grass, for example. These animals produce hides, bones, meat and milk as well as loads of manure. When slaughtered, they still could have about 50 kg of feed undergoing processing in their stomach system. Microbes participate in this process [Flint, 1997]. There have been selected a typical flora for the rumen [Watase and Takenouchi, 1978; Hobson and Wallace, 1982].

The rumen flora includes many obligate anaerobes, such as *Ruminococcus flavefacians, Ruminococcus albus, Bacteroides succinogenes, Butyrivibrio fibrisolvens* [Jayne-Williams, 1979; Rychlik and Russell, 2002]. In these conditions it becomes inevitable to see the remarkable potential of anaerobic metabolism demonstrated also in case of anaerobic organisms [Hakalehto *et al.*, 2007].

The ruminant cow with the aid of its microbes effectively exploits the chemical energies stored in the macromolecules and other substances in grass or in other feed. Particularly such chemical structures as cellulose or hemicelluloses are degraded by the rumen microorganisms and their energy content is liberated [Williams *et al.*, 1989]. Vast amounts of biomaterials are produced from the cattle nutrition by the ecosystem comprised of the animals and their microbes.

The high efficiency of the host-microbe systems is based on the existence of enhanced ecosystems (see Table 3). Often the distinctions or borderlines between the different types are somewhat blurred.

Research on them is made possible by the PMEU technologies. These ecosystems are relatively rare compared with the scattered or concentrated systems in which ecosystems the "laws on the microscale level " are totally different from the accelerated pace in the enhanced ecosystems. The latter ones do therefore give an explanation on several phenomena in the body and its organs.

In the circumstances of the alimentary tract especially the colon microflora is living under conditions where competition could often take place [Perunova *et al.*, 2010]. On the other hand, if the biofilm structures are to be formed, not much exclusive antimicrobial activity could be exhibited by any single strain or group of micro-organisms. Such examples as bacteriocins as antibiotics produced by gut bacteria have been presented in the literature [Rychlik and Russell, 2002; Rea *et al.*, 2011].

**Table 3. The system of microbial communities**

| Type | Nutrients | Diffusion limitation | Example |
|---|---|---|---|
| 1. Scattered ecosystem | Scarce and dispersed, sometimes occasionally available | Present | Common soil habitat |
| 2. Concentrated ecosystem | Highly abundant | Present | Manure, biofilms, colon |
| 3. Enhanced ecosystem | Frequently highly abundant, often in cycles | Removed or lowered | Duodenal epithelia |
| 4. Resting state, or restricted ecosystem | Variable | Insignificant or less determining | Hot or salt spring, very poor soil condition |

The effective protection against microbes is often based on extensive usage of antimicrobial substances [Syed *et al.*, 2010]. Emerging antibiotic resistant strains constitute a severe risk on the population and the safety in hospitals [Hakalehto, 2006; Visca *et al.*, 2011]. In order to struggle against disseminating hazardous antibiotics resistant strains, it is of utmost importance to understand and further investigate the microbiome as a whole [Cox and Popken, 2010; Hakalehto, 2012]. If the strategies for this battle are taking into account the protective means provided by the healthy normal flora, the outcome of the confrontation could be a more positive one. The PMEU cultivation is demonstrating the antibiotic effects of different substances in a very sensitive way. The different possible growth curves in the PMEU Spectrion® are presented in Figure 3.

The microbes in nature are often prepared for confrontations, where the antibiotic substances are the "molecular weapons" used by bacterial, mould and other microbial strains for conquering more space in the populated microbial communities. They are directed toward the competing other strains in soil, for example.

In medicine, these molecules of microbial origin have been used as healing agents since the discovery of penicillin. This has been facilitating the eradication of many devastating communicable diseases from our societies. However, if a particular pathogenic bacteria could produce resistant forms against otherwise effective antibiotic medicines, they again constitute an increased health risk [Hakalehto, 2006]. The bacterial strains in the intestines carrying the antibiotic resistance genes have been shown to transfer the corresponding genetic material and traits to other strains and even to other species [Hakalehto 2012].

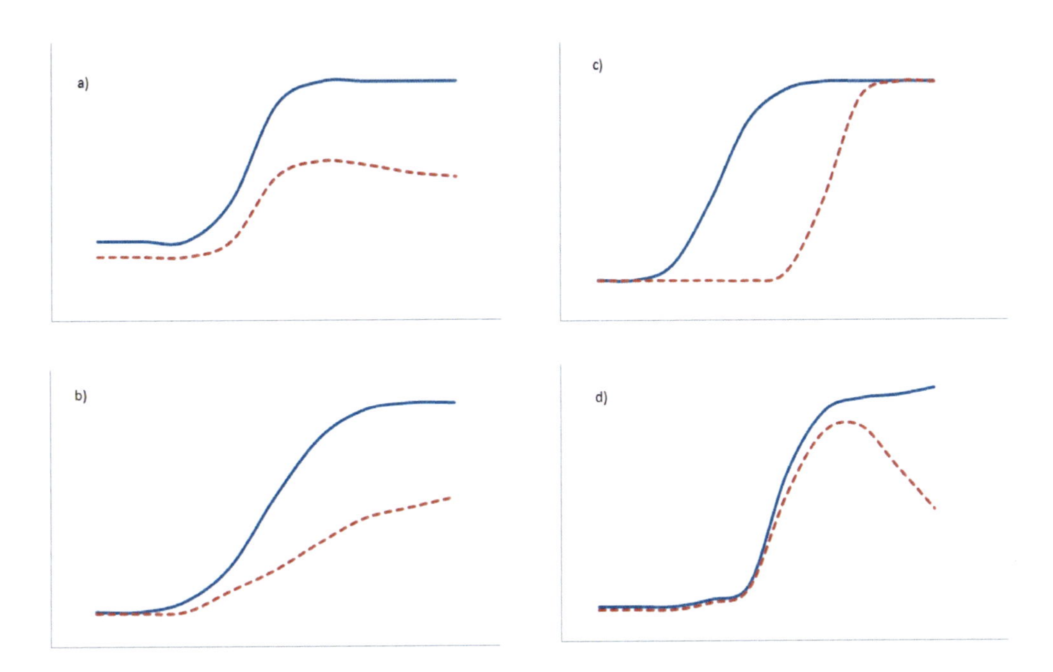


Figure 3. Model culture growth patterns in the PMEU Spectrion® as published first by Hakalehto (2012). The upper unbroken lines represent bacterial growth without added antibiotics; the lower ones illustrate the effect of different antibiotics on the growth. The antibiotic influence is directed toward a) the volume of growth, b) the growth rate, c) the onset of growth, or d) the accelerated cell death. These curves illustrate the potential of the PMEU method in evaluating the effects of antibiotics on the bacterial cultures in conditions simulating the active state cells in the human body.

Microbial reactions with antibiotics have been studied in the PMEU [Hakalehto *et al.*, 2010; Hakalehto, 2011, 2012]. Responses of the PMEU cultures of some bacteria to the clinically important antibiotic, cefuroximide, are presented in the Figure 4.

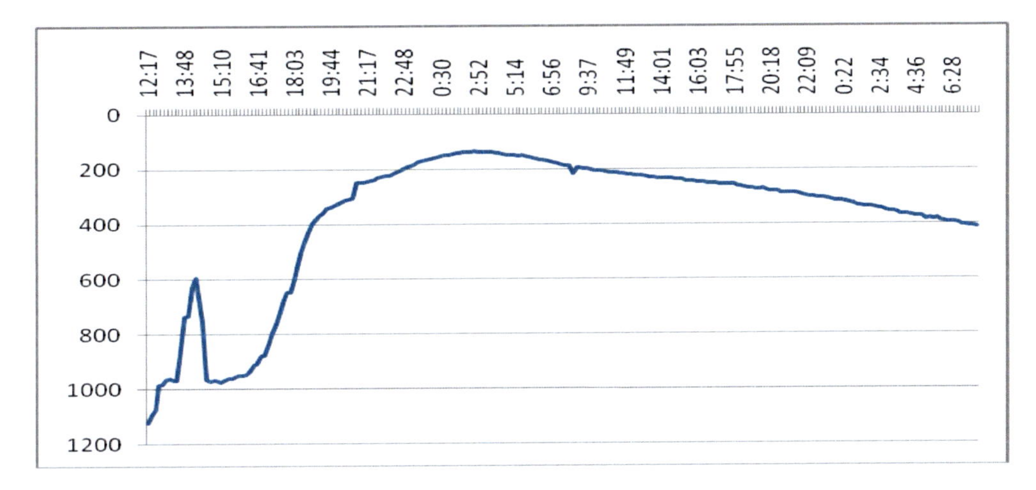


Figure 4a. Growth of *Staphylococcus aureus* in the PMEU Spectrion® without any antibiotic addition. The culture medium was the TYG (Tryptone, Yeast extract, Glucose) medium at 37 °C , and sterile air was applied for stirring and bubbling of the culture.

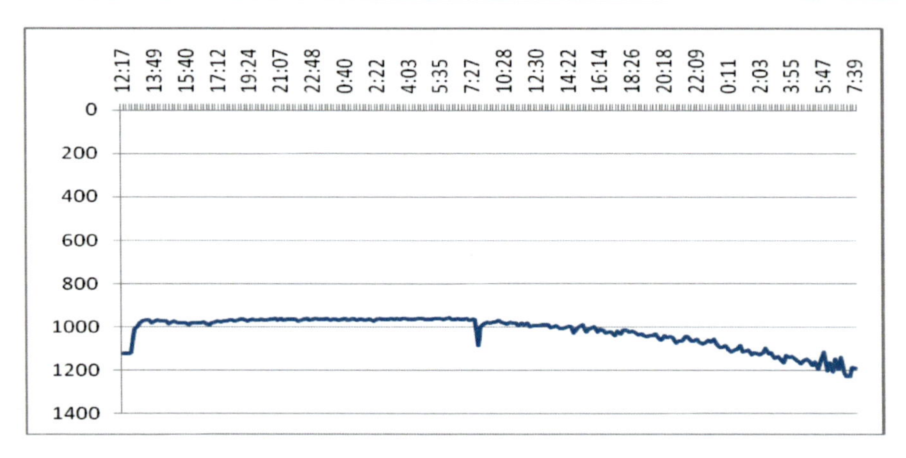


Figure 4b. Growth of *Staphylococcus aureus* in the PMEU Spectrion® with clinical concentration of cefuroximide (Zinacef™) at a final concentration of 0.025mg/ml. For other cultivation conditions, see Figure 4.a.

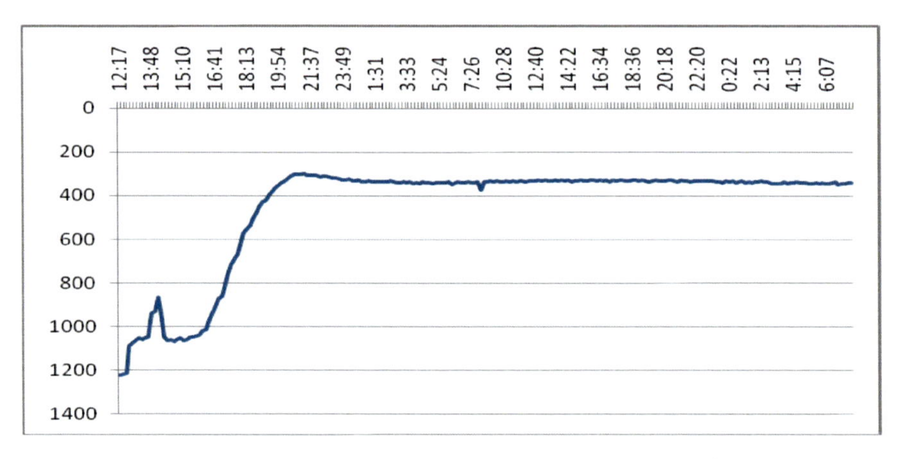


Figure 4c. Growth of multiresistant *Enterobacter cloaceae* in the PMEU Spectrion® without any antibiotic addition. For other cultivation conditions, see Figure 4a.

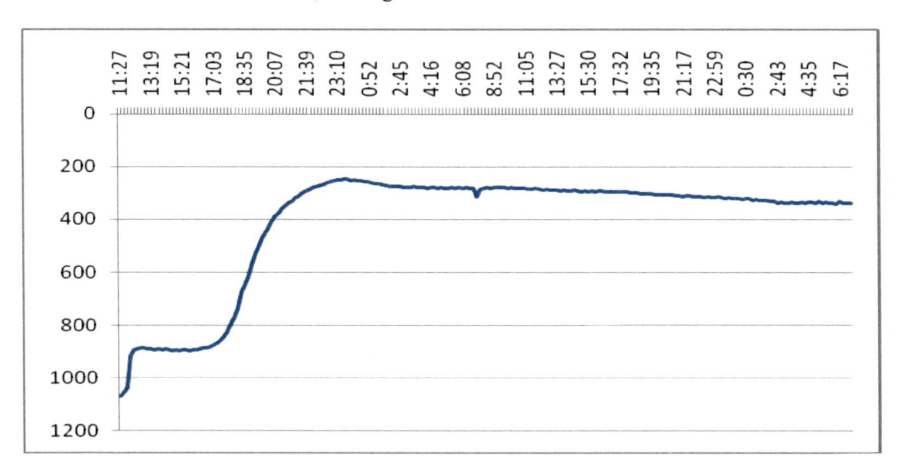


Figure 4d. Growth of multiresistant *Enterobacter cloaceae* in the PMEU Spectrion® with clinical concentration (0.025mg/ml) of cefuroximide (Zinacef™). In this case only the initiation of the growth had been delayed with a couple of hours, and no other clear effects of the antibiotic medication were seen. For other cultivation conditions, see Figure 4 a.

In this particular experiment all other clinical isolates were made of strains, which occurred mainly in the body, whereas the *Enterobacter cloacae* is typically a bacterium present in the equipment in the hospitals. It could be supposed that this strain in the neonatal intensive care unit, even though isolated from a newborn patient, was originating from some catheters or other instruments. It also was more stable against the antibiotic commonly used in this ward. The strains isolated from "pure" human origins had not acquired such resistance markers.

Note that these patients had lived only for several days, and their microflora had not been adopted to the antibiotic usage. In natural populations such adaptations inside the gut could hypothetically be less required from the bacterial point of view than in the harsh conditions inside the clinical unit and its equipments. Consequently, it could also be hypothesized that the most rapid causes for multiresistance lie in the hospital environment and the equipment where monocultures more easily are formed.

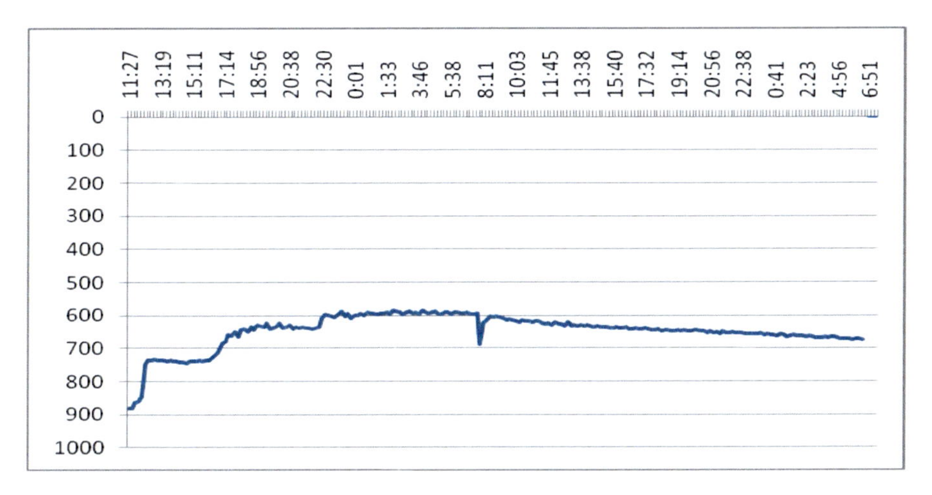


Figure 4e. Growth of *Sreplococcus agatilactis* in the PMEU Spectrion® without any antibiotic addition. The growth is very weak in these conditions. For other cultivation conditions, see Figure 4 a.

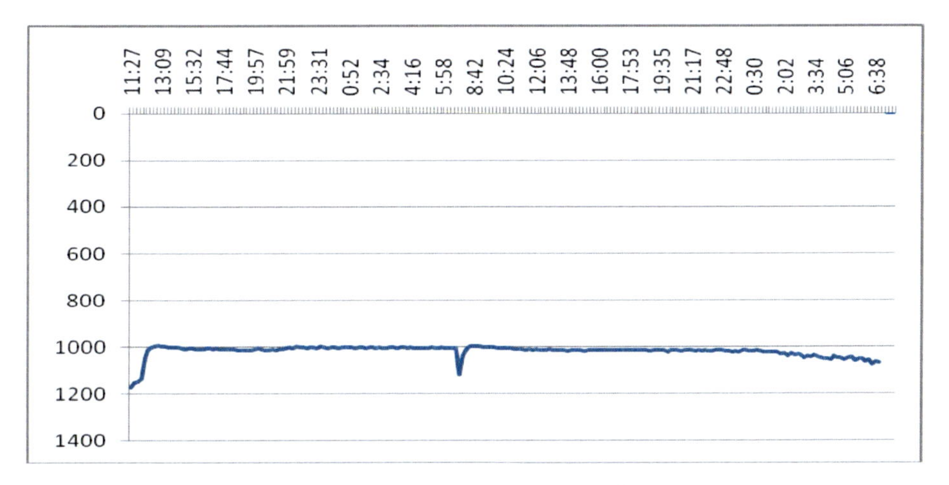


Figure 4f. Growth of *Sreplococcus agatilactis* in the PMEU Spectrion® with clinical concentration (0.025mg/ml) of cefuroximide (Zinacef™). All growth has ceased, or been prevented by the antibiotics. For other cultivation conditions, see Figure 4 a.

The antibiotic resistance factors are usually carried in extra-chromosomal genetic material, in plasmids. For example, in *E. cloacae* there have been found plasmid genes carrying multiresistance markers for producing resistance against five different antibiotics [Sato, 2002].

In the presence of the specific antibiotics, these genetic elements carrying out the resistance factors could spread from cell to cell. Consequently, the excessive use of antibiotics in the hospital environment is promoting the resistance genes in the bacterial population. The plasmids carrying the resistance are often able to transfer even across species boundaries, which phenomenon is one of the reasons for the extending risks of multiresistant epidemics. Therefore, the overall hygiene in the hospital areas and equipment where antibiotics are used is of utmost importance.

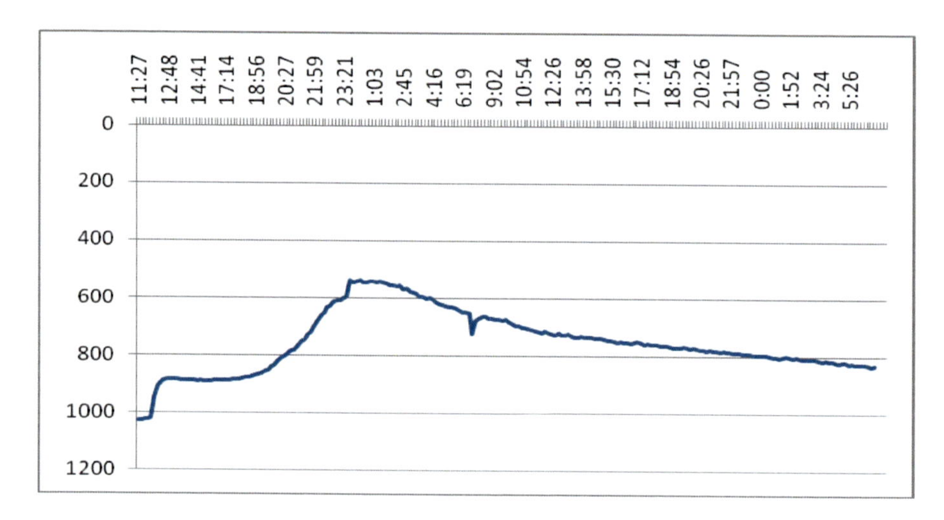


Figure 4g. Growth of a coagulase negative staphylococci in the PMEU Spectrion® without any antibiotic addition. The growth has started relatively well but then some impaired detection has occurred, perhaps due to the precipitation of the cells. For other cultivation conditions, see Figure 4 a.

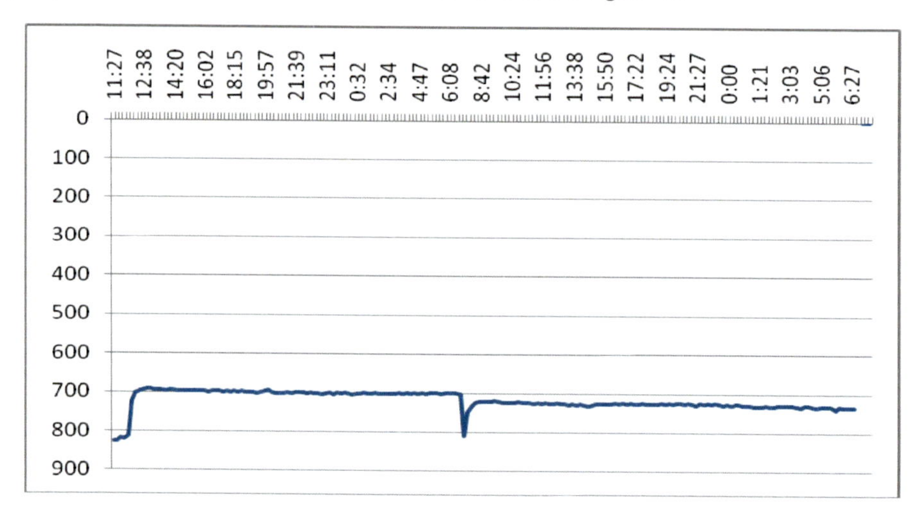


Figure 4h. Growth of a coagulase negative staphylococci in the PMEU Spectrion®. Absolutely no growth is occurring in the presence of a clinical concentration (0.025mg/ml) of cefuroximide (Zinacef™). For other cultivation conditions, see Figure 4 a.

The competition is not the primary task for the duodenal microbes [Hakalehto *et al.*, 2008]. As indicated below in the Figure 5.a.- c. the two facultative species of *Klebsiella mobilis* and *Escherichia coli* are benefiting from each other, The klebsiellas apparently utilize some compounds produced by the *E. coli*.

It is also possible to observe repeatedly in the PMEU Scentrion® that the mixed culture of these two bacteria is producing equal amounts of cells but about half of the volatile emission than the two pure cultures separately from a doubled amount of nutrients. This data is presented in the Figure 5 and in Table 4 below.

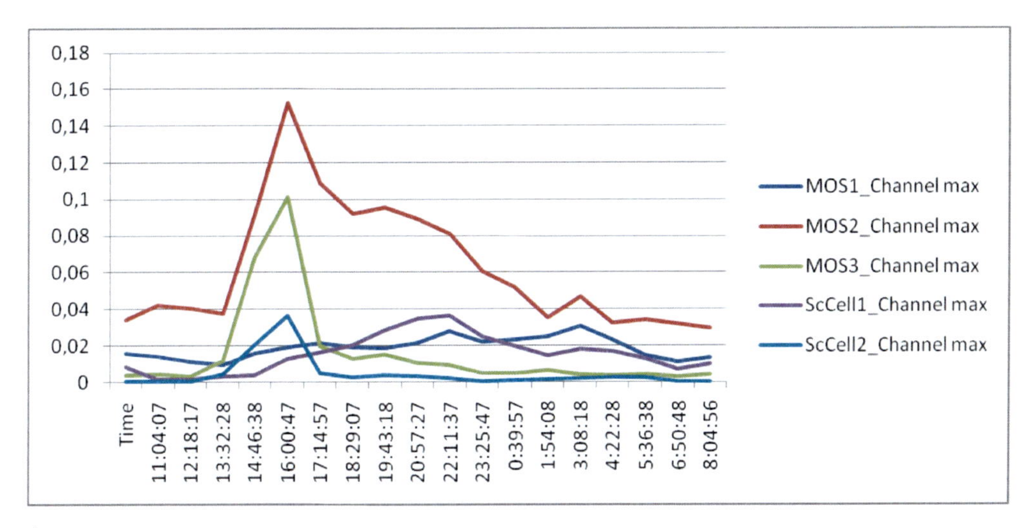


Figure 5a. Growth of *Klebsiella mobilis* in the PMEU Scentrion®in TYG broth (Tryptone Yeast Extract Glucose) at 37 °C. The X-axis shows the time elapse. On the Y-axis there is the relative drop in the transmission as measured by various Metal Oxide Sensors (MOS) and the Semiconductive Cells (ScC) of the device. For further explanations see previous publications regarding the PMEU Scentrion® [Hakalehto *et al.* 2008, 2009, 2010, 2011a].

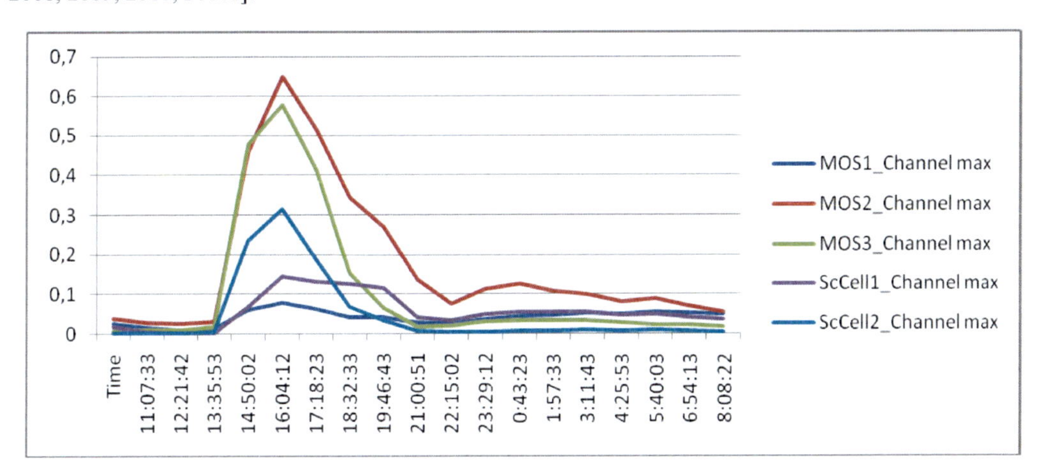


Figure 5b. Growth of *Escherichia coli* in the PMEU Scentrion® in TYG (Tryptone Yeast Extract Glucose) broth at 37 °C. For other conditions of the culture and the measurement, see Figure 5 a.

**Summa summarum: In specific conditions of "collapsed competition"** the microbes need not keep developing any molecular weapons for securing their survival. Instead, competition becomes somewhat counter-productive. It destroys the biodiversity, and thus lowers the potential of the microbial community to preserve as a whole. In the PMEU cultivation the multitude of strains could produce growth at maximal rate together as long as the limits of the culture media have not yet been reached as a community [Hakalehto *et al.*, 2010; Hakalehto, 2011]. They also get the nutrients exploited more efficiently by a joint metabolic activity, which has been demonstrated in the PMEU Scentrion® experiments. The various subcultures are during this phase maintaining the optimal growth rate without demonstrating any competitiveness. This situation differs totally from the conditions in many other natural environments, such as soil.

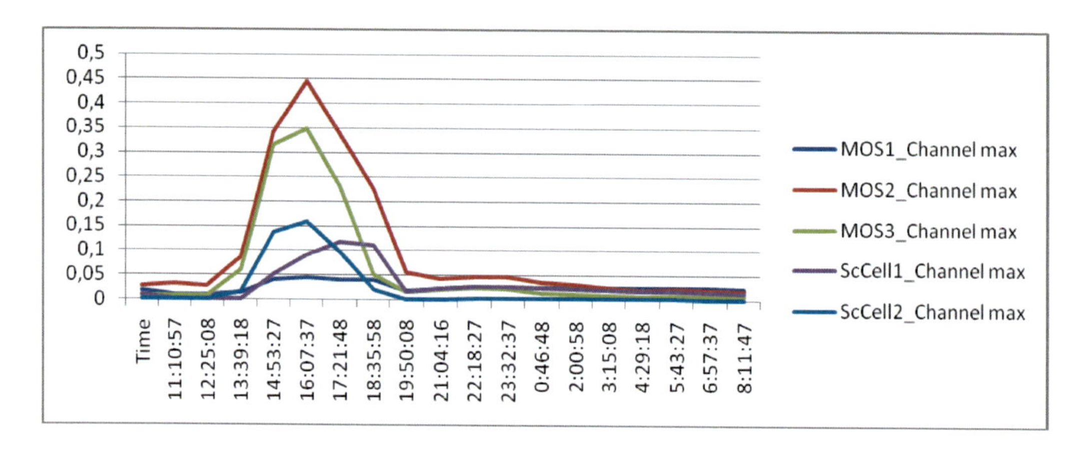


Figure 5c. Growth of both *K. mobilis* and *E. coli* in the PMEU Scentrion®in TYG (Tryptone Yeast Extract Glucose) broth at 37 °C. For other conditions of the culture and the measurement, see Figure 5 a. It is noteworthy that 1. the amplitude of the curves is at least 1/3 less than in the Figure 5 b. with the *E. coli* growing alone; 2. the "shoulder" of the curves is being removed (when compared to Figure 4.b.) indicating the consumption by *K. mobilis* of some substances in the mixed culture liberated by the *E. coli* subpopulation. As a net effect the joint culture of the two facultatives, they emit less volatiles than the the two pure culture emission summed up would be (about 50%) In Table 4, the growth figures of all three cultivations are presented (see below).

**Table 4. Bacterial colony counts at fully grown PMEU Scentrion® cultures after 10 hour enrichment in the TYG medium. Plate counts obtained from inoculated ChromAgar™ Petri dishes cultivated at 37 °C. Both strains were multiplying in nearly as high numbers in the mixed culture as in the pure cultures. The sizes of the inoculations were the same in all cultures. This result indicates that the two facultative strains could better exploit the nutrients of the medium together, as they both were able to grow equally well in the co-culture**

| Microbial culture | Colony count/ dilution -7 (cfu) |
|---|---|
| Pure *K. mobilis* | 85 |
| Pure *E. coli* | 250 |
| Mixed *E. coli* with *K. mobilis* | 117 *E. coli* / 78 klebsiellas |

# Biofilms Are Concentrated Ecosystems

Biofilms are in a way a sign of stagnation of a microbial ecosystem. These structures have been researched extensively during the last decades. The most developed biofilms in our body system could be found in the colon [Andrade *et al.*, 2011]. Understanding their function in important with respect to health, but these structures or microbial communities are in fact not effecting as much on the nutrient uptake, as the latent microflora of the duodenum or in other small intestines.

In the colon the biofilms participate in the formation of eg. some vitamins such as vitamin K2 [Vermeer and Braam, 2001]. In fact, perhaps the absorption of this vitamin can take place mostly in the proximal small intestine, where some colon bacteria are transported with some backward flowing of the intestinal fluids. Vitamin K2 (menaquinone) is reducing coronary calcification, thereby decreasing the risk of cardiovascular disease [Beulens *et al.*, 2009]. The adequate supply of this substance requires a functioning gut flora. However, as in the colon the waste materials of the body are removed, its major task is to maintain the cleanliness of the body system. In constipation the toxic compounds could easily get reabsorbed into the body. The microbes continue processing the remaining parts of the food, and constitute a network of metabolical reactions and relations. In order to understand their importance for the host nutrient uptake, more research is needed. Self-evidently, the proper mechanical functions of the colon ensure the entire proper action of the intestines. Delays in these could cause accumulation of toxin compounds.

If the colonic microflora is considered as a health factor for their human host, it is of utmost importance that the organisms belonging to that flora are decoding poisonous, mutagenic or carcinogenic substances instead of producing them. This is extremely important for the health in the long term, especially taking into account the excessive use of chemical substances belonging to our industrialized life-style. In chapter 5 of this book the absorption of medicines into the human body has been outlined. These substances could also be further processed by the gut microflora.

These processes could then cause some side-effects of the medical treatments. However, in the overall picture of the functions of the gastrointestinal tract, the colonic biofilms form a kind of "waste treatment unit". Consequently, the qualities of the gut flora or ecosystem influence on the disposal of substances of the body.

The colon is also taking part in the water circulation and balance of the body. Many microbial activities are dependent on the moisture content or substrate concentration. Therefore, constipation could trigger disadvantageous effect for both the alimentary microbiome, and the host intestines and entire body system. Moreover, diverticulosis may generate areas where some undesired strains could achieve dominance. This, in turn, builds up the conditions for disease development. If the water flows out of the tissues, or other cleaning mechanisms are not flushing away the poisonous substance, this may result in ill health, and serious consequences. In some cases, colonic flushes are recommended for keeping up adequate movements of the colonic contents. Also many host body functions regulate the water circulation in the body, and in the intestines.

Also in between the planktonic submerged cells and the surface associated cells there is some kind of balance. This matter could be of importance also with respect to pathogenesis, and also for the development of biofilms. Type 1 fimbriae are used for attachment by many bacterial strains [Abraham and Beachey, 1987]. Also other fimbrial structures are mediating attachment in various bacterial groups. For example, type IV pili provide *Clostridium* perfringens also with a gliding motility [Varga *et al.*, 2006]. It was shown that the cells rapidly loose these appendiges, if they are not needed in the culture [Hakalehto et al. 2000]. *Salmonella* species seem to express the structures more readily in the presence of oxygen then in complete anaerobiosis [Hakalehto *et al.*, 2007]. This is a sign the importance of oxidative conditions for the facultative bacteria.

The facultative bacteria form a big part of the intestinal flora especially in the upper regions. It could be possible that their activities are somewhat controlled by the blood circulation in the intestinal epithelia. The blood is transporting oxygen into the minor veins [Snyder, 1989]. This takes place also in the areas where the nutrients are taken up. Such pathogens as salmonellas search for these niches in order to root themselves into the alimentary tract. If it happens, this leads to pathogenesis.

## Host Body Participating in Defining the Microbiome

Human body possesses many control and signaling systems, which are directing the functions of various tissues all over. Specific organs themselves also transmit information, and the same can be said regarding individual cells. Some of the most important mechanisms involved in the overall regulation consist of:

1. neurological control and monitoring
2. hormonal functions
3. tissue hormones
4. growth factors
5. electric potentials across the cell membranes
6. electromagnetic emissions of the microtubules and other intracellular structures

All these body functions interact with the normal flora in our body, including the alimentary microbes. There has to exist many forms of adaptation from both sides. The local phenomena occur in small scale, but their influences are measurable in a larger context. For example, blood test result is the general change in particular parameters in the prevailing status of the information network inside the body*, including the microbiome.*

Any intrusion by hazardous or foreign microbes into the microbiological ecosystems could produce measurable effects on our body's physicochemical activities and functions. Physiology and medicine based on the electric signals produced by individual cells in our body were established already a long time ago [Stadler, 2011].

Since then these techniques, as well as the observations on the electromagnetic fields in the body have been further elaborated. Also different gases regulate the conditions in our body and in the microbiome. Oxygen concentrations in our tissues need to be on adequate levels for optimal health [van Beest *et al.*, 2011].

On the other hand, all micro-organisms have shown to require carbon dioxide for their growth initiation [Kendall, 1927]. Similarly, anaerobic clostridia have been documented to benefit from the $CO_2$ produced by other microflora [Hakalehto and Hänninen, 2012]. In fact, the clostridia were shown to grow well and in an accelerated pace in the analysis of clinical strains in the PMEU when 100% $CO_2$ was applied for gassing [Hell *et al.*, 2010].

## Microbial Coverage of the Epithelial Surfaces and Various Niches inside the Alimentary Tract

Bacterial surface structures are an essential part of their movement and attachment [Hakalehto, 2000]. The eubacterial flagellin and fimbrin molecules are synthesized on the cells in ribosomes adjacent to the cell membranes. They are transported out their site of action by different mechanisms. Bacterial flagellins (the actual structural proteins of the flagellar filaments) are transported within the tubular filament as linear molecules. The type 1 fimbriae have been shown to act in the formation of bacterial layers onto surfaces. They are not expressed at 20 °C but are readily produced at 37 °C by salmonellas.

These pathogens are capable of making these structures in a short period of time for attachment until. The N-terminal sequences of all eubacterial flagellins resemble very closely each other, which also is indicating their important roles in cell movements (Hakalehto *et al.*, 1997). The fimbriae, on the other hand, are shorter filaments which facilitate the attachment onto the mucosal membranes and then get deprived from them for liberating the cells from the surfaces. This mechanism is helping the pathogens (and also other microbes?) to colonize the entire intestines rapidly. What is then the ecological consequence of this?

Type 1 fimbriae are used for attachment by many bacterial strains, such as facultative salmonellas [Baek *et al.*, 2011]. It was shown that the cells rapidly loose these appendiges, if they are not needed in the culture [Hakalehto *et al.,* 2000]. *Salmonella* species seem to express the structures more readily in the presence of oxygen then in complete anaerobiosis [Hakalehto *et al.*, 2007]. This is a sign of the importance of oxidative conditions for the facultative bacteria.

These bacteria form a big part of the intestinal flora especially in the upper regions. It could be possible that their functions are somewhat controlled by the blood circulation in the intestinal epithelia. The blood is transporting oxygen into the minor veins in the areas where the nutrients are taken up. Such pathogens as salmonellas search for these niches in order to root into the alimentary tract. In case of the intrusion of fimbriated bacteria, these organelles act as like anchors onto the gut walls, in order to facilitate colony formation in the different areas.

Microbes are living things being incredibly fast in their use of their preferable nutrients. According to the PMEU studies and simulations the intestinal bacteria seem to collaborate with each other aiming for optimal use of available nutrients [Hakalehto *et al.*, 2008; Hakalehto *et al.*, 2010; Hakalehto, 2011]. This balance between different bacterial strains found in the PMEU experiments is somewhat antagonistic piece of evidence to the "nutrient-niche hypothesis", according to which every microbial strain has its own niche in the intestines determined by its abilities to exploit carbon sources, or simple sugars [Chang *et al.,*2004].

From the PMEU results we can see that different bacterial strains tend to build up a kind of balance where the metabolic end-products of one strain could be instantaneously consumed by the other thus fading the boundaries between niches [Hakalehto, 2010; Hakalehto *et al.*, 2010]. For instance, *E. coli* and klebsiellas are able to cooperate when living together in the small bowel. In the PMEU they produce equal concentrations of cells produced by each strain as in the pure cultures [Hakalehto *et al.*, 2008]. They also regulate the duodenal pH for the benefit of the host, and thus maintain the balance of the local flora and perhaps the rest of the intestines.

## Role of Bacterial Surface Structures in Establishment of Ecosystems and in Host-Microbe Interactions

In order to understand the bacterial behavior in the alimentary tract we need to pay further attention to their extracellular organelles and outer surfaces. These structures mediate the contacts between the bacterial cells, between them and the solutions and suspensions, or surfaces. Also the contacts with human cells, tissues and organelles take place with the aid of these molecular matrices or structures of our prokaryotic companions. Some 5-10% of protein molecules can be of flagellar origin [Macnab and Castle, 1987]. The N-terminal sequences of the eubacterial flagellar filament protein are rather similar from species to species [Hakalehto *et al.*, 1997]. This part of the molecule is in dissolved form when produced inside the cells, on the inner side of the membranes. Then the steric structures direct it toward the growing filament from the ribosomes attached to the membrane, and further into the tube-like growing filament. This filament can be expressed in a few seconds. They also can disappear in a moment. This could occur when the acidic conditions in the stomach have removed them from the cell surfaces, or when the bacterial cell has deliberately cut them off. The latter situation may be a result of need for quick detachment due to exhausting resources in the case of fimbriae [Hakalehto *et al.*, 2010], or alternatively due to flagellar or fimbrial phase variation intended for luring the host immunosystem [Rhen *et al.,* 1983]. If present, however, the flagellae are actively moving the bacterial cells toward beneficial niches and conditions. Their contribution to the orientation of the cells and to their relocation in the GI canal is of utmost importance for the bacteria. Consequently, it could be supposed that the motile strains could have broader probiotic influences than the non-motile organisms.

Fimbrial structures, and other molecular tools for attachment such as adhesions, are responsible for the rapid adhesion and adsorption of many bacteria onto the epithelial surfaces (Freter and Jones, 1983). As like flagellins, also fimbrial structures are synthesized within an extremely short period of time, in a few seconds [Abraham and Beachey, 1987; Farfan *et al.*, 2011]. The fimbriated bacteria are able to detach these fibrils at once. For example, the salmonellas express type 1 fimbriae during the exponential growth phase, and later give them up as the population density on the surface is increasing, and the resources are being exhausted [Hakalehto *et al.,* 2000]. This mechanism enables the rapid bacterial colonization of the small intestines together with the enhanced anaerobic metabolism of the facultatives [Hakalehto *et al.*, 2007]. If the living space or nutrients are deprived from the bacterial cells, they leave the fimbriae behind, and possibly consume them for their own novel biosynthesis. They also are then liberated from the previous location, and move into new sites on the gut walls. In order to avoid host immune defenses they may carry out flagellar or fimbrial phase variation [Rhen *et al.,* 1983; Chen *et al.*, 2010; Morowitz *et al.*, 2011]. This means in practice that the same strain is expressing an altering flagellin or fimbrin epitope or epitopes in order to mislead the immunosystem. The production of fimbrial subunits counts for 95% of the intracellular protein precursors at the time point of 1.5 hours in *E. coli* cultures [Abraham and Beachey, 1987].

Besides the antibodies against the above-mentioned surface organelles or their subunits, the host body is producing different types of antibodies against capsular polysaccharides (K antigens), or the O-specific side chains of the lipopolysaccharides (O antigens) [Hood *et al.*, 2004]. The so-called H antigens are the flagellar antigens. Some structures on the cell surfaces are universally present on the surfaces of a certain group of bacteria, such as the enterobacterial common antigen (ECA). Mutant *Salmonella* strains devoid of the ECA can cause tedious permanent infections [Gilbreath *et al.*, 2011].

When the expression of antigens against type 1 fimbriae was studied in the case of *Salmonella* sp. it was found that these immunological reactions were stronger and prevailed for a longer period of time in aerobiosis [Hakalehto *et al.*, 2007]. This finding could indicate the preferencies among salmonellas toward aerobic conditions also *in vivo.* They could get oxygen rich blood available when penetrating across the one host cell deep surface of the small bowel epithelium. Stronger expression of attachment fimbriae in the presence of oxygen in cultures with equal growth rates aerobically and anaerobically, could imply to active search for aerobic conditions by the facultatively anaerobic, pathogenic salmonellas. In turn, this preference is making the salmonellas basically also unfit as members of the intestinal microbial community. This very character is one reason behind their true nature as causative agents of contagious intestinal diseases.

## Anaerobic Metabolism of Bacteria and Its Consequences for Intestinal Function and Overall Health

In the bacterial metabolism there characteristically is a potential into extremely fast growth with increase in cell volume, with doubling time of about 20 min in optimal conditions. However, the joint volume of multiplied cells starting from one cell to divide with that rate would in 48 hours produce more cytoplasm than is the earth's volume [Russell and Setchell, 1992]. Therefore, it is obvious that "in the life of a bacterium any number of essential nutrients do soon become limiting"(modified from Recny *et al.* 1990). The PMEU environment provides most cells and strains an optimal milieu for some time [Hakalehto, 2010; Hakalehto, 2011]. In these conditions, various strains aim at utilizing the available nutrients at optimal speed.

However, not all chemical energy liberated by the microbe culture is utilized for growing cell number or biomass [Russell and Setchell, 1992]. When *Klebsiella aerogenes* (*K. mobilis*) was cultivated continuously limited by carbon, ammonium, sulphate, or phosphate, the rate of carbon source was always higher when carbon was available in excess [Neijssel and Tempest, 1976]. It also has been documented that the maintenance energy of the cells was rather small in comparison with the growth requirements in the actively growing cells [Pirt, 1965]. This gives an idea that the bacterial life in optimal conditions is almost exclusively directed toward growth. The idea is consistent with our finding that in the PMEU culture the optimal growth could be enhanced effectively, this enhancement of growth equaling in anaerobiosis with the aerobic conditions in case of *e.g.* facultatively anaerobic salmonellas [Hakalehto *et al.*, 2007]. This surprisingly high productivity of the oxygen-deprived metabolism is somewhat contradicting with the finding that in aerobic conditions the Krebs cycle could produce more stored chemical energy in the form of ATP. It can be further deduced that the pace of metabolic activity is again dependent on the amount of available carbon. Even this value is likely to have its optimum, because in case of the oversupply of carbon sources, the *Klebsiella mobilis* cells needed to use some kind of overflow metabolism in order to relieve the osmotic pressure and sinking pH [Hakalehto *et al.*, 2012]. This reaction was carried out in anaerobic conditions, and could therefore not utilize the Krebs cycle or any other form of oxidation, such as the methylglyoxal bypass pathway [Tempest *et al.*, 1983] Because the bacteria are unicellular, their maintenance is not precluding such keeping up of electric fields as the multicellular body and its organs do generate. Therefore, in a non-growing state the bacteria are probably less likely to overcome the propulsion of the epithelial and other living surfaces inside the body. This could explain the finding that in the stationary phase (7 h in the PMEU) the *Salmonella* sp. cells threw away the type 1 fimbrial attachment bodies which had been expressed strongly in the exponential phase (3,5 h culture time in the PMEU) [Hakalehto et al. 2000].

In different tissues, the neurons have usually a positive charge, whereas the bacteria are negatively charged. This could lead to increased affinities to the neurons of the micro-organisms. Maybe this also explains some of the tedious inflammations in the neural passageways. Although the central nervous system is relatively well protected against infections, the peripheral nerves are probably more easily reachable. In case of newborn babies the neural infections are more often caused by *E. coli* or group B streptococci than by such respiratory pathogens as *Neisseria meningitidis*, *Haemophilus influenzae* or *Streptococcus pneumoniae* [Nester *et al.*, 1982]. This implies to a route of infection via the intestines, and more precisely from the duodenal tract, bile duet, and liver. It is possible that the less developed immune system of a neonate is allowing the intrusion and neural inoculation by the intestinal flora. Some results supporting this hypothesis have been obtained from studies with patient strains from Kuopio University Hospital neonatal intensive care unit [Hakalehto *et al.*, 2009]. The PMEU Scentrion® units have been recently validated in this institute and the department for children tumor and blood diseases of the same hospital [Pesola and Hakalehto, 2011]. The strategies for maintaining hygienic levels in hospitals using the PMEU technologies have been surveyed earlier [Hakalehto, 2006; Hakalehto, 2010].

In case of septic infection the pathogens invade the body and circulation from an infected area in the urinary tract, intestines, respiratory organs, or via other routes. In case of Gram-negative infections, the lipopolysaccharides (LPS, endotoxin) from the outer leaflet of the outer membranes are detached, and cause damage to the blood vessel walls, whose ability to contract is seriously disturbed [Nester *et al.,* 1982]. Dilution in blood vessels then causes a drop in blood pressure, which result in inadequate supply of oxygen in vital organs. This may have fatal consequences even though the infection gets under control by *e.g.* an antibiotic treatment. This is one reason for the Gram-negative septic infections being generally more serious ones than those caused by molds or the Gram-positives. The most common causative agents of the Gram-negative infections are such coliformic rod-shaped bacteria as *E. coli*, *Enterobacter aerogenes*, *Serratia marcescens,* and *Proteus mirabilis*, all of which could be derived from the GI tract. All these species are facultatively anaerobic bacteria. Also obligate anaerobic intestinal strains, such as *Bacteroides* sp., are often causing septic infections. In all these cases endotoxic shock could cause fatal consequences. Therefore, it is of utmost importance to get the septic infection treated fast enough. The PMEU system provides the hospital personnel with information on the infective organism in a few hours time [Hakalehto *et al.*, 2009; Pitkänen *et al.*, 2009]. This could be achieved simultaneously with information on the antibiotic resistances of the bacterial isolates [Hakalehto 2011, 2012].

As indicated earlier also in another chapter of this book (Chapter 5), the intestinal bacteria seem to strive for cooperation in their natural environment, in the gut epithelia. However, elsewhere in the body they may cause severe pathogenic inflammations. The hazardous conditions may also be caused by aerobic environmental bacteria, such as *Pseudomonas aeruginosa,* which is a common Gram-negative bacterium causing both wound and septic infections [Nester *et al.*, 1982]. It has been estimated that about 10% of us humans are carrying *Ps. aeruginosa* in our intestines. Also members of the obligate anaerobic genus *Clostridium* are quite common opportunistic pathogens being responsible for many dangerous illnesses, such as gas gangrene. Both Pseudomonas sp. and many species of the genus Clostridium use intruders from a "different microbial world", that of the scattered ecosystems with high levels of competitions. In the GI tract these anaerobes could also cause extensive damage. *Cl. difficile* is an agent often spreading in the intestines after antibiotic medication which would have weakened the normal flora (and the immunosystem). *Cl. perfingens* or *Cl. botulinum* may cause serious food poisonings. If a clostridial infection is spreading in the tissues, it could sometimes be cured with hyperbaric oxygen treatment.

If we consider the finding that in the anaerobiosis a particular bacterium may have such metabolic speed, which is equalling with the extreme, fastest growth rates of the aerobes, this observation opens up new windows to the anaerobic pathogenesis. Their role in normal digestive functions has also been underestimated. This also explains the devastating speed and effect of these infections in many cases. In the intestinal dysbiosis, or in case of septic infections, or in clostridial or streptococcal infections, the anaerobic bacteria rapidly take over the space, producing toxins and other substances invalidating the body defenses. The fastidious microaerobic streptococci have traditionally been grouped with the strict anaerobes [Finegold *et al.*, 1986]. When the author was acting as a visiting scientist in the biotechnology unit of the Hebrew University of Jerusalem in 1989, the large scale fermentation cultivations of strictly anaerobic *Pectinatus* bacteria revealed that these organisms multiplied with a speed equalling to the growth speed of *E. coli* (results not shown here). These Gram-negative bacteria have so far been isolated from the beer making processes only [Lee *et al.* 1978]. Their LPS chemical composition turned out to be partially unique [Helander *et al.*, 1983], and was documented at least as toxic as the one from *E. coli*. The LPS endotoxins are causing inflammation on the mucosal membranes of the gut walls [Fukuoka *et al.*, 2011].

In fact, the conquest of the intestinal tract by the obligate anaerobic pathogens seem to be for a great extent a failure of the ecosystem to maintain the Bacterial Intestinal Balance (BIB) [Hakalehto, 2011]. From the intestines or from the other parts of the body including the teeth and the oral cavity, these pathogens may readily spread all over the body causing severe infections. These include brain abscesses, actinomycosis, gas gangrene and other soft tissue infections, chronic osteomyelitis, tetanus and botulism [Finegold *et al.*, 1986]. Note that many of these conditions are caused by spore-forming anaerobic clostridia which originate from soil. Hence they are derived from a scattered ecosystem (see Table 3). In these conditions all strains are taking part in harsh competition, which often takes place by using the antibiotics as molecular weapons [Hakalehto, 2012]. Consequently, *Clostridium* sp. strains can sustain antimicrobials quite well. This explains also the outcome of *Cl. difficile* infestions and epidemics in the hospital system [Hell *et al.,* 2010]. In the PMEU studies, we have been able to track one of the potential mechanisms by which the alimentary ecosystem is keeping the intruders under control. Namely, the gas production of lactic acid bacteria and other members of the normal flora in the cecum, for instance, is boosting the clostridia, thus preventing their sporulation and persistent take over of the digestive track [Hakalehto and Hänninen, 2012]. The implementation of some types of bacteremia such as the one caused by *Cl. septicum,* although usually mild, is a sign of a serious underlying malignant disease of the cecum or the colon [Finegold *et al.*, 1986]. This is an example of the tight interdependence between the microbial ecosystems and the host condition.

In case of septic infections, the most common causative agent among the strict anaerobes is the genus *Bacteroides* [Finegold *et al.*, 1986] .The Gram-negative species of this genus belong to the gut normal flora but they also can induce septic infections with high mortality rates (15-35% in anaerobic bacteremia). This underlines the importance of keeping the gut anaerobes in a non-pathogenic mode. In *Bacteroides* bacteremia, the mortality rate was 25% in one study, when the most appropriate antibiotics were used, and rose up to 48% if the medication was chosen on erroneous grounds. It could be stated that the urgent determination of the antibiotic resistances of any pathogens is required both in intestinal and systemic infections. Moreover, the gastrointestinal tract was found to be the most common route of entry for all bacteremias (62%) [Marcoux *et al.*, 1970]. Slightly above 50% of all anaerobic infections occurred in the GI area [Finegold *et al.*, 1986]. In these regions, the anatomical defects or abnormalities (also possibly caused by infections) could build up favorable conditions for the pathogenesis. One example is diverticulosis (Figure 6).

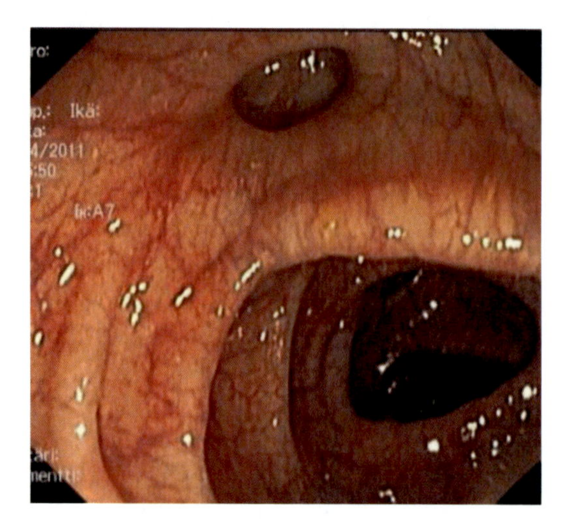


Figure 6. Intestinal diverticle (up in the picture). This kind of structure is giving space for separate microbial ecosystem in the gut, which could eventually cause pathogenic alterations.

The inflammation can also start from very small areas which have got irritated or where the intrusions have been attempted by some hostile microbes (Figure 7).

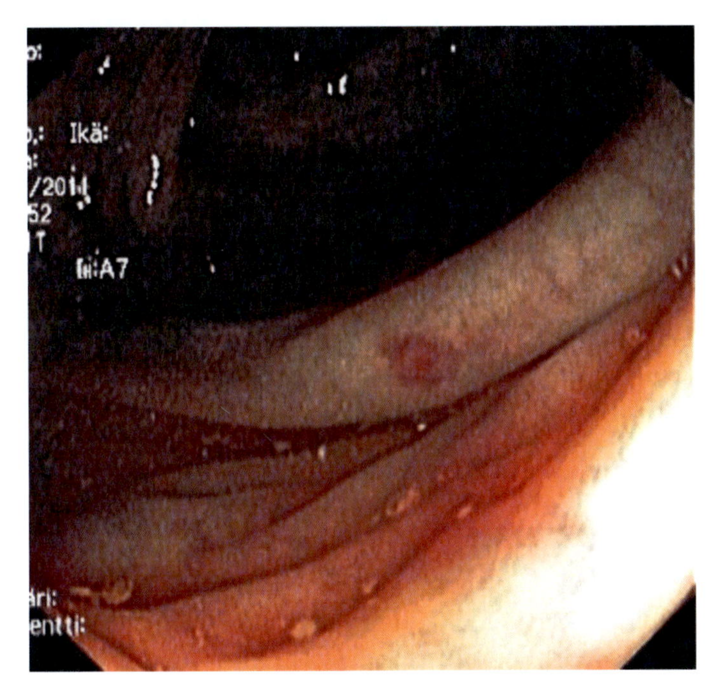


Figure 7. Site of inflammation in the colon. The red spot in the center representates mucosal haemorrhage as a starting point of a potential infection site.

Problematic formations could also include fistulas, stagnation of bowel contents caused by blind loops, hindrances in blood circulation into the intestines, and infected biliary tract and pancreatic fluid entering the duodenum. Partial prevention in the blood access to any of the gut areas may cause diminished nutrient uptake, and also alterations in the microflora. In order to have a positive influence in the patient´s condition, the treatment then often needs to include operation, laparotomy, or other such methods for correcting the misfunctioning gut. It should be emphasized that the problems on the organismal level usually provokes problems on the microbiotic organism, and *vice versa.*

Intra-abdominal infections are an important manifestation of the disturbed intestinal flora. In 73 of such cases detected in 62 patients, the bacterial isolates consisted of strict anaerobes (about 52%) and facultative strains (about 48%) [Finegold *et al.*, 1986]. The isolated strains per infections averaged 4.5 per patient case. Out of the facultatives, *E. coli* represented 15% of all infections, and group D streptococci around 10%. These groups were the dominating ones among the bile tract isolates but with a still higher incidence rate [Hakalehto *et al.*, 2010]. The causative agent in the intra-abdominal infections also reflected the site of disease process or surgery [Finegold *et al.*, 1986]. The lower in the alimentary tract the infection located, the more likely the strain was to be of an obligate anaerobic origin. Nearly a half of these were caused by *Bacteroides* sp. the situation being almost parallel to the one with bacteremias. One third of the strict anaerobic, and one sixth of all causes of the intra-abdominal infections were induced by both *Clostridium* sp. and *Peptostreptococcus* sp. In any event, the bacterial presence outside the GI tract in the abdomen is an indication of severe pathogenic condition. This condition usually develops from a diseased, or imbalanced, gut region. The disturbed intestines are studied more closely in Chapter 7 of this book. One indirect proof of the "organismal" nature of the alimentary microbiome could be the search for novel body areas as a consequence of bypass surgery. This strive of the micro-organisms could cause such infections as bypass enteropathy, rupture of gas-filled blebs, proctitis, and protein-losing enteropathy. The intestinal stagnation may initiate bacterial overgrowth in the suppressed bowel, which then has the tendency of spreading the inflammation and disease.

# Conclusion

The microbiological world consists of countless interactions on the molecular level, which is sometimes called as "molecular communication". In detecting these microscopic events in the alimentary tract ecosystems we always need to simplify. The PMEU method offers means for simplifying these phenomena. Microbial cultures in the enrichment syringes are in a fiercely metabolizing homogenous state, thus reflecting and representing an individual cell on the mucosal membranes. In these ecosystems all reactions occurring in the vicinities of various cells generally take place extremely fast. Therefore, the entire idea of microbial ecology in these environments is different forms that of the more dispersed communities in the soil, for example. Consequently, the relations between strains become different, instead of needing to battle with each other for example by by extensive secretion of antibiotics, the bacteria search for cooperation between strains.

Besides avoiding the direct competition, the bacterial species in the duodenum are collaborating with each other in making the conditions bearable for the microbial community as a whole. This is advantageous also for the host body, because in so doing the microbial strains belonging to the normal flora are maintaining the conditions stable in the region. The studies with the PMEU equipment are indicating the metabolic cooperation between *Escherichia coli* and *Klebsiella mobilis*, for example. These two bile tolerant members of the small intestinal flora manage to maintain the pH in a suitable range during and in between the meals. These facultative bacteria set up the basis for the establishment of the intestinal microbial ecosystems. The multitude of strains in the cecum and in the colon are exploiting the left over and digested food materials after the host AND the microflora in the small bowel. The latter is a key component in our health. These bacteria could also exploit the overflow metabolism in stabilizing the intestinal milieu. This stabilization is a prerequisite for such vital functions as the host food uptake and bile circulation.

In the understanding of intestinal diseases it is essential to have an idea about the nature of microbial ecosystems and the behavior of bacterial communities. The tendencies for co-operation protect them against intruders. On the other hand, if a pathogenic strain starts overpowering, the balance may become counter productive. Therefore, in any disease the maintenance and recovery of the alimentary microbiome and its ecosystems is of crucial importance.

# References


Abraham, SN; Beachey, EH. Assembly of a chemically synthesized peptide of *Escherichia coli* type 1 fimbriae into fimbria-like antigenic structures. *J. Bacteriol.,* 1987; 169, 2460-2465.

Andrade, JA; Freymuller, E; Fagundes-Neto, U. Adherence of enteroaggregative *Escherichia coli* to the ileal and colonic mucosa: an *in vitro* study utilizing the scanning electron microscopy. *Arq. Gastroenterol.,* 2011; 48, 199-204.

Baek, CH; Kang, HY; Roland, KL; Curtiss, R,3rd. Lrp acts as both a positive and negative regulator for type 1 fimbriae production in *Salmonella enterica* Serovar Typhimurium. *PLoS One,* 2011; 6, e26896.

Beulens, JW; Bots, ML; Atsma, F; Bartelink, ML; Prokop, M; Geleijnse, JM; Witteman, JC; Grobbee, DE; van der Schouw, YT. High dietary menaquinone intake is associated with reduced coronary calcification. *Atherosclerosis,* 2009; 203, 489-493.Bures, J; Horak, V; Duben, J. Importance of colicinogeny for the course of acute bacillary dysentery. *Zentralbl. Bakteriol. Orig. A.,* 1979; 245, 469-475.

Chang D-E; Smale, DJ; Tucker, DL; Leatham, MP; Norris, WE; Stevenson, SJ; Anderson, AB; Grissom, JE; Laux, DC; Cohen, PS; Conway, T. Carbon nutrition of *Escherichia coli* in the mouse intestine. *PNAS,* 2004; 101, 7427-7432.

Chen, Q; Decker, KB; Boucher, PE; Hinton, D; Stibitz, S. Novel architectural features of *Bordetella pertussis* fimbrial subunit promoters and their activation by the global virulence regulator BvgA. *Mol. Microbiol.,* 2010; 77, 1326-1340.

Cox, LA,Jr; Popken, DA. Assessing potential human health hazards and benefits from subtherapeutic antibiotics in the United States: tetracyclines as a case study. *Risk. Anal.,* 2010; 30, 432-457.

Dalton, H; Stirling, DI. Co-metabolism. *Philos. Trans. R. Soc. Lond. B. Biol. Sci.,* 1982; 297, 481-496.

Davelos, AL; Kinkel, LL; Samac, DA. Spatial variation in frequency and intensity of antibiotic interactions among *Streptomycetes* from prairie soil. *Appl. Environ. Microbiol.,* 2004; 70, 1051-1058.

Farfan, MJ; Cantero, L; Vidal, R; Botkin, DJ; Torres, AG. Long polar fimbriae of enterohemorrhagic *Escherichia coli* O157:H7 bind to extracellular matrix proteins. *Infect. Immun.,* 2011; 79, 3744-3750.

Finegold, SM; Lance-George, W; Mulligan, ME, editors. *Anaerobic infections. A Disease-a-month classic.* Year Book Medical Publishers. Chicago, USA, 1986.Flint, HJ. The rumen microbial ecosystem--some recent developments. *Trends Microbiol.,* 1997; 5, 483-488.

Floch, MH. Intestinal microecology in health and wellness. *J. Clin. Gastroenterol.,* 2011; 45 Suppl, S108-10.

Freter, R; Jones, GW. Models for studying the role of bacterial attachment in virulence and pathogenesis. *Rev. Infect. Dis.,* 1983; 5 Suppl 4, S647-58.

Fukuoka, S; Richter, W; Howe, J; Andra, J; Rossle, M; Alexander, C; Gutsmann, T; Brandenburg, K. Biophysical investigations into the interactions of endotoxins with bile acids. *Innate Immun.,* 2011.

Gilbreath, JJ; Colvocoresses Dodds, J; Rick, PD; Soloski, MJ; Merrell, DS; Metcalf, ES. Enterobacterial common antigen mutants of *Salmonella enterica* Serovar Typhimurium establish a persistent infection and provide protection against subsequent lethal challenge. *Infect. Immun.,* 2011.

Guedon, E; Desvaux, M; Payot, S; Petitdemange, H. Growth inhibition of *Clostridium cellulolyticum* by an inefficiently regulated carbon flow. *Microbiology,* 1999; 145, 1831-1838.

Hakalehto, E. Characterization of *Pectinatus cerevisiiphilus* and *P. frisingiensis* surface components. Use of synthetic peptides in the detection of some Gram-negative bacteria. Kuopio, Finland: Kuopio University Publications C. Natural and Environmental Sciences 112; 2000.

Hakalehto, E. Semmelweis' present day follow-up: Updating bacterial sampling and enrichment in clinical hygiene. *Pathophysiology,* 2006; 13, 257-267.

Hakalehto E, inventor. Method and apparatus for concentrating and searching of microbiological specimens. US Patent No. 7,517,665. 2009.

Hakalehto, E. Hygiene monitoring with the Portable Microbe Enrichment Unit (PMEU). *41st R3 -Nordic Symposium. Cleanroom technology, contamination control and cleaning. VTT Publications 266.* Espoo, Finland: VTT (State Research Centre of Finland); 2010.

Hakalehto, E. Simulation of enhanced growth and metabolism of intestinal *Escherichia coli* in the Portable Microbe Enrichment Unit (PMEU). In: Rogers MC, Peterson ND, editors. *E. coli infections: causes, treatment and prevention.* New York, USA: Nova Publishers; 2011.

Hakalehto, E. Antibiotic resistant traits of facultative *Enterobacter cloacae* strain studied with the PMEU (Portable Microbe Enrichment Unit). In Press. In: Méndez-Vilas A, editor. *Science against microbial pathogens: communicating current research and technological advances. Microbiology book series Nr. 3.* Badajoz, Spain: Formatex Research Center; 2012.

Hakalehto, E; Hänninen, O. Lactobacillic CO2 signal initiate growth of butyric acid bacteria in mixed PMEU cultures. Manuscript in preparation. 2012.

Hakalehto, E; Santa, H; Vepsalainen, J; Laatikainen, R; Finne, J. Identification of a common structural motif in the disordered N-terminal region of bacterial flagellins--evidence for a new class of fibril-forming peptides. *Eur. J. Biochem.,* 1997; 250, 19-29.

Hakalehto, E; Hujakka, H; Airaksinen, S; Ratilainen, J; Närvänen, A. Growth-phase limited expression and rapid detection of *Salmonella* type 1 fimbriae. In: Hakalehto E. *Characterization of Pectinatus cerevisiiphilus and P. frisingiensis surface components. Use of synthetic peptides in the detection fo some Gram-negative bacteria.* Kuopio, Finland: Kuopio University Publications C. Natural and Environmental Sciences 112; 2000. Doctoral dissertation.

Hakalehto, E; Pesola, J; Heitto, L; Narvanen, A; Heitto, A. Aerobic and anaerobic growth modes and expression of type 1 fimbriae in *Salmonella. Pathophysiology,* 2007; 14, 61-69.

Hakalehto, E; Humppi, T; Paakkanen, H. Dualistic acidic and neutral glucose fermentation balance in small intestine: Simulation *in vitro. Pathophysiology,* 2008; 15, 211-220.

Hakalehto, E; Pesola, J; Heitto, A; Deo, BB; Rissanen, K; Sankilampi, U; Humppi, T; Paakkanen, H. Fast detection of bacterial growth by using Portable Microbe Enrichment Unit (PMEU) and ChemPro100i((R)) gas sensor. *Pathophysiology,* 2009; 16, 57-62.

Hakalehto, E; Hell, M; Bernhofer, C; Heitto, A; Pesola, J; Humppi, T; Paakkanen, H. Growth and gaseous emissions of pure and mixed small intestinal bacterial cultures: Effects of bile and vancomycin. *Pathophysiology,* 2010; 17, 45-53.

Hakalehto, E.; Heitto, A.; Heitto, L.; Humppi, T.; Rissanen, K.; Jääskeläinen, A.; Paakkanen, H.; Hänninen, O. Fast monitoring of water distribution system with portable enrichment unit – Measurement of volatile compounds of coliforms and *Salmonella* sp. in tap water. *Journal of Toxicology and Environmental Health Sciences,* 2011a; 3, 223-233.

Hakalehto, E; Vilpponen-Salmela, T; Kinnunen, K; von Wright, A. Lactic Acid bacteria enriched from human gastric biopsies. *ISRN Gastroenterol,* 2011; 2011b, 109183.

Hakalehto, E; Tiainen, M; Laatikainen, R; Paakkanen, H; Humppi, T; Hänninen, O. Rapid bacterial metabolic activity by the production of ethanol and 2,3 -butanediol without population growth protects intestinal flora against osmotic stress. *Manuscript in preparation,* 2012.

Helander, I; Hakalehto, E; Ahvenainen, J; Haikara, A. Characterization of lipopolysaccharides of *Pectinatus cerevisiophilus. FEMS Microbiol. Lett.*, 1983; 18, 223-226.

Hell, M; Bernhofer, C; Huhulescu, S; Indra, A; Allerberger, F; Maass, M; Hakalehto, E. How safe is colonoscope-reprocessing regarding *Clostridium difficile* spores? *J. Hosp.Inf.*, 2010; 76, 21-22.

Hobson, PN; Wallace, RJ. Microbial ecology and activities in the rumen: part 1. *Crit. Rev. Microbiol.,* 1982; 9, 165-225.

Hood, DW; Randle, G; Cox, AD; Makepeace, K; Li, J; Schweda, EK; Richards, JC; Moxon, ER. Biosynthesis of cryptic lipopolysaccharide glycoforms in Haemophilus influenzae involves a mechanism similar to that required for O-antigen synthesis. *J. Bacteriol.,* 2004; 186, 7429-7439.

Jayne-Williams, DJ. The bacterial flora of the rumen of healthy and bloating calves. *J. Appl. Bacteriol.,* 1979; 47, 271-284.

Johnston, I; Nolan, J; Pattni, SS; Walters, JR. New insights into bile acid malabsorption. *Curr Gastroenterol. Rep.,* 2011; 13, 418-425.

Kendall, J. The Abuse of Water. *Science,* 1927; 66, 610-611.

Khalil, R. Evidence for probiotic potential of a capsular-producing Streptococcus thermophilus CHCC 3534 strain. *Pol. J. Microbiol.,* 2009; 58, 49-55.

Kutsch, WL; Merbold, L; Ziegler, W; Mukelabai, MM; Muchinda, M; Kolle, O; Scholes, RJ. The charcoal trap: Miombo forests and the energy needs of people. *Carbon Balance Manag,* 2011; 6, 5.

Lyte, M. The microbial organ in the gut as driver of homeostasis and disease. *Medical Hypotheses,* 2010; 74, 634-638.

Lyte, M; Li, W; Opitz, N; Gaykema, RP; Goehler, LE. Induction of anxiety-like behavior in mice during the initial stages of infection with the agent of murine colonic hyperplasia Citrobacter rodentium. *Physiol. Behav.,* 2006; 89, 350-357.

Macnab, RM; Castle, AM. A variable stoichiometry model for pH homeostasis in bacteria. *Biophys. J.,* 1987; 52, 637-647.

Marcoux, JA; Zabransky, RJ; Washington, JA,2nd; Wellman, WE; Martin, WJ. Bacteroides bacteremia. *Minn. Med.,* 1970; 53, 1169-1176.

Marks, CG; Hawley, PR; Peach, SL; Drasar, BS; Hill, MJ. The effects of phthalylsulphathiazole on the bacteria of the colonic mucosa and intestinal contents as revealed by the examination of surgical samples. *Scand. J. Gastroenterol.,* 1979; 14, 891-896.

Morowitz, MJ; Denef, VJ; Costello, EK; Thomas, BC; Poroyko, V; Relman, DA; Banfield, JF. Strain-resolved community genomic analysis of gut microbial colonization in a premature infant. *Proc. Natl. Acad. Sci. USA,* 2011; 108, 1128-1133.

Neijssel, OM; Tempest, DW. The role of energy-spilling reactions in the growth of Klebsiella aerogenes NCTC 418 in aerobic chemostat culture. *Arch .Microbiol.,* 1976; 110, 305-311.

Nester EW, Pearsall NN, Roberts JB, Roberts CE. *The Microbial Perspective.* Philadelphia, PA, USA. CBS College Publishing. 1982.

Peach, S; Lock, MR; Katz, D; Todd, IP; Tabaqchali, S. Mucosal-associated bacterial flora of the intestine in patients with Crohn's disease and in a control group. *Gut,* 1978; 19, 1034-1042.

Perunova, NB; Ivanova, EV; Bukharin, OV. Microbial regulation of biological characteristics of bacteria in human gut microsymbiocenosis. *Zh Mikrobiol Epidemiol Immunobiol,* 2010; (6), 76-80.

Pesola, J; Hakalehto, E. Enterobacterial microflora in infancy - a case study with enhanced enrichment. *Indian J Pediatr,* 2011; 78, 562-568.

Pirt, SJ. The maintenance energy of bacteria in growing cultures. *Proc. R Soc. Lond B Biol. Sci.,* 1965; 163, 224-231.

Pitkänen, T; Bräcker, J; Miettinen, I; Heitto, A; Pesola, J; Hakalehto, E. Enhanced enrichment and detection of thermotolerant *Campylobacter* species from water using the Portable Microbe Enrichment Unit (PMEU) and realtime PCR. *Can. J. Microbiol.,* 2009; 55, 849-858.

Rea, MC; Dobson, A; O'Sullivan, O; Crispie, F; Fouhy, F; Cotter, PD; Shanahan, F; Kiely, B; Hill, C; Ross, RP. Effect of broad- and narrow-spectrum antimicrobials on Clostridium difficile and microbial diversity in a model of the distal colon. *Proc. Natl. Acad. Sci. USA,* 2011; 108 Suppl 1, 4639-4644.

Recny, MA; Neidhardt, EA; Sayre, PH; Ciardelli, TL; Reinherz, EL. Structural and functional characterization of the CD2 immunoadhesion domain. Evidence for inclusion of CD2 in an alpha-beta protein folding class. *J. Biol. Chem.,* 1990; 265, 8542-8549.

Repuske, R; Clayton, MA. Control of *Escherichia coli* growth by $CO_2$. *Journal of Bacteriology,* 1968; 135, 1162-1164.

Rhen,M., Mäkelä,P.H.; Korhonen, TK. P-fimbriae of *E.coli* are subject to phase variation. *FEMS Microbiol. Lett.,* 1983; 19, 267-271.

Russell, DW; Setchell, KD. Bile acid biosynthesis. *Biochemistry,* 1992; 31, 4737-4749.

Rychlik, JL; Russell, JB. Bacteriocin-like activity of Butyrivibrio fibrisolvens JL5 and its effect on other ruminal bacteria and ammonia production. *Appl .Environ. Microbiol.,* 2002; 68, 1040-1046.

Sanchez, PA; Bandy, DE; Villachica, JH; Nicholaides, JJ. Amazon basin soils: management for continuous crop production. *Science,* 1982; 216, 821-827.

Sato, Y. Neonatal bacterial infection. *Nihon Rinsho,* 2002; 60, 2210-2215.

Snyder, JR. The pathophysiology of intestinal damage: effects of luminal distention and ischemia. *Vet. Clin. North Am. Equine Pract.,* 1989; 5, 247-270.

Stadler, M. Biophysical double-lives, 1939-1946. Or: spaces of boredom. On 'information discourse' and (dis)continuities in the life sciences. *Ber. Wiss,* 2011; 34, 27-63.

Syed, MA; Manzoor, U; Shah, I; Bukhari, SH. Antibacterial effects of Tungsten nanoparticles on the Escherichia coli strains isolated from catheterized urinary tract infection (UTI) cases and Staphylococcus aureus. *New Microbiol.,* 2010; 33, 329-335.

Tempest, DW; Neijssel, OM; Teixeira De Mattos, MJ. Regulation of metabolite overproduction in *Klebsiella aerogenes. Riv. Biol.,* 1983; 76, 263-274.

van Beest, P; Wietasch, G; Scheeren, T; Spronk, P; Kuiper, M. Clinical review: use of venous oxygen saturations as a goal - a yet unfinished puzzle. *Crit. Care,* 2011; 15, 232.

Varga, JJ; Nguyen, V; O'Brien, DK; Rodgers, K; Walker, RA; Melville, SB. Type IV pili-dependent gliding motility in the Gram-positive pathogen Clostridium perfringens and other Clostridia. *Mol. Microbiol.,* 2006; 62, 680-694.

Vermeer, C; Braam, L. Role of K vitamins in the regulation of tissue calcification. *J Bone Miner Metab.,* 2001; 19, 201-206.

Visca, P; Seifert, H; Towner, KJ. Acinetobacter infection - an emerging threat to human health. *IUBMB Life,* 2011.

Vissers, MM; Driehuis, F; Te Giffel, MC; De Jong, P; Lankveld, JM. Minimizing the level of butyric acid bacteria spores in farm tank milk. *J. Dairy Sci.,* 2007; 90, 3278-3285.

Watase, H; Takenouchi, T. Bacterial flora in the digestive tract of cattle. I. Comparison of nonselective culture medium and changes in fecal bacterial flora with age. *Natl. Inst. Anim. Health Q (Tokyo),* 1978; 18, 143-154.

Westerhoff, HV; Juretic, D; Hendler, RW; Zasloff, M. Magainins and the disruption of membrane-linked free-energy transduction. *Proc. Natl. Acad. Sci. USA,* 1989; 86, 6597-6601.

Williams, AG; Withers, SE; Strachan, NH. Postprandial variations in the activity of polysaccharide-degrading enzymes in microbial populations from the digesta solids and liquor fractions of rumen contents. *J. Appl. Bacteriol.,* 1989; 66, 15-26.

Wolfaardt, F; Taljaard, JL; Jacobs, A; Male, JR; Rabie, CJ. Assessment of wood-inhabiting Basidiomycetes for biokraft pulping of softwood chips. *Bioresour. Technol.,* 2004; 95, 25-30.

In: Alimentary Microbiome: A PMEU Approach
Editor: Elias Hakalehto ISBN: 978-1-61942-692-4
© 2012 Nova Science Publishers, Inc.

Chapter VII

# Intestinal Imbalances and Diseases of the Digestive Tract

***Kaarlo Jaakkola*[1] *and Elias Hakalehto*[2]**
[1] Helsinki Antioxidant Clinic, Helsinki, Finland
[2] Department of Biosciences, University of Eastern Finland, Kuopio, Finland

## Abstract

The imbalanced microbiological condition of the alimentary tract is often a consequence and an indication of a general disease. For example, weakened immune defense could result a manifestation of microbial overgrowth in the intestines, or is opening up an access for a pathogenic intrusion into the tissues. On the other hand, any minor effects in the microbiome could easily accumulate, and its disturbances then reflect to our health. Therefore, also any improvements in repairing the malfunctioning of the gastrointestinal tract could also be a consequence of a sum effect of several factors. The nutrition plays an important role in this process. Individual nutritional therapy with micronutrients is a potential tool for preventing and curing the intestinal disturbances. Thus small changes could influence remarkable cure by several accumulated impacts to the positive direction.

It is of great importance to recognize the fast bacterial or other microbial growth inside the gastrointestinal channel, which is not necessarily evoking the innate immunity or other biochemical defenses in all cases. Even if the microbial propagation does not lead to intrusion into the body system as such, fierce microbial metabolism inside the alimentary tract contributes to different symptons. For example, irritable bowel syndrome (IBS) could at least partially be a consequence of acid and gas production by the microbes inside the gut. It is remarkable, that these organisms could even belong to the host normal flora, or are considered as commensal by nature. In order to have an overall view to the various problems of the digestive tract from the microbiome perpective, various dysfunctions are communicated in this chapter.

# Functional Disorders of the Gastrointestinal Tract

Gastrointestinal tract disorders are common, often annoying and downright invalidating. The patient has often been told that nothing serious is in question but most patients do not accept this kind of response. In many cases the explanation for the symptoms could be found in microbiological or immunological incompatibilities, or imbalances. Often the neurological or humoral control is somewhat mispaired, too.

In the case of the so called functional disorders the question is about several mechanisms, such as:

1. intestinal abnormalities in the physical functions
2. hypersensitivity of internal organs
3. microbe infections
4. inflammatory conditions
5. disorders of the autonomic nervous system
6. disorders of hormonal regulation
7. irritable bowel syndrome.

These functional disorders cause sensational occurrence in the brain and also of psyche. Functional disorders of the gastrointestinal tract include swallowing disorder, non-cardiogenic chest pain, functional dyspepsia, biliary contractions disorders, functional constipation, functional diarrhea and in practical medicine a significant irritable bowel syndrome (IBS).

The purpose of the digestive system is first to break down food with the help of enzymes and then to transfer nutrients through the intestinal mucosa into the blood circulation to be used by the organism. This process is somewhat integrated with the microbial metabolism. The wall of the intestine also acts as a selective "gate" by denying entry from insoluble or badly dissolved proteins and different toxins, bacteria, yeasts and parasites. The function of the intestines is controlled by the nervous system of the intestines [Gershon, 1999]. Its function is mainly independent from the brain but coplex communication mechanisms do occur between the different nervous systems [Van Oudenhove *et al.*, 2011]. Also the so-called peristaltic motion is controlled by many factors. With the peristaltic motion the muscles of the walls of the stomach and intestines maintain a progressive slow movement to transport the food mass from the stomach to the small intestine and forward to the colon and rectum. Any of the below mentioned disorders or malfunctions could possibly be caused by the involvement of the micro-organisms.

# Breakdown of Functional Disorders

- Esophagus disorders
- Stomach and duodenum disorders
- Intestines disorders
- Stomach ache syndrome
- Gall bladder and bile duct closure disorders

- Anus and rectum disorders
- Disorders of newborns and breastfeeding
- Disorders of children and adolescents

# A Functional Entity

The digestive tract and liver must be reviewed as a functional entity. Together they operate as the protection of the organism against harmful substances. Micronutrients are crucial factors in maintaining our health [Prasad, 2011]. The damaging of the intestinal mucosa can lead to the dissolving of harmful substances in the contents of the intestine. The task of the liver's detoxification system is to render harmless the chemicals and the metabolic products that are alien to the organism [Santoro *et al.*, 2007]. The liver has a big role in the detoxification system. Disorders of it overload the antioxidant system. Also, on the contrary, the oxidative stress in the tissues is stressing the liver. This results in chronic oxidative stress i.e. a spoiling process and toxic load that shorten the life span of cells and that cause changes of function and structure of several organs in the intestines, liver and almost everywhere in other organs also. Nowadays the organisms toxic load is increased by stress, medication, allergic reactions, infections, stimulants and drugs and also toxins developed inside the organism that strain and often damage both the intestines and liver and increase the permeability of the intestines. This develops into a vicious circle that further aggravates the stress on the liver and increases the toxic load of the organism and worsens the oxidative stress of the entire organism.

# The Wall of the Intestine, the Size of a Tennis Field, as a Permeable Filter

The length of the small intestine is about 5 metres [Fritscher-Ravens and Swain, 2002]. The surface area of the mucous membrane is increased by transverse, less than one centimeter high, circular folds and villi. The latter are protrusions in the epithelium, which enlarge the total surface area. One square millimetre has 20-40 villi. The surface area is further increased by the fluffy border ridge of the epithelial cell, the so called microvilli, through which most of the absorbing of nutrients takes place. Each epithelial cell contains 2000-4000 microvilli, so the surface area of the small intestine is about 300m$^2$. The intestine mucous membrane, the epithelium, has only one layer of cells. The billions of specialized cells in the villi recur once every 3-7 days [Lawson, 1989]. New epithelial cells are born in the tubular small intestine glands, the so called Lieberkühn glands, that are located between the villi. From there they migrate to the tip of the villi where the release into the intestine cavity takes place. This delicate system is also highly vulnerable, and its proper function requires protection by the immune system, and other defense mechanisms.

Cell membranes are formed of phospholipid bilayers. The most important phospholipids of the cell membrane are the choline containing phosphatidylcholine and sphyngomyelin [Spengler *et al.*, 2008]. The third largest group is the aminophospholipids; phosphatidylserine and phosphatidylinositol. Cholesterol is one of the cell membrane building blocks. It controls the solubility of cell membranes in situations where the cell tries to multiply too much. Such instances are for example the effect of alcohol or medication to cell membranes. Cell membranes contain small amounts of glycolipids. Their "carbohydrate tails" act among others as antibody receptors. Glycolipids are almost always located on the outside surface of cell membranes.

As the functioning of the cell changes, phospholipids will organize themselves again. Cell membranes can contain over 100 different proteins. They act as enzymes, transport proteins and receptors of hormones and neurotransmitters. In the small intestinal surface areas, in between the various formations, are remaining some bacterial flora that constitutes the normal flora of this region. They start functioning vigorously, as the food materials arrive in pulses [Hakalehto *et al.*, 2008]. The rapid growth, and overall metabolism of the normal flora in the small intestines is synchronized with the host metabolism. Both parties of this somewhat symbiotic relationship take advantage of the joint process, as the food is forwarded in the gut. The microbes are transported within the flow of material into the cecum, where they merge into the microbe population there. Inside the colon, human excretion is influencing the microbe community which is a sum of the intestinal fluid with the incoming strains of duodenal origin, and the "local" population residing in the even more strictly anaerobic colon areas. In recent studies of the cecum microflora, there has been found a correlation between the obesity and the composition of the microbiota in the region [Turnbaugh and Stintzi, 2011]. According to our studies, the indication of increased amount of butyric acid bacteria in the cecum of obese individuals could be derived from 1. slightly different composition of the flora, and 2. of its metabolic activities in the upper small intestines [Hakalehto *et al.*, 2008]. If the neutral acid substances producing strains of the *Klebsiella/Enterobacter* type of the facultatives dominate in the duodenum, it causes the increased production of ethanol which is absorbed fast, and contains a high energy content. The 2,3-butanediol is then transported in the chime into the colon where it is being metabolized by the local flora. In turn, if the mixed acid fermentation is converting an increasing portion of the food substances into lactic and acetic acid, and other organic acids. These molecules are absorbed somewhat less readily than the ethanol, and also cause growing acidity. This, in turn, could influence the microflora composition down to the colon.

Lipophilic and especially low molecular weight compounds are able to permeate the cell membrane based on their solubility [Staines *et al.*, 2000]. Hydrophilic compounds have different ways of solution that normally require energy when permeating cell membranes. The mucous membrane of the digestive tract must function partly as a permeable filter so that it allows the absorption of small nutrients but prevents the absorption of large alien molecules. When the mucous membrane of the intestines is healthy, water and nutrients will permeate it. A healthy mucous membrane can prevent the absorption of large molecules such as antibodies and many toxins.

The selective permeability of the epithelial layer is controlled by several mechanisms of the organism. The surface of the epithelial cells of the small intestine are covered by a porous surface layer that contains the enzymes needed in the dispersing of proteins and carbohydrates and receptor proteins and carrier molecules needed in the absorption of nutrients. Epithelial cells have an effective ability to tightly close the space between cells. Tight junctions prevent carrier protein movements to the wrong part of the cell membrane. Therefore in a healthy mucous membrane there is very little absorption through the spaces.

Nutrients are absorbed primarily through the cell. Mucus excreted by the epithelium forms a barrier to the adhesion of harmful and pathogenic bacteria and the absorption of macromolecules. Some pathogens exploit specific mechanisms and different routes for intrusion in the body through the damage of epithelial surfaces [Kabaroudis *et al.*, 2003].

The surface cells, i.e. epithelial cells of a healthy mucous membrane and an effective immune system are key issues to the functioning of the selective permeability of the intestines. Current nutrition therapy uses products that can protect and repair the mucous membrane, such as silicon and cellulose products, animal lecithin products (including all neurolipids needed in human cells), new generation probiotic products, linseed components such as fibres and mammalian lignin, vitamin A and carotenoids, vitamin E and tocopherols, vitamin C and bioflavonoids, several medicament molecules that protect mucous membranes and several miniature nutritional factors that are indispensable for the organism [Farhadi *et al.*, 2006].

## Leaky Gut Syndrome

The digestive tract must function as a partly permeable filter so that it allows the absorption of small nutrients but prevents the absorption of toxic compounds. *Several microbes are adsorbed and attached onto the membranes, and their metabolic functions are associated with host functions on the epithelial surfaces.* When the mucous membrane of the intestines is healthy, water and nutrients permeate it but a healthy mucous membrane also prevents the absorption of antibodies and toxins. Many of the organism's mechanisms control this selective permeability.

The integrity of the intestines mucous membrane is affected by many factors, especially genetic and nutritional ones, and the general toxic burden of the organism. Nutrients and their ability to absorb affect the intestinal mucous membranes. The recurrence of the epithelium is important to the functioning of the intestines and the prerequisite to this is a good balance of miniature nutritional factors. Exotoxins, toxins from the ambient environment, and endotoxins, toxins that are born inside an organism, affect the condition and functionality of the digestive tract. The main exotoxins are food additives, food allergy causing compounds, chemical medication and environmental toxins.

When the surface tissue and villi structures of the intestines are damaged, this can cause the absorption of both endotoxins and exotoxins. This is called the leaky gut syndrome [Maes and Leunis, 2008]. Many common drugs, such as NSAIDs, many antibiotics, cytostatics, i.e. cytotoxins and alcohol can increase intestine permeability. Also many other harmful matters such as smoking, food allergies, dysbiosis, and some infections cause the same. This leads to the hyperactivity of the immune defense and can cause allergies, hypersensitivity and autoimmune diseases.

## The Origin of Leaky Gut Syndrome

With the increase of oxidative stress, the tissue damage of mucous membranes in the inflammatory area expands. This is how cytokines, free radicals and other oxidative substances are born. As a chain reaction they increase the inflammation also in surrounding tissues. These are fire-like damage processes that extensively weaken the defense mechanisms of the intestines. The probiotic bacteria, that are important to the normal function of the intestines, decrease making room for many harmful bacteria and fungus. The damage of intestine walls advances and leads to increase permeability of the intestine. This condition is called leaky gut syndrome [Rapin and Wiernsperger, 2010].

When a patient has increasing bowel symptoms, also different allergic reaction, hypersensitivities, diarrhea, stomachaches, bloody stool and different acid reflux symptoms will appear. In this case there is powerful metabolic degradation in the organism (catabolism), in which case the organism starts to use its own tissues increasingly as a nutrient. The neural regulation is an important factor on the level of mucosal surfaces [Manicassamy and Pulendran, 2011]. The patient's weight will decrease. Approximately with one third of patients with irritable bowel syndrome the dominating harmful intestinal microbe is *Candida* fungus. This is, however, not a classic microbial fungus disease but instead a candida prevailing dysbiosis in the intestines.

## The Intestines Immune System and Its Systemic Influences

The contents of the intestines are 1/3 unabsorbed food material, 1/3 dead cells of the intenstine mucous membranes and 1/3 in the intestines autonomously living microbiota. In order to prevent toxins of the fecal mass to get into the organism, the mucous membrane must be not only intact and protected, but also effectively functioning [Indrio *et al.*, 2011]. Otherwise, the food may be only partially absorbed, which then could lead to growth of the problematic microbes. In any kinds of situations, the immune cells in the intestines should be able to identify pathogenic microbes from those that are useful or harmless.

It has been estimated that the majority of the human body's lymphatic system is in the digestive tract area being an important protective and maintenance mechanism for our health [Reis and Mucida, 2012]. Almost 70% of the antibodies needed in the cleansing of the organism are produced in the intestines. The intestinal flora, i.e. the so called good bacteria, have a powerful effect to the environment's lymphatic system and prevention of mucous membrane damage. In the intestines there is continuous interaction between the microbiota and the immune system. Microbes have significant roles in this interaction. They can affect the immune system in such a way that the harmful factors of nutrition do not too easily cause immune defense reactions. Good bacteria that are generally called lactid acid bacteria are therefore important in the maintenance of the intestines mucous membranes and immune system.

The gut microbes have been shown to influence the systemic innate immunity by affecting the neutrophils in the blood circulation of mice [Philpott and Girardin, 2010;Clarke *et al.*, 2010]. It also seems important for health that the innate immunity is reaching on some level of activity. Mice detective of the Nod 1 protein by nutrition were more susceptible to pneumococcal sepsis than the wild type mice. While the immune system defends against pathogens and growth of excessive intestinal flora, it has to maintain simultaneously a state of immune tolerance to the components of the normal flora [Reis and Mucide, 2012]. Many bacteria produce antigens into their cytoplasm, which then are expressed very rapidly onto the surfaces [Abraham and Beachey, 1987]. Therefore the immune system needs to monitor and keep up immune homeostasis. On the contrary, the microbes are influencing the host immune development, immune responses, and the susceptibility to diseases such as IBD (intestinal bowel disease), diabetes mellitus, and obesity. Intestinal epithelial, and both innate and adaptive immune systems exploit pattern recognition receptors (PRPs). The recognition of microbes by the PRPs initiate numerous defenses, such as cytokine and chemokine signaling pathways, antimicrobial proteins, phagocytosis, autophagy and reactive oxygen and nitrogen species. The formation of the latter is initiating oxidative stress whose compensation requires adequate and continuous replacement of antioxidants. There are plentiful of mechanisms, which are used by the host to control the microflora, and to keep it on a moderate level. These include enzymatic modifications of the LPS (lipopolysaccharides), which suitable are called as endotoxins. They constitute the outer leaflet of the Gram-negative cell walls, and thus regularly get in contact with the host mucosal and cell surfaces. The host functions, such as intestinal production of defensins (peptides with antimicrobial activities produced by the host body), are formative factors in regulating the microbial ecology in the gut [Salzman et al. 2010]. The activities of defensins have been recorded in the PMEU to be 100-1000 fold in comparison with the antibiotics [Hakalehto. 2011; Hakalehto *et al.*, 2011b]. On the other hand, microbial symbiosis is preventing the intestinal inflammatory diseases [Mazmanian *et al.*, 2008].

# The Brain – Intestines Connection

According to professor Michael Gershon, the nervous system of the intestines is so well developed and independent in structure and function, that it can be considered the organism's "second brain" [Gershon, 1999]. The intestinal nervous system functions mainly independently from the brain, but it has many common features with the brain. In the human small intestine alone there are more than 1200 million nerve cells which is at least as much as in the spinal cord [Kramer *et al.*, 2011]. As in the brain, more than 30 influencing neurotransmitters have up to now been found in the intestines.

The human intestine is actually one long tube that is surrounded by a multilayer wall. Nerve cells are located on the inner surface of the intestines and nerve branches go to the surface of the mucous membrane. These branches collect information on the content of the intestine. Other nerve cells observe when to stretch the intestine wall so that the food mass can move forward [Bassotti *et al.*, 2005]. Any disturbances in the movement could accumulate the toxic substances, and the microbial balance (BIB, bacterial intestinal balance) could get distorted. Cells that observe the content of the intestine secrete a lot of serotonin neurotransmitter that is found also in the brains [Banerjee *et al.*, 2007]. According to studies published during the past years, over 95% of all the serotonin in the organism is formed in the intestines [Gershon, 2004]. Moods and feelings of a person have been found to correlate with the amount of serotonin in the brain.

A mechanical or chemical stimulus in the intestine can cause the release of serotonin. Irritable bowel syndrome is one of the most common stomach problems in the western world. Its cause is not known. According to one theory it could be caused by a deficit of serotonin in the intestines. According to the above mentioned background information, a vague or lost balance in the microflora could be the causative agent in these symptoms. Irritation by microbial acid or gas production could be sensed by the nerve endings on the gut walls. Also the attachment of hazardous bacteria, and their possible intrusions, could produce defensive reactions by the neural actions. For example, the salmonellas could cause the development of the diseased gut and diarrhea by further active offenses on the epithelia, which is initiated by an attachment [Hakalehto *et al.*, 2007]. It is inevitable that the neural tissues also get irritated.

The intestinal nerve cells also react to signals from elsewhere in the organism. When we are excited, stressed, happy, or in love, hormones and other informational material, that affect also the nerve cells of the digestive tract, are released into the organism. The communication system between the brain and the intestines has an important role in most vital functions. The central nervous system coordinates the different functions of digestion, the peristaltics of the intestines (the advancing contraction wave that moves the contents forward). Similarly, signals from the intestines can control the function of the central nervous system. An acute inflammatory condition of the intestines can cause the central nervous system to sensitize and further cause *e.g.* mental reactions.

Therefore, a lipid substitute containing all neurolipids is needed for traumatized human cells.

According to current understanding, serotonin is the most significant neurotransmitter. A sudden decrease of its production can increase pain and cause depressed mood. Clinical treatment studies have shown serotonin treatment to relieve the symptoms in part of the patients.

## Dysbiosis Causes Dysfunction

A disturbed balance of intestinal microbial flora (dysbiosis) causes dysfunctions of the gastrointestinal tract. The amount of intestinal bacteria alone in the colon is more than 100 billion. The amount is larger than the total of all the cells in the human body. When lactic acid bacteria, probiotics, are well represented in the intestines, they maintain the immune defense system of the intestines. A healthy microbiota can change the chemical composition of industrial food and drugs, produce and break down vitamins, break down food toxins and prevent growth of pathogenic bacteria and yeasts. Healthy microbes boost the immune system of the intestines. The administering of probiotics in time is an important principle in the treatment of the intestines.

The disturbed alimentary interactions between man and his microbiota could be divide into three categories, namely:

- microbial overgrowth,
- inflammation, and
- intrusion

Although both situations are usually manifestations of disturbed BIB (Bacterial Intestinal Balance), they differ in the fundamental matter of nature of the pathological action. In both cases the condition could develop into dysbiosis [Floch, 2011]. The first category involves the microbes using the alimentary tract as a compartment for their uncontrolled growth. In this case one or several strains have escaped out of the control; they do not strive for balance, but simply spread in the intestines without provoking much direct host resistance. This is due to the fact that in the first place the microbes do not try to spread on the epithelia or into the tissues. The immunological defenses have thus not awakened. This makes it considerably difficult for the host to eradicate the disturbing micro-organism.

The second type of imbalances includes the increased coverage of the intestinal or other surfaces by the opportunistically pathogenic strains. These could promote many host defenses, such as the occurrence of diarrhea symptoms, by which the tissues flush the intestinal organ [Bosseckert, 1983]. In such cases, micro-organisms actively seek for attachment and usually possess cellular appendixes for that adsorption. For example, salmonellas and many enteric bacteria apply the type 1 fimbriae for quick anchoring to the cell surfaces onto mannose containing side chains of the epithelial cells [Hakalehto *et al.*, 2007]. In these cases the pathogens are more likely to challenge the immune system and other potent host defenses. At least the innate immunity as the first line of defenses is wakened up [Harrison and Maloy, 2011].

Most severe form of pathogenic action is the penetration into the mucosal membranes by the disease causing bacterium. It usually contains more specific means of attachment and intrusion, such as the P fimbriae in the uropathogenic *E.coli* [Narciso *et al.*, 2011]. *Salmonella* sp. also typically represents such an organism with considerable abilities to interfere with the host digestion and other vital functions [Agbor and McCormick, 2011]. Salmonellas sometimes could develop intracellular pathogenesis, which capability they share with *e.g.* yersinias [Torres *et al.*, 2011].

Any microbial function that provokes the host immune system involves the invasion onto or into the surfaces. The third type of intervention activates both the innate and the adaptive immunity. All involvement of the immunosystem includes the inflammatory reactions, which struggle for keeping the body clear of pathogens but often cause problems by overreacting. Therefore, the "fire of the disease" needs to be repaired and this inflammation to be settled down in order to avoid disturbances and such consequences as autoimmune reactions. This reparation requires good nutrition, acquisition of individually important micronutrients. The approach should belong to the future treatment arsenal of personalized natural medicines.

Better understanding of the immune system has confirmed the different stages in the pathogenic microbial actions also inside the gastrointestinal tract. However, it is noteworthy that not all unusual microbial actions is being noticed or causes immunological responses. In the intestines it is possible to develop distressing conditions which involve bacterial or other microbial strains that are not intruding inside the body system. Still they produce plentiful of gases and acids, and cause remarkable irritation. These forms of microbial interference is usually neither detected in the blood test nor recognized by the standard medicine.

## Systemic Effects of the Intestinal Disorders

Even though a microbiologically alarming situation occurs in one part of the alimentary tract, it is extending its effects into the entire body system. The oxidative stress could be compared with a fire that generates inflammation. The reparation requires a plentiful of antioxidative nutritional substances. Proper nutrition, in turn, is maintaining a healthy microflora. That enables the support for healthy normal flora, and also the prevention of hostile intruders, such as soil bacteria.

The free radicals in the body cause an oxidative stress, which is producing inflammation and eventually influencing entire immunosystem. Most medicinal treatments increase the stress on the body system, which could be prevented by antioxidant treatment. This is influencing the membranes which surround the cells. There the sphingomyelins and cholesterol form lipid platforms [Silva *et al.*, 2009]. The former substances are converted into a signal substance, ceramide [Romanowicz and Bankowski, 2010]. This in turn takes part in many important series of reactions, such as cell growth, differentiation, aging, movement, attachment, and apoptosis.

Recently, there has been proclaimed "the European fight against malnutrition" [Ljungqvist *et al.*, 2010]. The concept includes a campaign for improved nutritional care in all types of care facilities, such as hospitals, care homes, as well as the in the community. The basic idea is to promote awareness of the relevance of the nutritional status for the patient′s health status. It is an essential point that all patients should be provided with diets rich in protective elements. In the campaign there is an important goal to learn to detect the malnutrition in all its forms at an early stage [Muscaritoli *et al.*, 2010]. Sarcopenia can be observed at any age as a consequence of inflammatory disease, malnutrition, disuse or endocrine disorders. Malnutrition in this case could also mean a qualitative loss of vital components in nutrition. Accordingly, the prerequisite for recovery from any inflammation is proper antioxidant level in order to avoid degenerative effects of the oxidative stress caused by accumulation of free radicals.

In 1900, the first organic free radical, triphenylmethyl radical, was identified by Moses Gomberg in the University of Michigan, USA [Lafont, 2007]. The free radicals are derived from oxygen or nitrogen in humans being atoms, molecules, or ions with impaired electrons. They are highly reactive, causing oxidative stress to any living organism. The aerobically metabolizing cells utilizing oxygen for cell respiration need for survival an antioxidant system. These molecular mechanisms protect the cells against damages produced by free radicals.

Any inflammatory reaction is causing the above-mentioned oxidative stress in the tissues [Escobar *et al.*, 2011]. The consequences of such a prolonged sequence of reactions can have a severe impact on our health. Therefore, reparation by continuous corrective nutrition is required and is essential for the recovery from the microbial diseases, too. Any of these illnesses can produce lifelong track in the cell memory, and has its traces in our metabolism. Our body requires oxygen but excessive free radicals are disastrous for our health.

The continuous inflammation of the mucosal membranes and the different tissues, which could be a consequence of partial recovery of an illness or of any imbalance in the microbiome, is depriving the body from resources. The alarming situation needs to be stopped with adequate compensation of micronutrients [Prasad *et al.*, 2001]. Unfortunately, the pathogens often attempt to create conditions where the disturbed body functions prevail for the benefit of the harmful microbes. This kind of development is detrimental also for the normal flora. For example, yeasts can produce substances, which deflect, deactivate, or diminish the neurological or immunological functions [Kumamoto and Pierce, 2011]. This, in turn, is altering the circumstances inside the gut and on the membranes into a direction of a monoculture system. Such negative progress is then invaliding the nutritional uptake mechanisms.

The microbiological condition of the alimentary tract is often a consequence of a general disease. For example, weakened immune defense could result in a manifestation of microbial overgrowth in the intestines, or opening up an access for a pathogenic intrusion into the tissues [Cerutti *et al.*, 2011]. Any minor effect in the microbiome could easily accumulate. Therefore, also any improvements on diseased conditions, or in repairing the malfunctioning of the gastrointestinal tract could also be a consequence of a sum effect. Thus small changes could influence remarkable cure by several accumulated impacts on the positive direction. It is of great importance to recognize the fast bacterial or other microbial growth inside the gastrointestinal channel, which is not necessarily evoking the immunity.

Some pathogens, such as *Mycobacterium tuberculosis*, HI-virus, or intracellular pathogen bacterium *Borrelia burgdorferi* can hinder the functions of the VDR (Vitamin D Receptor) sites [Haussler *et al.*, 2011]. It has been found that up to 98% of the receptors sites were blocked by the living borrelias that cause the Lyme disease [Marchal *et al.*, 2011]. An extract from this bacterium diminished the number of these sites to 1/8. This kind of action prevents the vitamin D from entering the cells. This is a serious condition, because this vitamin plays a key role in maintaining the innate immunity system [Bischoff-Ferrari *et al.*, 2011]. Besides, this vitamin is central in human cell division and differentation. Moreover, the bile acid lipocacheic acid is normally binding to the same receptors [Castillo *et al.*, 2011]. As the fatty acids are assimilated from micelles comprised of also bile salts, the defect in the contact of cells and intestinal surfaces with the micelles could hamper the the uptake of essential fatty acids, too. If this function is deprived or diminished from the cells, it could lock the anabolic metabolism to a great extent, because of the prevention of the membrane synthesis, and propagation of any cells. This is an example of a mechanism, by which the pathogens are able to seriously damage the overall health of an individual.

## Microbial Overgrowth

Even if the microbial propagation does not necessarily lead to intrusion into the body system as such, fierce microbial metabolism inside the alimentary tract contributes to different symptoms. For example, irritable bowel syndrome (IBS) could at least partially follow from acid and gas production by the microbes inside the gut. It is remarkable, that these organisms could ever belong to the host normal flora, or are not at least commensal by nature.

Irritable bowel syndrome (IBS) could perhaps be seen also as like a condition where unusual microbial activities interfere with the host health more or less seriously, but without definite signs of pathogenesis. The causative organisms are often not identified. However, they easily build up physical pressure in the gastrointestinal tract by gas production. The body reacts accordingly, trying to remove the problem, for example by intestinal movements. If the gas pressure is elevated in the stomach, it might inhibit the natural mechanisms for feeling the hunger, and thus disturb body functions related to proper nutrition [Deenichin *et al.*, 2010]. If the excessive microbial activities lead to continuous acid production, for example, this could irritate the mucosal membranes and initiate inflammation. Also in the gastric areas the microbial overgrowth may induce hydrochloric acid overproduction. Our defense system in the stomach is not necessarily able to distinct the intruding microbes from the intestines from those derived from the food and drink. Therefore, the problems are becoming more serious when the body attempts to remove the harmful agents.

Other disturbed body functions caused by the dysbiosis include allergic reactions toward the the microbial strains. Also in this case, the body is striving for balance, but the natural immunological defenses cannot remove the trouble making microbe strain, because it remains active in the intestinal fluid. All counterproductive host reactions only worsen the situation in the disturbed alimentary tract. The best way for healing the diseased gut and the IBS is in many cases to reconstruct the BIB (Bacterial Intestinal Balance) [Hakalehto, 2011]. This requires various components of the healthy microflora to overcome the troublesome monocultures or pathological balances of undesired strains. The transfer on inoculation of feces has been used in some hospitals for regaining the balance [Härkonen, 1996]. In any case, the better treatment and cure of IBS requires more research on the interactions of the alimentary microbiome. The uncomfortable symptoms of the IBS could be avoided or overcome by adjusting the BIB.

## Inflammatory Intestinal Diseases

Inflammatory intestinal diseases include ulcered colon infection (colitis ulcerosa), chronic rectum infection (proctitis) and Crohn's disease (CD). The cause of these diseases is unknown. The incidence of inflammatory bowel disease (IBD) in Finland is about 20/100.000 inhabitants and the prevalence is 300-400/100.000 inhabitants. Of these, 2/3 suffers from colitis ulcerosa. IBD normally appears at age 20-40 years and Crohn's disease with slightly younger people than ones with colitis ulcerosa. IBD is a somewhat unusual disease group in the sense that a general practitioner does not get adequate experience in treating it. Patients are often young and the disease is socially restrictive. The course of the disease is often variable, which makes the diagnostics difficult. In fact, the various disorders of the intestines are related to the microbial imbalances which need to be repaired in order to achieve complete healing from the illness. In this case it is desirable that not much permanent damage had been caused to the intestinal mucosa. Nevertheless, the antioxidant therapy could provide the body system the substances required for the recovery.

## Nausea and Vomiting

There are many causes for acute nausea and vomiting: general infections, gastric and intestinal infections, drugs, hepatitis, pancreatitis, cholelithiasis, post-surgical conditions, irritation of the organ of equilibrium, migraine attacks, radiotherapy, urinary poisoning and ketoacidosis, hypertension crisis, increase of intracranial pressure, poisoning etc.

Causes of chronic nausea and vomiting are functional disorder of the esophagus, tumor or infection, gastric ulcer disease and gastric tumor or functional emptying disorder, bowel obstruction and bowel functional disorder, early pregnancy, increase of intracranical pressure, severe metabolic diseases, nutritional disorders and mental reasons, not to forget the overgrowth of the microbial strain in the intestinal compartment. Nausea and vomiting are fairly common, however, literature does not present exact incidence figures. The numerous disturbances causing the various symptoms indicate the sensitivity of the alimentary tract on nutritional or microbiological malfunctioning. A general consequence of these is the stress and accumulation of toxic compounds in the intestines. Their elimination requires extra resources from the body. This is emphasized in cases when the normal microflora has been weakened and cannot fully act as a cleaning body. Any disturbances in this vital function could be then experienced as nausea.

# Diarrhea

Diarrhea means that the excrement becomes loose and liquid. When defecation is needed 3 times a day, one can talk of diarrhea. The mechanisms and factors behind diarrhea are very variable. In fact, the diarrhea could be considered as some kind of cleaning procedure in the intestines. An acute bloodless diarrhea normally relates to an infection caused by microbes and the source of infection is often spoiled food or water. Based on the infection environment the condition is characterized as tourist diarrhea or domestic food poisoning. Any such illness is challenging our:

- immunosystem including the innate and adaptive immunity
- other body defenses including the cytokines
- repair mechanisms for damages on tissue, organ and cellular level

These detrimental effects of an inflammatory disease are often directed toward the mucosal membranes of the alimentary tract.

Most common causes of tourist diarrhea are especially *E. coli*. The reason, length and treatment of regular diarrhea (length over 3 weeks) depend on its cause and birth mechanism. The origins of the disease lay again on the disturbed, or imbalanced, gut flora. It is of crucial importance for the human body to maintain healthy normal flora which also is a protection against diseases.

Diarrhea in these acute cases, when the intestinal microbiome meets incomers, could reflect the efforts of the gut to normalize the flora. This could be logical especially for the reason that the normal flora if persisting, would be favored by *e.g.* the immune system.

Causes of functional chronic diarrhea are irritant bowel syndrome and so-called functional diarrhea [Niemelä, 2009]. If diarrhea has begun in connection with antibiotic medication or after it, the possibility of the aggressive bacteria *Clostridium difficile* should be clarified using endoscopy of the colon. In this case a serious monoculture event of spore-forming clostridia has been taking over the alimentary tract. The antibiotics may have been used in excess or extensively.

As a result, the anaerobic clostridia, being relatively sturdy against such effects (perhaps due to their soil origins) could compete out the other flora. More detailed studies on *Cl.difficile*, as well as on the antibiotic usage can be found elsewhere in this book. Bile acid diarrhea can also be connected with irritant bowel syndrome without a local organic disease. In this case in question is the agitated motor function of the small intestine causing the too fast passage of bile acids through the small intestine into the colon without adequate time for absorption.

The severe excretion function disorder of the pancreas, especially in chronic pancreatitis, relates to the decrease of the break-up of fats and absorption disorder of fats.

## Diseases Causing Diarrhea

An acute diarrhea is normally an infectious disease caused by a virus. It spreads often as epidemics and is contagious. Antibiotics can also cause diarrhea as in addition to killing the bacteria that are harmful to the system they often also kill bacteria that are advantageous to the intestines thus causing an imbalance of the micro-organism of the intestines.

The most common symptom of diarrhea is loose excrement several times a day. Often increased temperature, vomiting and temporary mild stomach ache are connected with it. The treatment is mainly the treatment of the general condition. Diarrhea heals normally by itself in three days. There is a risk of dehydration of the person's system because of the loss of fluids and salts especially if there is high fever and vomiting in connection with the diarrhea. Most important in the treatment of diarrhea is to drink plenty of liquids. Depending on age and size at least 1.5 to 2 liters of liquids should be drunk per day, small amounts at a time. Water, diluted juice and diluted tea are recommended. Sweet soft drinks are not recommended since too much sugar can increase the risk of drying. If one has appetite, food can normally be eaten during diarrhea, however, fatty, fried and spicy foods should be avoided.

Recent studies have added important new information to our understanding of the pathogenesis and ethiology of diarrheal disease [Neter, 1982]. *Vibrio cholerae* produces a heat-labile enterotoxin, affecting cyclic AMP. A very similar heat-labile enterotoxin is produced also by certain strains of *Escherichia coli*, as well as by *Citrobacter, Klebsiella*, and *Aeromonas. E. coli* may also produce a heat-stable enterotoxin, stimulating guanylate cyclase activity. In order to produce the pathologic effects, *E. coli* first attaches to epithelial cells of the intestinal tract by means of fimbriae (pili) or surface antigens [Hakalehto et al. 2000]. Enterotoxin can be demonstrated by both *in vivo* and *in vitro* tests, but none are yet suitable for routine diagnostic laboratories. A third mechanism whereby *E. coli* causes diarrheal disease consists of enteroinvasiveness.*Campylobacter, Yersinia*, and *Clostridium difficile* have been added to the list of enteric pathogens of man.

## Children's Diarrhea

A common disease in children is diarrhea in connection with infectious diseases. In developing countries this is still the most common cause of infant mortality. Microbes that cause diarrhea produce toxins, so-called enterotoxins that bind to the mucous membrane cells of the bowel causing disorder of the normal function of the bowel. In this case instead of absorbing normally, salts and water excrete strongly.

If a child has symptoms of dehydration, restlessness, fatigue and a decrease of the excretion of urine, it is recommended to drink over-the counter liquids available at pharmacies using instructions provided on packaging. These are designed to correct dehydration and are suitable for both children and adults. A doctor should be consulted if the diarrhea has lasted for over 5 days or the diarrhea is substantial together with vomiting, the general condition is very weak, fever is high or the child cannot drink adequate amounts of liquid. Serious dehydration should be treated in hospital. Nowadays probiotics are recommended in the treatment of diarrhea.

In Finland the diarrhea of infants is normally caused by rotavirus. Acute diarrhea occurs mainly in the beginning of the year. In a 1996 survey in the maternity and infant clinics of the city of Espoo, under 5 year old children had on average of 0.5 diarrhea episodes per year [Rautanen *et al.*, 1998]. Studies show that for example the use of foods containing a probiotic, Lactobacilli GG has shortened the length of diarrhea in rotavirus infections. A randomized comparative study found that lactic acid bacteria of Lactobacilli GG was the most effective probiotic product in this survey. The use of it decreased significantly the risk of getting a diarrhea lasting over 3 days. Based on new studies it is obvious that probiotic products are useful in the treatment of infectious diarrhea for both infants and adults, especially if the diarrhea is caused by rotavirus. Novel strategies and views for introducing the probiotic strains are presented in the Chapter 12 of this book. In any case, the maintenance of a specific strain on the epithelia requires continuous delivery.

## Chronic Diarrhea

For practical reasons diarrhea can externally be split into bloody and non-bloody diarrhea based on appearance. Based on anamnesis and survey of the excrement one can distinguish between watery, slimy and fatty forms of diarrhea. When diarrhea has lasted for less than 3 weeks it is acute diarrhea and a diarrhea that has lasted over 3 weeks is considered chronic.

The occurrence of chronic diarrhea is approximately 5% of the population. In the background are tens of different causes of which most are uncommon. Chronic diarrhea is gradually increasing. Fever does not normally relate with it. Gastrointestinal inflammation (gastroenteritis) and inflammatory bowel disease (enteritis) caused by microbes normally start suddenly. In the background of a non-bloody diarrhea is often either irritable bowel disease, celiac disease, lactose intolerance or other food intolerance or food allergy.

Most patients with chronic diarrhea can be examined in general health care system. Irritable bowel disease is the most common cause of non-bloody diarrhea with adolescents and others with a generally good overall condition. It can occur after acute gastroenteritis and trouble for a long time. If the patient is over 50 years old with chronic watery diarrhea, endoscopy of the colon should be done and samples taken to find microscopic colitis. At the same time this will be an exclusion examination for colon cancer.

## Microscopic Colitis

About ten per cent of those with chronic diarrhea have been diagnosed with microscopic colitis. When performing an endoscopy of the colon the mucous membrane appears normal but a microscopic examination of the biopsies finds inflammatory changes. Tissue doctrinal changes refer either to so-called collagen colitis or to lymphocytic colitis.

Diseases that relate to microscopic colitis are celiac disease, rheumatoid arthritis, thyroid diseases, asthma and both types of diabetes. This chronic inflammation in question is most often an autoimmune mediated disease. 40% of those with collagen colitis have been diagnosed with one or more associated diseases.

Also joint and muscle pains occur. Quite uncommon associate diseases are different systematic connective tissue diseases, "collagenoses" such as Sjögren syndrome, SLE, Raynaud`s disease, ankylosing spondylitis, chronic atrophic gastritis, sarkoidiosis and psoriasis.

Patients' predisposition to autoimmune diseases, the majority being women and the efficacy of cortisone to symptoms refer to auto immune origin mechanisms. The process can be triggered by bacteria or its antigen, bacteria toxin or a viral infection. Microscopic colitis may start suddenly similarly to infectious diarrhea. At least in the beginning antibiotics are often efficient against symptoms. Patient data shows that those diagnosed with microscopic colitis use more anti-inflammatory drugs than the general population. The long-term use of painkillers has been found to increase the permeability of intestines which quite obviously increases the access of the cause of the inflammation from the bowel into the mucous membrane. Also the secretion of hydrochloric acid and proton pump inhibitor drugs have been found to cause microscopic colitis which has healed after ending the use of the drug.

As to the clinical picture, collagen colitis and lymphocytic colitis resemble each other. Both diseases cause chronic watery diarrhea but bloody excrement do not fit into the clinical picture. Microscopic colitis are even less common than IBD. If a biopsy has not been taken some of them are diagnosed as irritable bowel syndrome or functional diarrhea. In mild cases a symptom based treatment of diet, diarrhea drugs and nutritional therapy is often adequate.

The causes and birth mechanisms of irritable bowel syndrome, inflammatory bowel disease and microscopic colitis are still largely unknown. Despite the fact that many symptoms are partly similar, the character of these diseases, treatment possibilities and complication risk are different. Therefore the diagnostic separation of these conditions is important. Only irritable bowel syndrome can be diagnosed based on anamnesis and clinical examination.

# Causes of Food Poisoning

In Finland food poisoning epidemics have been systematically followed since 1975. After that there have been altogether 1.800 epidemics in which about 70.000 people have fallen ill according to the statistics of the Finnish National Institute for Health and Welfare (Interview of Prof. Anja Siitonen in Finnish *Tiede-magazine* ("Science") 8/2009). Each year about 1.000-2.000 people get ill by food poisoning. Most likely this is only the tip of the iceberg as all diarrheas are not reported to the authorities.

In the 1950's, food poisoning was caused in practice by three different bacterial species: *Salmonella enterica, Staphylococcus aureus* and *Clostridium perfringens*. Presently there is a multiple amount of microbes that cause food poisoning, besides them also viruses and protozoa. In these cases pathogenic microbes have been increasing their numbers in improperly stored or prepared food. In many cases also water serves as the route for contamination. In order to maintain good status of general hygiene, more research should be directed toward the behaviour of these pathogens in the environment. They could be detected in very low numbers [Hakalehto *et al.*, 2011a], which enables effective research on their distribution. Any epidemics or outbreaks of food poisoning require rapid action with research capabilities on site of the problem.

The bacteria *Yersinia enterocolica* is capable of multiplying in refrigerator temperature and in anaerobic conditions. The increased frequency of cold storage of meat and vacuum packing, have caused problems. During the next decade the cause of food poisoning was often *Campylobacter* that grew in a warm environment within chicken bowels. Epidemics increased as the consumption of chicken meat grew. Although slaughter hygiene has improved and meat storage times shortened, both *Yersinia* and *Campylobacter* cause more food poisonings than *Salmonella*. In the developing countries we have investigated the spreading of *Yersinia* and *Campylobacter* strains in the irrigation water [Heitto *et al.* 2011]. In these studies in Burkina Faso, Africa, these pathogens were outnumbered in waste waters by other strains. However, they remained a threat for the population even though their detection required such specific methods as the use of the PMEU. The water monitoring could be automated by using the ASCS (Automated Sample Collection System) with the PMEU. This procedure has also been tested in several water departments in Finland.

In the 2000's food contamination was caused by the soil bacteria *Bacillus cereus* which multiplies especially when the food is prepared on the day before consuming. The bacteria tolerate heating better than average. The detection techniques of *Bacillus cereus* and other aerobic bacilli with the PMEU instrument has been introduced earlier (Mentu *et al.*, 2009).

If food is heated carelessly, other bacteria die, but *B. cereus* remains alive and multiplies and produces a very powerful cereulide toxin. Reportedly cereulide poisoning has been found on about 400 patients. Even if the food is heated again prior to serving, the toxin will not disappear from the food and results in intense vomiting. Also rice and pasta may be the source of cereulide poisoning. Soil bacteria can also multiply in damp building insulations in eg. mineral wool causing cereulide poisoning to the inhabitants. Also the increased use of catering services has influenced in the increase of *B.cereus* epidemics. Nowadays it is one one of the most common causes of food poisoning in Finland.

Current food poisonings are international and affect thousands of people around the world. One example is ice cream contaminated with *Salmonella* in the United States in the 1990's. One quarter of a million people got ill. The epidemic was noticed only after a long delay. The raw materials for the ice cream production batch had been transported in a tank that had held eggs. Also viruses can cause food poisoning. An example is a Danish epidemic in which raspberries grown in Eastern Europe made over 1.000 people ill. The cause was irrigation water to the raspberries that was contaminated with norovirus. A similar case occurred in Kuopio in Finland some years ago. Then the imported strawberry jam contained also norovirus agents which derived from the irrigation water.

Also food poisonings caused by protozoa have been reported. One of these is *Cryptosprodium parvum* which is a single-celled protozoan that is spread by water contaminated with excrements when the water has been used to irrigate vegetables, berries or fruit. It causes a fierce diarrhea, fever and nausea. In North America and Britain thousands of people have fallen ill in epidemics caused by protozoa. Another single-celled protozoan *Cyclospora cayetanensis* has spread through food like salad and raspberries that have been irrigated with contaminated water.

Many food poisonings are caused by waterborne pathogens. Autumn 2007 there occurred a serious outbreak of several diseases in the Finnish town of Nokia where the waste water was erroneously mixed with the household water. Since then the PMEU technologies have been developed for maintaining high standard of cleanliness in all water distribution systems in communities, industries, hospitals etc. [Heitto *et al.* 2009; Hakalehto, 2010; Wirtanen and Salo, 2010; Hakalehto, 2011].

The fast detection of coliformic or pathogenic bacteria could take place in about ten hours also in cases where just a single cell was contaminating the water. However, in some real life situations the prolonged bacterial lag phase in cultures was delaying the verification. Even in these circumstances, the PMEU was promoting the recovery of strains. For example, in cases with research on neonatal intestinal flora, it was documented that the PMEU method improved the recovery of environmentally stressed cells manifold in the culture experiments with faecal samples [Pesola *et al.*, 2009; Pesola and Hakalehto, 2011].

It is worth also mentioning a bacteria that is related to cholera, *Vibrio vulnificus*, that can be obtained from fish, seafood and octopus that are from tropical waters and poorly cooked. The consequence is a fierce infection that can lead to blood poisoning and even death. It is also possible that the bacterial strains behind the pathogenesis of common diseases change their characteristics during some period of time. For example, the causative agent of cholera, *Vibrio cholerae* has changed its mode of metabolism connected with the epidemics. Earlierly this bacterium was metabolizing glucose by mixed acid fermentation, but lately all epidemics have been provoked by strains capable of carrying out the 2,3-butanediol fermentation [Yoon and Mekalanos, 2006].

Changes in the food supply chain have also increased the appearance of new disease causing microbes. In the global food production chain the distance from the producer to the consumer is often long. The farther the food is produced and the more processed it is and the more intermediaries it travels through, the more vulnerable we are.

## Constipation

The bowel normally empties every 8 to 72 hours. Constipation means the irregular, too seldom happening emptying of the bowel. The decrease in the number of defecations, difficulty in defecating and often the hardness and sparseness of the excrement (less than 35g per day) relate to constipation. The prevalence of constipation has been studied in the United States.

About 15% of over 18 year olds and almost 20% of 30 to 64 year olds reported that they suffer constipation. This increases with age and is more common in women. In healthy adults constipation occurs in 1 to 6% and in the elderly with limited mobility the number is up to 80%. Constipation is thus a common symptom with people who do limited exercise and are on a low fiber diet. If a bowel that has functioned regularly begins to function much more infrequently, it is advisable to consult a doctor, especially if at the same time the person has stomach aches, blood in the stool or if constipation and diarrhea alternate.

Low fiber content of nutrition is the most common cause for constipation. To function to its full potential the bowel needs a certain amount of solid contents. If the content is minimal, the peristalsis, or the wavelike contraction of the bowel will bit function properly to empty the bowel.

Our daily diet should include adequate amounts of nutrition fiber. Also limited exercise and insufficient intake of liquids promote constipation. Some of those with irritable bowel syndrome also suffer from constipation.

# Western Food Predisposes to Dysbiosis

In our opinion, the change in the normal flora of the colon, so that there is hyperplasia of harmful microbes, causes at least to some extent similar intestinal dysfunctions as described with irritable bowel syndrome. Plenty of physiological statuses occur, which are either pathological or somewhat bearable depending on the condition of the individual. Essentially the health is dependable on the degradation of the food substances in the gut. It is said that "any undissociated food becomes poison in the middle of the small intestines". Additionally, when the normal anaerobic bacteria and enterobacteria of the intestines disappear and their growth place is taken by different harmful bacteria and yeasts, this is called dysbiosis [Floch, 2011]. The human body and its microbes are indeed interdependent.

The colon of an adult person contains at least hundreds of different microbe strains. In addition to antibiotics, the composition of normal flora is affected by several factors such as the quality of nutrition, lifestyles, age and hormone activity. Industrially produced food used in the western world predisposes to dysbiosis [Hillilä *et al.*, 2007; Kajander *et al.*, 2008]. As a result of many factors there may be plentiful growth of harmful microbes in which case the amount of useful bacteria has decreased. The metabolism of harmful bacteria can cause large amounts of toxic substances that preserve the colon irritation and if this continues for long it can lead to serious illnesses such as colon cancer and inflammatory bowel disease.

# Dysbiosis Definition

*Dys* means flaw and *bios* means life and growth. Thus dysbiosis means flawed life. The medical term bowel dysbiosis is often understood to describe the harmful changes of the intestines bacteria flora. A few years ago little was known about the human bacteria but the amount of knowledge is quickly increasing. It has been shown that microbe population has a great meaning to health.

Human microbes consist mainly of bacteria. It is estimated that human intestines contain 100.000 billion microbes [O'Hara and Shanahan, 2006]. They contain 10 times more bacterial cells than own cells and up to 1000 times more microbe genes than own genes. Most of the microbiota is located in the large intestines. If this part of the alimentary tract is sterile, an individual gets diseased. We need our symbiotic partners. The interactions with them are extremely complicated. The microbe population can contain even more bacteriophages than bacteria. A recent new estimate is that our bacteria consist of an average of 100 bacterial genus, 800 species and 2500-7000 bacterial strains.

A part of the intestines bacteria lives there permanently but others enter with food and travel and exit through the digestive tract. Some of the bacteria are useful to human but many harmful, so called pathogenic bacteria can cause diseases. Our basic flora is established during the early years [Oien *et al.*, 2006]. The term “intestinal microbes” covers all the bacteria, fungus, viruses and other microbiota in the intestines. Several microbes are called commensal which equals to some kind of indifference with respect to host functions. In reality, many microbes are taking part in the nutritional network in the intestines, and even many pathogenic organisms remain latent within the intestinal ecosystem. The bacteria of an adult intestine are fairly stable if the diet remains the same. Often in dysbiosis the question is of the increase of harmful microbes and the essential decrease of useful bacteria. If the physiological condition, nutritional status or other essential parameter changes, it will eventually lead to some reformulation of the gut flora. In any case, the basic composition remains usually unaltered. If it will collapse, that leads to dysbiosis.

The intestine of a newborn is totally free of bacteria [Forchielli and Walker, 2005]. The colonization of the intestine, i.e. the settlement of microbes into the intestines takes place with food. The microbiota of a child’s intestine is stabilized only about two years after birth. Only some of the bacteria that travel through the digestive tract during the early age stay permanently in the intestines as the organism is able to choose microbes that can stay in the intestine for the duration of life. Every person has one’s own individual intestinal microbiota.

## The Significance of Intestine Microbiota

The European Union has strongly started to support research projects that clarify how useful bacteria can be used to promote the health and lifestyle of aging population. Lactid acid bacteria, of which most belong to the probiotics group, are the objects of intensive research in Europe [Lebeer *et al.*, 2011]. In Finland, the *Lactobacillus GG* has been extensively studied, and many poritive health effects have been documented [Luoto *et al.*, 2011]. However, besides the LAB (lactic acid bacteria) also other organisms have essential function in protecting our health.

The whole alimentary tract microbiome, when in balanced condition, works for the benefit of its particular individual host. Although the LAB strains are rather persistent, and have been found to survive for long periods even in the gastric region [Hakalehto *et al.*, 2010], already in the duodenum starts another type of population growth associated with pulsed food arrival into the tract [Hakalehto *et al.*, 2008; Hakalehto *et al.*, 2010]. This view on the microbial interactions is confirmed in numerous laboratory experiments in Finnoflag Oy`s laboratory.

The intestinal microbes affect the functions of the organism both biochemically and immunologically [Mitra *et al.*, 1998]. Therefore it is important to take care of the intestinal microbe flora so that the amount of probiotic bacteria is sufficient [Giorgi, 2009]. Probiotics, according to the present view this means practically the bacteria that produce lactic acid, together with the organism´s own microbe population, are capable to repair dysfunctions that are caused by unsufficient food, stress and other burdens, environmental toxins and different chemical drugs such as antibiotics, chemical hormone products and NSAIDs (non-steroidal anti-inflammatory drugs) [Bezirtzoglou and Stavropoulou, 2011].

The intestinal microbe flora affects the immune system and controls the organism's ability to clear itself of alien substances. The bacteria population apparently controls lipid metabolism.

In apple-shaped obese people some inflammation tracers have increased and so-called large cell carcinoma of the immune system have been found in increasing numbers [Eckl *et al.*, 2011]. Intestinal microbes are part of the organism's toxin removal system. A microbiota that is in balance and optimal for health helps in the absorbing of many micro nutrients such as minerals, trace elements, vitamins and amino acids.

# Battle for Living Space

A case of a potential disease can be seen when the normal anaerobic bacteria and enterococci of the intestines disappear simultaneously as different harmful bacteria and fungus take their place. A continuous "battle" for living space is going on in the intestines between useful and harmful microbes. In other words, the BIB (Bacterial Intestinal Balance) keeps the battle out, and if that is shaken, there starts the struggle. The metabolism of harmful microbes can produce large amounts of toxic substances that maintain colon irritation and have an effect to the entire gastrointestinal tract. Symptoms are often similar to irritable bowel syndrome. Food quality, lifestyles, age and hormone action have an effect on the composition of intestine normal flora. Chronic dysbiosis may be a result of many chronic underlying diseases, such as diabetes and imflammatory bowel diseases (Crohn's disease or ulcerative colitis). Crisis that burden the organism, such as infections and their antibiotic treatments, major operations, radiotherapy, severe social stress, problematic pregnancy and difficult delivery, are situations that can cause the development of acute dysbiosis.

## Acute Dysbiosis

Acute dysbiosis is a quick and significant decrease of lactid acid bacteria and other beneficial (or balanced) flora in the small intestines and colon. At the same time harmful microbes, like *Clostridium difficile* bacteria and/or *Candida albicans* fungus, increase strongly in the intestines. Most common causes for this type of negative and imbalanced development are antibiotic treatment, food poisoning, gastric and intestinal inflammation, anorexia, malnutrition and significant deficits of micronutritional factors. Gastroenteritis, i.e. the acute inflammation of the stomach and intestines, is caused by eating food that is contaminated by pathogenic microbes or toxins produced by them.

## Chronic Dysbiosis

Subtypes of gastrointestinal tract dysbiosis:

- Lack of sufficient useful bacteria.
- Overgrowth of harmful microbes in which case the amount of bacteria and fungus in the intestine has increased. Bacteria that cause problems include

*Helicobacter pylori, Klebsiella pneumonia, Proteus mirabilis, Pseudomonas eruginosa, Staphylococcus aureus, Staphylococcus epidermis, Streptococcus pyogenes, Group A Streptococcus, Citrobacter freundii, Candida albicans and other fungi (yeasts or molds)*

- Immunosupressive dysbiosis, when there are microbes in the intestines that produce toxins that weaken immunity.
- Hypersensitivity or allergy causing dysbiosis, when normal yeast or bacteria cause exaggerated immune reactions.
- Inflammatory dysbiosis and reactive arthritis.
- Amoebas, protozoa and other parasites, such as *Dientamoeba fragilis, Entamoeba histolytica* and *Giardia lamblia* may cause chronic symptoms.

In dysbiosis the digestion may be disturbed to the extent that more than the usual amounts of undigested alien substances, such as badly digested food proteins, begin to form in the intestines. Some of the most significant harmful food proteins are in the so called gluten grains (wheat, rye and barley). Gluten degradation products, peptides, are able to activate the immune cells of the intestine walls and start inflammations in which neurotransmitters that cause inflammation are released into the organism [Mochizuki *et al.*, 2010]. This is a defensive reaction that can become too powerful and will not focus on the substance that caused the reaction. As a consequence, there is a strong immunological inflammatory reaction in the cell mucous membrane. Mucous membrane damage weakens the handling of alien substances. Thus a vicious circle has been created. The intestines are not able to handle immunologically active liver enzymes and harmful food components anymore. One potential problem is related with impartial digestion of food substances in the upper small intestines. If the material is undigested when arriving the cecum that has adverse consequencies for health. It could also give an opportunity to microbes to overgrow at the expense of unused nutrients.

## What Diseases May Follow from Dysbiosis?

The meaning of intestinal microbiota to the health of the alimentary canal is always not known in detail. However, many diseases have a clear connection with the gastrointestinal tract's microbial flora. According to researcher Steele, at least in the origins of the following diseases there is relevance with unfavorable gastrointestinal tract microbes nutrient [Steele, 2010].

*Hyperplasia of small intestine bacteria*: According to studies of the amounts of bacteria in the small intestine, control patients have 20% hyperplasia but irritable bowel syndrome patients have 84% and fibromyalgia patients no less than 100% [Maes and Leunis, 2008]. A correct antimicrobial drug treatment decreases symptoms. According to a new study the amount of bifidobacteria in the microbe flora has decreased [Bosscher *et al.*, 2009]. About 20% of the population suffers from *irritable bowel syndrome*. The intestines' colibacteria, lactobacilli and bifidobacteria have often decreased in these cases.

*Inflammatory Bowel Disease* (IBD): About 20 million people worldwide suffer from Crohn's disease and ulcerative colitis nutrient [Prince *et al.*, 2011]. Patients often have low amounts of lactobacilli and bifidobacteria in their intestines and large amounts of cocci-like anaerobic and sulphate reducing bacteria.

*Colon cancer*: This is believed to be, at least partially, a disease caused by harmful bacteria. The metabolic products of many bacteria can also provoke cancer or damages to genetic material. Such products are nitrosoamines, secondary bile acids, heterocyclic amines, polycyclic aromatic hydrocarbons, azo compounds and ammonia nutrient [Roberfroid *et al.*, 2010]. Many bacteria produce these harmful substances. According to research findings, it seems that bacteria that produce short-chained fatty acids can prevent the enzyme activities of cancer causing substances. Bifidobacteria and lactobacilli prevent the effects of harmful bacteria.

*Gastroenteritis*: Acute gastro-intestinal inflammation is caused by several pathogenic bacteria such as *Listeria, Yersinia, Campylobacter, Escerichia coli, Vibrio* and *Clostridium perfringens* and also by shigellas and salmonellas [Luca *et al.*, 2011]. The microbe flora of a healthy intestine with ample bifidobacteria and lactobacilli forms a strong protection against these bacteria.

*Necrotizing enterocolitis* (a bowel inflammation that kills and destroys bowel cells) is a fierce illness in newborns that have a low birth weight [Ganguli and Walker, 2011]. Symptoms include a very powerful increase of gas volume, a strong increase of pathogenic bacteria in the intestines and mucous membrane damages. It leads to the death of every tenth child with the illness.

*Pseudomembranous colitis* is also known as antibiotic colitis [Berman *et al.*, 2008]. The most frequent cause of diarrhea (antibiotics associated) is *Clostridium difficile* bacteria that produces two strong toxins. Especially number of patients with severe complications has increased for patients over 64 years of age [Lyytikäinen *et al.*, 2009]. Probiotics have been used in treatment of severe *Clostridium difficile* infection. The benefit and exact role of probiotics is still open and further studies are needed [Na and Kelly, 2011].

*Urinary tract infections*: Before menopause, the normal bacterial flora of women's urinary and genital organs, contain for example lactobacilli. Some of these bacterial strains form protection against infections caused by pathogenic bacteria [Cadieux *et al.*, 2009].

The PMEU method has been successfully applied for the quick verification and diagnosis of the alimentary and urinary tract infections, as well as the septic diseases [Hakalehto *et al.*, 2007; , 2009; Hakalehto *et al.*, 2011b]. In these culture conditions (in the PMEU) the bacterial and other microbial growth takes place in such a fashion that more closely resembles that of the *in vivo* situation.

This approach is important in understanding the behavior of the strains, as well as for getting an early warning of the infection. During menopause, when menstruation comes to an end and due to hormonal changes, the bacterial flora changes so that its protection against infections weakens. Thus vulnerability towards urinary tract infections increases. Probiotic medication has been used to try to prevent urinary tract infections. Results are promising in the treatment and prevention of recurrent urinary tract infections.

*Colon diverticulitis*: Diverticula appear throughout the entire gastrointestinal tract but by far the most in the colon. Most diverticula are symptomless. Currently the disease burdens the health care system significantly financially. It is impossible to determine the absolute incidence of colon diverticulitis. It increases continuously with age. The risk to get diverticula at the age of 60 is about 50% and at the age of 80 about 75%. Most diverticula are found accidentally while diagnosing different bowel disorders. Approximately about 10% of diverticula patients will develop symptoms. Three out of ten will develop serious complications.

## Bacterial Interference Causing by Intrusions from the Alimentary Tract

As a general rule we could assume that the micro-organisms would prefer staying in the alimentary tract as an integral part of this complex ecosystem. In this purpose they also form symbiotic relationships between different strains and with the host. Any singular bacterial or other microbial strains that break this rule could become a true pathogen. Consequently, it also is escaping the regulation from both the host body and the other microbial strains, and the entire community. In many cases, due to some neurological, hormonal or nutritional malfunction, some strain or strains make the balance (BIB; Bacterial Intestinal Balance) to collapse [Hakalehto, 2011]. This situation is both detrimental to the host and offensive against the regular members of the microflora. If the newcoming intruders, or trouble-makers within the normal flora contain several members, the situation could develop into a mixed infection. In this case the microflora may seek for a new balance which, however, is less beneficial to the host.

The interactions between commensal organisms, opportunistic pathogens, and disease causing microbes could be studied in the PMEU device by simulating the cocultures in the intestines [Hakalehto *et al.*, 2010]. Some invasive pathogens, such as salmonellas start penetrating the intestinal membranes [Hakalehto *et al.*, 2007]. The molecular mechanisms of this attack, as well as the host defense option, are surveyed in this chapter. However, for the balanced and healthy function of a human body, it would be important to maintain the BIB rather than to try to repair the damage. This means that the body has to struggle against an invading strain which rapidly disturbs the balance, converting it towards less a supportive, or stabilized, status.

In order to eradicate permanently the pathogens, often an antibiotic medication is required. In the clinical practices it is of utmost importance to stabilize the body with parenteral infusion therapy in order to help the body system to sustain the infection, and to develop the immunological responses for it. Some methods, such as novel probiotics, or passive immunisation, could be attempted for returning the BIB, which is the best option for getting rid of the pathogenic function. In fact, it is often possible that the once pathogenic strain has “escaped” the control of the human organs as well as the microbial community. Consequently, the re-establishment of the balance could return the opportunistically pathogenic strain into commensal status, For example, the species *Klebsiella pneumoniae* is found in the intestines of many individuals where it is a common part of the ecosystem. However, in some undesired circumstances the same bacterium could find its way to the respiratory tract and turn into a pathogen there [Parm *et al.*, 2011].

## Intracellular Pathogens

One of the most severe forms of pathogenecity is the intracellular intrusion. Such pathogens as *Chlamydia* sp., *Borrelia burgdorferi*, *Treponema pallidum*, some yersinias and salmonellas could penetrate through human cell membranes and cause serious infections, which could also include latent periods [Guttman *et al.*, 2006]. These diseases are difficult to get diagnosed, and their treatments are often very complicated. For example, the *B. burgdorferi* IgAG antibodies could be detected from human serum with such rapid molecular methods as dot blot, MarDx, and VIDAS enzyme immunoassays [Jespersen *et al.*, 2002]. The two former methods were found to require a Western Blot confirmation due to their low specificities. One big challenge in researching this type of pathogens which penetrate the cells, is the sampling thresold. When can we be sure of having a reliable specimen material, without the risk of the pathogens continuously hiding inside the human cells?

As many bacteria are able to colonize the host cell intracellularly, there is a need for a mechanism of ontophagy for the destroyal and removal of these intruding pathogens such as *Staphylococcus aureus*, for example [Amano *et al.*, 2006]. Such intestinal microbes as *Escerichia coli* and *Candida albicans* and *Streptococcus pneumoniae* influence influence the complement cascade depriving it from the full activity [Zipfel *et al.*, 2007]. Regardless of the multitude of the mechanisms which are exploited by the host, and by the microbial strains, to control each other, the decisive studies on the microbial behavior in the alimentary ecosystems remain to be elucidated. Since it is extremely tedious to monitor the events inside the gut, for example, effective means are required for the simulations of these effects *in vitro*. The PMEU method has been developed in order to facilitate precise mapping of the relations between micro-organisms, as well as their reactions to various host parameters, and actions [Hakalehto. 2010; Hakalehto, 2011].

Besides the commonly known intracellular pathogens, it has been suggested that intracellular forms may explain such phenomena as persistence of pathogenic *E.coli* infections against antibiotic treatments [Kerrn *et al.*, 2005]. It seems possible that some organisms may penetrate the epithelium and persist there during antibiotic treatments. In these cases, flushing the pathogens out of the epithelium with some intracellular cAMP (cyclic adenosine monophosphate) increasing drugs could be a useful strategy in the cure of disease [Bishop *et al.*, 2007]. The cAMP is an important mediator in the regulation of the intracellular effects of several hormones. It is also evident that for proper functions of gastrointestinal tract or the urinary tract it is of utmost importance to keep the intracellular bacterial strains or viruses out of the cell interior. Therefore, it can be deduced that in the microfloral development it is essential to find microbial ecosystems which would be capable of overgrowing any of these intruding strains. Also, the innate immunity is needed for maintaining such a flora, and for supporting it.

The mechanisms of severe infections often include taking over of the host body, or directing its actions. For example, the borreliosis as like some other serious pathogens, such as HIV virus and tuberculosis bacteria (*Mycobacterium tuberculosis*) are blocking the vitamin D receptors [Spector, 2011; Khoo *et al.*, 2011]. These, in turn, are key factors in building up the innate immunity, which is the first line defense against infections. This is due to the central role of vitamin D in the immunity [Gleisner *et al.*, 2011]. Moreover, the same receptors function also as receptors for lipocacheic acid which is one of the important bile acids [Steiner *et al.*, 2011]. By preventing the binding of the vitamin D and the bile salt the pathogen is regenerating the defenses both immunologically, and by weakening the fatty acid dissimilation, which in turn is damaging the cell membranes.

Many bacteria, such as *Escherichia coli* and *Bacteroides* sp. in the large intestine, can synthesize vitamin $K_2$ (menaquinone-7), but not vitamin $K_1$ (phylloquinone) [Vermeer and Braam, 2001]. In these bacteria, menaquinone will receive two electrons from small molecules, such as lactate, formate, or NADH, in a process called anaerobic respiration [Jones *et al.*, 2011]. The menaquinone, with the help of another enzyme, will in turn in another enzymatic reaction transfer these electrons to an oxidant, such fumarate or nitrate. Adding two electrons to fumarate or nitrate will convert the molecule to succinate or nitrite + water, respectively. These reactions resemble the generation of cellular energy by eukaryotic cell aerobic respiration, except that the final electron acceptor is not molecular oxygen, but fumarate or nitrate, for example. Thus the vitamin K2 synthesis offers some intestinal bacteria to carry out oxidative respiration in the anaerobiosis. This provides them with more option in their energy metabolism.

In the host body, the lack of vitamin K2 produced by bacteria may cause building up of plaques to the coronary arteries. Therefore, calcium and vitamin D supplementation needs to be completed with an adequate vitamin K2 supply. Vitamin K2 activates osteocalcin hormone, produced by osteoblasts, which is needed to bind calcium into the bones. If this activation is inadequate, the calcium may end up into the arteries. The vitamin K2 also improves the bone health in our body. This example is an indication of the continuous interdependence between man and the microbiome.

## Trends in Nutritive Therapies

The micronutrients are important for the proper functioning of the immune system, in repairing tissue damages, wounds and fractures, as well as in the production of new cells, and in the toxin removal. Clinical experiences have taught in practise that therapies with one or a few supplements are seldom satisfactory, or fulfilling the patient´s expectations. The results from the treatments are remarkably improved when numerous substances needed for supporting the antioxidant and immune system, and for repair and regeneration of the cells, are used simultaneously. These substances with selected herbal medications need to be given according to an individual plan for every patient. The treatments always include also the balancing of the intestinal microflora with probiotics.

The nutritional treatments given to problem patients should be as often as possible based on profound laboratory tests giving a view to the individual nutritional status of a patient. Doctors at the Finnish antioxidant clinics have had at their disposal the results from vitamin, trace mineral and fatty acid determinations [Laakso *et al.*, 1990; Laakso *et al.*, 1990; Mahlberg *et al.*, 2007]. These methods have been developed and implemented since 1981 at the mineral laboratory Mila Oy [Jaakkola *et al.*, 1983; Jaakkola *et al.*, 1986a; Jaakkola *et al.*, 1986b]. Besides the traditional tests of clinical chemistry repertoire, there are 14 analysis techniques for finding out the levels of minerals and trace elements from blood and its fractions, such as sera, isolated eryhtrocytes and whole blood. Also the measurements of 10 vitamins and vitaminlike cofactor Q10 have been carried out in this laboratory. In addition to traditional lipid analysis, there are performed determinations for the molarities of 16 fatty acids, and the propotions of different lipid groups.

Tests analysing the balances of micronutrients have a great information value for a practitioners, who has got familiar with the biochemical background of the laboratory methods and the nutrition therapy. On the basis of several test results on micronutrients, the doctor is able to more effectively diagnose the patient´s condition, and to repair imbalances and deficiencies in the nutritional status. Decades lasting research efforts support diet advises to be based on laboratory results. Together with practitioners professional experience in the Finnish clinics, they give the the individual micronutrient supplement therapies. Because the treatments are most often given to severely ill patients, it is of crucial importance to apply the medications in pharmacologically individualized portions [Jaakkola *et al.*, 1992; Jaakkola *et al.*, 1992; Mahlberg *et al.*, 2006]. The wide spectrum nutrition therapy based on laboratory studies is often initiating a quite rapid period of recovery during which time the GI microflora is also often getting balanced. Also the other potentially occuring diseases of the patient can be more safely carried out when the patient is under the replacement therapy. This also is diminishing the risk of hospital infections due to the more rapidly improving condition of the patient.

## Conclusion

The intestinal balance function requires a balanced microbiome as well as its smooth cooperation with host functions. This symbiosis is reciprocally beneficial interaction, which could justify the concept of "microbiome" in the sense of having an extra organ inside the body in the form of alimentary microbes. From this very basis, it is understandable and logical to suppose that the micro-organisms participate in human nutrition by ways. This indeed is the case, and malfunction in the gut microbial interactions are causing health problems for us, and *vice versa.* Most intestinal disease have a microbial link. In order to repair the oxidative and inflammative damages caused by pathogens or by the microbial overgrowth, we need to get "outside help". This "rescue service" for our body system consists of 1. correct replacement therapy with trace elements, vitamins, and other micronutrients; 2. corrected nutrition; and 3. carefully chosen microbial probiotics. The third part is essentially including the safe-quarding of our normal flora as the basis for our well-being, as well as its balacing when needed. This gives our body system the basis to combat against all kinds of contagious diseases as well as to maintain the bacteriological and other balances inside the alimentary tract.

## References

Abraham, SN; Beachey, EH. Assembly of a chemically synthesized peptide of *Escherichia coli* type 1 fimbriae into fimbria-like antigenic structures. *J. Bacteriol.,* 1987; 169, 2460-2465.

Agbor, TA; McCormick, BA. *Salmonella* effectors: important players modulating host cell function during infection. *Cell Microbiol.,* 2011; 13, 1858-1869.

Amano, A; Nakagawa, I; Yoshimori, T. Autophagy in innate immunity against intracellular bacteria. *J. Biochem.,* 2006; 140, 161-166.

Banerjee, S; Akbar, N; Moorhead, J; Rennie, JA; Leather, AJ; Cooper, D; Papagrigoriadis, S. Increased presence of serotonin-producing cells in colons with diverticular disease may indicate involvement in the pathophysiology of the condition. *Int. J. Colorectal. Dis.,* 2007; 22, 643-649.

Bassotti, G; de Roberto, G; Castellani, D; Sediari, L; Morelli, A. Normal aspects of colorectal motility and abnormalities in slow transit constipation. *World J. Gastroenterol.,* 2005; 11, 2691-2696.

Berman, L; Carling, T; Fitzgerald, TN; Bell, RL; Duffy, AJ; Longo, WE; Roberts, KE. Defining surgical therapy for pseudomembranous colitis with toxic megacolon. *J. Clin. Gastroenterol.,* 2008; 42, 476-480.

Bezirtzoglou, E; Stavropoulou, E. Immunology and probiotic impact of the newborn and young children intestinal microflora. *Anaerobe,* 2011; 17, 369-374.

Bischoff-Ferrari, HA; Dawson-Hughes, B; Stocklin, E; Sidelnikov, E; Willett, WC; Orav, EJ; Stahelin, HB; Wolfram, S; Jetter, A; Schwager, J; Henschkowski, J; von Eckardstein, A; Egli, A. Oral supplementation with 25(OH)D(3) versus vitamin D(3) : effects on 25(OH)D levels, lower extremity function, blood pressure and markers of innate immunity. *J. Bone Miner. Res.,* 2011; 25, [Epub ahead of print].

Bishop, BL; Duncan, MJ; Song, J; Li, G; Zaas, D; Abraham, SN. Cyclic AMP-regulated exocytosis of *Escherichia coli* from infected bladder epithelial cells. *Nat. Med.,* 2007; 13, 625-630.

Bosscher, D; Breynaert, A; Pieters, L; Hermans, N. Food-based strategies to modulate the composition of the intestinal microbiota and their associated health effects. *J. Physiol. Pharmacol.,* 2009; 60 (Suppl 6), 5-11.

Bosseckert, H. Clinical aspects and differential diagnosis of malabsorption. *Dtsch. Z. Verdau Stoffwechselkr.,* 1983; 43, 27-32.

Cadieux, PA; Burton, J; Devillard, E; Reid, G. *Lactobacillus* by-products inhibit the growth and virulence of uropathogenic *Escherichia coli. J. Physiol. Pharmacol.,* 2009; 60 (Suppl 6), 13-18.

Castillo, HS; Ousley, AM; Duraj-Thatte, A; Lindstrom, KN; Patel, DD; Bommarius, AS; Azizi, B. The role of residue C410 on activation of the human vitamin D receptor by various ligands. *J. Steroid Biochem. Mol. Biol.,* 2012; 128, 76-86.

Cerutti, A; Chen, K; Chorny, A. Immunoglobulin responses at the mucosal interface. *Annu. Rev. Immunol.,* 2011; 29, 273-293.

Clarke, TB; Davis, KM; Lysenko, ES; Zhou, AY; Yu, Y; Weiser, JN. Recognition of peptidoglycan from the microbiota by Nod1 enhances systemic innate immunity. *Nat. Med.,* 2010; 16, 228-231.

Deenichin, GP; Kristev, AD; Mollov, VV; Turiiski, VI. Effect of elevated intra-abdominal pressure on the contractile activity and reactivity of smooth muscle tissue from rat gastrointestinal tract to galantamine and drotaverine (No-Spa). *Folia Med. (Plovdiv),* 2010; 52, 31-36.

Eckl, J; Buchner, A; Prinz, PU; Riesenberg, R; Siegert, SI; Kammerer, R; Nelson, PJ; Noessner, E. Transcript signature predicts tissue NK cell content and defines renal cell carcinoma subgroups independent of TNM staging. *J. Mol. Med. (Berl.),* 2012; 90, 55-66.

Escobar, J; Pereda, J; Arduini, A; Sandoval, J; Moreno, ML; Perez, S; Sabater, L; Aparisi, L; Cassinello, N; Hidalgo, J; Joosten, LA; Vento, M; Lopez-Rodas, G; Sastre, J. Oxidative and nitrosative stress in acute pancreatitis modulation by pentoxifylline and oxypurinol. *Biochem. Pharmacol.,* 2012; 83, 122-130.

Farhadi, A; Keshavarzian, A; Ranjbaran, Z; Fields, JZ; Banan, A. The role of protein kinase C isoforms in modulating injury and repair of the intestinal barrier. *J. Pharmacol. Exp. Ther.,* 2006; 316, 1-7.

Floch, MH. Intestinal microecology in health and wellness. *J. Clin. Gastroenterol.,* 2011; 45 Suppl, S108-10.

Forchielli, ML; Walker, WA. The role of gut-associated lymphoid tissues and mucosal defence. *Br. J. Nutr.,* 2005; 93 (Suppl 1), S41-8.

Fritscher-Ravens, A; Swain, CP. The wireless capsule: new light in the darkness. *Dig. Dis.,* 2002; 20, 127-133.

Ganguli, K; Walker, WA. Probiotics in the prevention of necrotizing enterocolitis. *J. Clin. Gastroenterol.,* 2011; 45 Suppl, S133-8.

Gershon, MD. Review article: serotonin receptors and transporters -- roles in normal and abnormal gastrointestinal motility. *Aliment. Pharmacol. Ther.,* 2004; 20 (Suppl 7), 3-14.

Gershon, MD. The enteric nervous system: a second brain. *Hosp. Pract. (Minneap.),* 1999; 34, 31-2, 35-8, 41-2 passim.

Giorgi, PL. Probiotics. A review. *Recenti Prog. Med.,* 2009; 100, 40-47.

Gleisner, MA; Rosemblatt, M; Fierro, JA; Bono, MR. Delivery of alloantigens via apoptotic cells generates dendritic cells with an immature tolerogenic phenotype. *Transplant. Proc.,* 2011; 43, 2325-2333.

Guttman, JA; Li, Y; Wickham, ME; Deng, W; Vogl, AW; Finlay, BB. Attaching and effacing pathogen-induced tight junction disruption *in vivo*. *Cell Microbiol.,* 2006; 8, 634-645.

Hakalehto E, inventor. Method and apparatus for concentrating and searching of microbiological specimens. US Patent No. 7,517,665. 2009.

Hakalehto, E. Hygiene monitoring with the Portable Microbe Enrichment Unit (PMEU). *41st R3 -Nordic Symposium. Cleanroom technology, contamination control and cleaning. VTT Publications 266.* Espoo, Finland: VTT (State Research Centre of Finland); 2010.

Hakalehto, E. Simulation of enhanced growth and metabolism of intestinal *Escherichia coli* in the Portable Microbe Enrichment Unit (PMEU). In: Rogers MC, Peterson ND, editors. *E. coli infections: causes, treatment and prevention.* New York, USA: Nova Publishers; 2011.

Hakalehto, E; Heitto,L; Heitto,A; Humppi,T; Rissanen, K; Jääskeläinen, A; Paakkanen, H; Hänninen, O. Fast monitoring of water distribution system with portable enrichment unit – Measurement of volatile compounds of coliforms and *Salmonella* sp. in tap water. *JTEHS,* 2011a; Vol 3(8), 223-233.

Hakalehto, E; Hell, M; Bernhofer, C; Heitto, A; Pesola, J; Humppi, T; Paakkanen, H. Growth and gaseous emissions of pure and mixed small intestinal bacterial cultures: Effects of bile and vancomycin. *Pathophysiology,* 2010; 17, 45-53.

Hakalehto, E; Hujakka, H; Airaksinen, S; Ratilainen, J; Närvänen, A. Growth-phase limited expression and rapid detection of *Salmonella* type 1 fimbriae. In: Hakalehto E. *Characterization of Pectinatus cerevisiiphilus and P. frisingiensis surface components. Use of synthetic peptides in the detection fo some Gram-negative bacteria.* Kuopio, Finland: Kuopio University Publications C. Natural and Environmental Sciences 112; 2000. Doctoral dissertation.

Hakalehto, E; Humppi, T; Paakkanen, H. Dualistic acidic and neutral glucose fermentation balance in small intestine: Simulation *in vitro*. *Pathophysiology,* 2008; 15, 211-220.

Hakalehto, E; Pesola, J; Heitto, L; Närvänen, A; Heitto, A. Aerobic and anaerobic growth modes and expression of type 1 fimbriae in *Salmonella*. *Pathophysiology,* 2007; 14, 61-69.

Hakalehto, E; Vilpponen-Salmela, T; Kinnunen, K; von Wright, A. Lactic acid bacteria enriched from human gastric biopsies. *ISRN Gastroenterol,* 2011b; 2011, 109183.

Härkönen, N. Recurrent pseudomembranous colitis treated with the donor feces (in Finnish). *Duodecim,* 1996; 112, 1803-1804.

Harrison, OJ; Maloy, KJ. Innate immune activation in intestinal homeostasis. *J. Innate Immun.,* 2011; 3, 585-593.

Haussler, MR; Jurutka, PW; Mizwicki, M; Norman, AW. Vitamin D receptor (VDR)-mediated actions of 1alpha,25(OH)2vitamin D3: genomic and non-genomic mechanisms. *Best Pract. Res. Clin. Endocrinol. Metab.,* 2011; 25, 543-559.

Heitto, L; Heitto, A; Hakalehto, E. Tracing wastewaters with faecal enterococci. (Poster). *Second European Large Lakes Symposium.* Norrtälje, Sweden; 2009.

Hillilä, MT; Siivola, MT; Färkkilä, MA. Comorbidity and use of health-care services among irritable bowel syndrome sufferers. *Scand. J. Gastroenterol.,* 2007; 42, 799-806.

Indrio, F; Riezzo, G; Cavallo, L; Mauro, AD; Francavilla, R. Physiological basis of food intolerance in VLBW. *J. Matern Fetal. Neonatal. Med.,* 2011; 24 (Suppl 1), 64-66.

Jaakkola, K. Long-term assessment of zinc status. First Meeting of the International Society for trace Element Research on Humans; 8.-12.12.1986; USA; 1986a.

Jaakkola, K. The selenium status of Finnish population. First Meeting of the International Society for Trace Element Research in Humans; 8.-12.12.1986; USA; 1986b.

Jaakkola, K; Lähteenmäki, P; Laakso, J; Harju, E; Tykkä, H; Mahlberg, K. Treatment with antioxidant and other nutrients in combination with chemotherapy and irradiation in patients with small-cell lung cancer. *Anticancer Res.,* 1992; 12, 599-606.

Jaakkola, K; Tummavuori, J; Pirinen, A; Kurkela, P; Tolonen, M; Arstila, AU. Selenium levels in whole blood of Finnish volunteers before and during organic and inorganic selenium supplementation. *Scand. J. Clin. Lab. Invest.,* 1983; 43, 473-476.

Jespersen, DJ; Smith, TF; Rosenblatt, JE; Cockerill, FR,3rd. Comparison of the *Borrelia* DotBlot G, MarDx, and VIDAS enzyme immunoassays for detecting immunoglobulin G antibodies to *Borrelia burgdorferi* in human serum. *J. Clin. Microbiol.,* 2002; 40, 4782-4784.

Jones, SA; Gibson, T; Maltby, RC; Chowdhury, FZ; Stewart, V; Cohen, PS; Conway, T. Anaerobic respiration of *Escherichia coli* in the mouse intestine. *Infect. Immun.,* 2011; 79, 4218-4226.

Kabaroudis, A; Papaziogas, B; Koutelidakis, I; Kyparissi-Kanellaki, M; Kouzi-Koliakou, K; Papaziogas, T. Disruption of the small-intestine mucosal barrier after intestinal occlusion: a study with light and electron microscopy. *J. Invest. Surg.,* 2003; 16, 23-28.

Kajander, K; Myllyluoma, E; Rajilic-Stojanovic, M; Kyrönpalo, S; Rasmussen, M; Järvenpää, S; Zoetendal, EG; de Vos, WM; Vapaatalo, H; Korpela, R. Clinical trial: multispecies probiotic supplementation alleviates the symptoms of irritable bowel syndrome and stabilizes intestinal microbiota. *Aliment. Pharmacol. Ther.,* 2008; 27, 48-57.

Kerrn, MB; Struve, C; Blom, J; Frimodt-Moller, N; Krogfelt, KA. Intracellular persistence of *Escherichia coli* in urinary bladders from mecillinam-treated mice. *J. Antimicrob. Chemother.*, 2005; 55, 383-386.

Khoo, AL; Chai, LY; Koenen, HJ; Oosting, M; Steinmeyer, A; Zuegel, U; Joosten, I; Netea, MG; van der Ven, AJ. Vitamin D(3) down-regulates proinflammatory cytokine response to *Mycobacterium tuberculosis* through pattern recognition receptors while inducing protective cathelicidin production. *Cytokine,* 2011; 55, 294-300.

Kramer, K; da Silveira, AB; Jabari, S; Kressel, M; Raab, M; Brehmer, A. Quantitative evaluation of neurons in the mucosal plexus of adult human intestines. *Histochem. Cell Biol.,* 2011; 136, 1-9.

Kumamoto, CA; Pierce, JV. Immunosensing during colonization by *Candida albicans*: does it take a village to colonize the intestine? *Trends Microbiol.,* 2011; 19, 263-267.

Laakso, J; The effect of daily consumption of cod liver oil on blood pressure and plasma lipids. *Ravitsemussymposium*; 19.-20.10.1990; Helsinki; 1990.

Lafont, O. Life and death of free radicals. *Rev. Hist. Pharm. (Paris),* 2007; 54, 475-478.

Lawson, HH. The duodenal mucosa in health and disease. A clinical and experimental study. *Surg. Annu.,* 1989; 21, 157-180.

Lebeer, S; Claes, IJ; Vanderleyden, J. Anti-inflammatory potential of probiotics: lipoteichoic acid makes a difference. *Trends Microbiol.,* 2011; Oct 24, [Epub ahead of print].

Ljungqvist, O; van Gossum, A; Sanz, ML; de Man, F. The European fight against malnutrition. *Clin. Nutr.,* 2010; 29, 149-150.

Luca, CM; Nemescu, R; Teodor, A; Fantanaru, R; Petrovici, CM; Dorobat, C. Etiological aspects of acute gastroenteritis--a ten-year review (1.01. 2001-31.12.2010). *Rev. Med. Chir. Soc. Med. Nat. Iasi,* 2011; 115, 712-717.

Luoto, R; Laitinen, K; Nermes, M; Isolauri, E. Impact of maternal probiotic-supplemented dietary counseling during pregnancy on colostrum adiponectin concentration: A prospective, randomized, placebo-controlled study. *Early Hum. Dev.,* 2011; Sep 24, [Epub ahead of print].

Lyytikäinen, O; Turunen, H; Sund, R; Rasinperä, M; Könönen, E; Ruutu, P; Keskimäki, I. Hospitalizations and deaths associated with *Clostridium difficile* infection, Finland, 1996-2004. *Emerg. Infect. Dis.,* 2009; 15, 761-765.

Maes, M; Leunis, JC. Normalization of leaky gut in chronic fatigue syndrome (CFS) is accompanied by a clinical improvement: effects of age, duration of illness and the translocation of LPS from gram-negative bacteria. *Neuro. Endocrinol. Lett.,* 2008; 29, 902-910.

Mahlberg, K; Changes in the whole blood seleniem levels in Finland from 1987 to 2003. Phytopharm 2007, 11th International Congress.; 27.-29-6-1007; Leiden, Netherlands; 2007.

Mahlberg, K; Safety and efficacy of antioxidants in the treatment of atopic eczema. The 10th International Phytopharm Conference 2006; 27.-30.6.2006; St. Petersburg, Russia; 2006.

Manicassamy, S; Pulendran, B. Dendritic cell control of tolerogenic responses. *Immunol. Rev.,* 2011; 241, 206-227.

Marchal, C; Schramm, F; Kern, A; Luft, BJ; Yang, X; Schuijt, T; Hovius, J; Jaulhac, B; Boulanger, N. Antialarmin effect of tick saliva during the transmission of Lyme disease. *Infect. Immun.,* 2011; 79, 774-785.

Mazmanian, SK; Round, JL; Kasper, DL. A microbial symbiosis factor prevents intestinal inflammatory disease. *Nature,* 2008; 453, 620-625.

Mentu, JV; Heitto, L.; Keitel, HV; Hakalehto, E. *Rapid Microbiological Control of Paper Machines with PMEU Method.* Paperi ja Puu / Paper and Timber, 2009, 91, 7-8.

Mitra, R; Saha, PK; Basu, I; Venkataraman, A; Ramakrishna, BS; Albert, MJ; Takeda, Y; Nair, GB. Characterization of non-membrane-damaging cytotoxin of non-toxigenic *Vibrio cholerae* O1 and its relevance to disease. *FEMS Microbiol. Lett.,* 1998; 169, 331-339.

Mochizuki, M; Shigemura, H; Hasegawa, N. Anti-inflammatory effect of enzymatic hydrolysate of corn gluten in an experimental model of colitis. *J. Pharm. Pharmacol.,* 2010; 62, 389-392.

Muscaritoli, M; Anker, SD; Argiles, J; Aversa, Z; Bauer, JM; Biolo, G; Boirie, Y; Bosaeus, I; Cederholm, T; Costelli, P; Fearon, KC; Laviano, A; Maggio, M; Rossi Fanelli, F; Schneider, SM; Schols, A; Sieber, CC. Consensus definition of sarcopenia, cachexia and pre-cachexia: joint document elaborated by Special Interest Groups (SIG) "cachexia-anorexia in chronic wasting diseases" and "nutrition in geriatrics". *Clin. Nutr.,* 2010; 29, 154-159.

Na, X; Kelly, C. Probiotics in *Clostridium difficile* infection. *J. Clin. Gastroenterol.,* 2011; 45 Suppl, S154-8.

Narciso, A; Nunes, F; Amores, T; Lito, L; Melo-Cristino, J; Duarte, A. Persistence of uropathogenic *Escherichia coli* strains in the host for long periods of time: relationship between phylogenetic groups and virulence factors. *Eur. J. Clin. Microbiol. Infect. Dis.,* 2011; Oct 12, [Epub ahead of print].

Neter, E. Enteropathogenicity: recent developments. *Klin. Wochenschr.,* 1982; 60, 699-701.

Niemelä, S. When to perform gastroscopy for a patient with upper abdominal pain? (in Finnish) *Duodecim.,* 2009; 125, 155-158.

O'Hara, AM; Shanahan, F. The gut flora as a forgotten organ. *EMBO Rep.,* 2006; 7, 688-693.

Oien, T; Storro, O; Johnsen, R. Intestinal microbiota and its effect on the immune system--a nested case-cohort study on prevention of atopy among small children in Trondheim: the IMPACT study. *Contemp. Clin. Trials.,* 2006; 27, 389-395.

Parm, U; Metsvaht, T; Sepp, E; Ilmoja, ML; Pisarev, H; Pauskar, M; Lutsar, I. Risk factors associated with gut and nasopharyngeal colonization by common Gram-negative species and yeasts in neonatal intensive care units patients. *Early Hum. Dev.,* 2011; 87, 391-399.

Pesola, J; Hakalehto, E. Enterobacterial microflora in infancy - a case study with enhanced enrichment. *Indian J. Pediatr.,* 2011; 78, 562-568.

Pesola, J; Vaarala, O; Heitto, A; Hakalehto, E. Use of portable enrichment unit in rapid characterization of infantile intestinal enterobacterial microbiota. *Microb. Ecol. Health Dis.,* 2009; 21, 203-210.

Philpott, DJ; Girardin, SE. Gut microbes extend reach to systemic innate immunity. *Nat. Med.,* 2010; 16, 160-161.

Prasad, KN. *Micronutrients in in health and disease.*2011; Boca Roton, FL, USA. CRC Press. Taylor & Francis Group.

Prasad, KN; Cole, WC; Kumar, B; Prasad, KC. Scientific rationale for using high-dose multiple micronutrients as an adjunct to standard and experimental cancer therapies. *J. Am. Coll. Nutr.,* 2001; 20, 450S-463S; discussion 473S-475S.

Prince, A; Whelan, K; Moosa, A; Lomer, MC; Reidlinger, DP. Nutritional problems in inflammatory bowel disease: the patient perspective. *J. Crohns. Colitis.,* 2011; 5, 443-450.

Rapin, JR; Wiernsperger, N. Possible links between intestinal permeablity and food processing: A potential therapeutic niche for glutamine. *Clinics (Sao Paulo),* 2010; 65, 635-643.

Rautanen, T; Halme, S; Vesikari, T. Community-based survey of paediatric diarrhoeal morbidity and home treatment practices in Finland. *Acta Paediatr.,* 1998; 87, 986-990.

Reis, BS; Mucida, D. The role of the intestinal context in the generation of tolerance and inflammation. *Clin. Dev. Immunol,.* 2012; 2012, 157948.

Roberfroid, M; Gibson, GR; Hoyles, L; McCartney, AL; Rastall, R; Rowland, I; Wolvers, D; Watzl, B; Szajewska, H; Stahl, B; Guarner, F; Respondek, F; Whelan, K; Coxam, V; Davicco, MJ; Leotoing, L; Wittrant, Y; Delzenne, NM; Cani, PD; Neyrinck, AM; Meheust, A. Prebiotic effects: metabolic and health benefits. *Br. J. Nutr.,* 2010; 104 (Suppl 2), S1-63.

Romanowicz, L; Bankowski, E. Altered sphingolipid composition in Wharton's jelly of pre-eclamptic newborns. *Pathobiology,* 2010; 77, 78-87.

Salzman, NH; Hung, K; Haribhai, D; Chu, H; Karlsson-Sjöberg, J., Amir, E; Teggatz, P; Barman, M; Hayward, M; Eastwood, D; Stoel, M; Zhou, Y; Sodergren, E; Weinstock, GM; Bevins, CL; Williams, CB; Bos, NA. Enteric defensins are essential regulators of intestinal microbial ecology. *Nat. Immunol.,* 2010; 11, 76-82.

Santoro, A; Mancini, E; Ferramosca, E; Faenza, S. Liver support systems. *Contrib. Nephrol.,* 2007; 156, 396-404.

Silva, LC; Futerman, AH; Prieto, M. Lipid raft composition modulates sphingomyelinase activity and ceramide-induced membrane physical alterations. *Biophys. J.,* 2009; 96, 3210-3222.

Spector, SA. Vitamin D and HIV: letting the sun shine in. *Top. Antivir. Med.,* 2011; 19, 6-10.

Spengler, MI; Bertoluzzo, SM; Catalani, G; Rasia, ML. Study on membrane fluidity and erythrocyte aggregation in equine, bovine and human species. *Clin. Hemorheol. Microcirc.,* 2008; 38, 171-176.

Staines, HM; Rae, C; Kirk, K. Increased permeability of the malaria-infected erythrocyte to organic cations. *Biochim. Biophys. Acta,* 2000; 1463, 88-98.

Steele, KE. Guest editorial: special focus on the pathology of biological select agents and toxins in animals and research challenges in biological defense. *Vet. Pathol.,* 2010; 47, 772-773.

Steiner, C; Holleboom, AG; Karuna, R; Motazacker, MM; Kuivenhoven, JA; Frikke-Schmidt, R; Tybjaerg-Hansen, A; Rohrer, L; Rentsch, KM; von Eckardstein, A. Lipoprotein distribution and serum concentrations of 7alpha-hydroxy-4-cholesten-3-one and bile acids: Effects of monogenic disturbances in high density lipoprotein metabolism. *Clin. Sci. (Lond.),* 2011.

Torres, R; Swift, RV; Chim, N; Wheatley, N; Lan, B; Atwood, BR; Pujol, C; Sankaran, B; Bliska, JB; Amaro, RE; Goulding, CW. Biochemical, Structural and Molecular Dynamics Analyses of the Potential Virulence Factor RipA from *Yersinia pestis*. *PLoS One,* 2011; 6, e25084.

Turnbaugh, PJ; Stintzi, A. Human health and disease in a microbial world. *Front. Microbiol.,* 2011; 2, 190.

Van Oudenhove, L; McKie, S; Lassman, D; Uddin, B; Paine, P; Coen, S; Gregory, L; Tack, J; Aziz, Q. Fatty acid-induced gut-brain signaling attenuates neural and behavioral effects of sad emotion in humans. *J. Clin. Invest.,* 2011; 121, 3094-3099.

Vermeer, C; Braam, L. Role of K vitamins in the regulation of tissue calcification. *J. Bone Miner. Metab.,* 2001; 19, 201-206.

Wirtanen G, Salo S. PMEU-laitteen validointi koliformeilla (Validation of the PMEU equipment with coliforms). 2010; Report VTT-S-01705-10, Statement VTT-S-02231-10.

Yoon, SS; Mekalanos, JJ. 2,3-butanediol synthesis and the emergence of the *Vibrio cholerae* El Tor biotype. *Infect. Immun.,* 2006; 74, 6547-6556.

Zipfel, PF; Mihlan, M; Skerka, C. The alternative pathway of complement: a pattern recognition system. *Adv. Exp. Med. Biol.,* 2007; 598, 80-92.

In: Alimentary Microbiome: A PMEU Approach
Editor: Elias Hakalehto
ISBN: 978-1-61942-692-4
© 2012 Nova Science Publishers, Inc.

Chapter VIII

# Importance of Intestinal Microbiota on Immunocompromised Pediatric Patients

***Jouni Pesola[1] and Elias Hakalehto[2]***
[1] Clinic of Children and Adolescents, Kuopio University Hospital, FI, Kuopio, Finland
[2] Department of Biosciences, University ofEastern Finland, FI, Kuopio, Finland

## Abstract

The relationship between the human body and intestinal microbiota is essential for health. Development of intestinal microbiota takes place during the first years of life. On the other hand, the protection against serious infections, if they are to come, is often based on heavy medications.

In a healthy situation a symbiotic balance exists between the host and the intestinal microbiota. Malfunctioning of the immune system makes the balance more vulnerable in the early childhood. Different situations and diagnoses related to the immunosuppression of children are prematurity, neonatal period, prolonged infections, malnutrition, specific immunodeficiency syndromes, malignancies and different treatments of cancer.

Immunosuppression makes it easier for harmful microbes to overcome the defense mechanisms of the body. This increases the risk of severe morbidity and mortality related to infections. Certain characteristics unique for different clinical situations are reviewed. In the clinical verification of infections, any sign of inflammation is to be taken seriously in the case of severely vulnerable patients.

# Introduction

## Methods for Studying the Intestinal Microbiota

The microbiota of the esophagus, ventricle and duodenum can be studied when esophago-gastro-duodenoscopy is performed. In clinical practice only samples for the detection of *Helicobacter pylori* are taken frequently. Although no other microbiological sample is routinely taken during the endoscopies of the alimentary tract, it is possible to get samples from different parts by this method. More active sampling and microbiological surveying of the upper GI tract of healthy and diseased individuals could give an improved view on the pathogenesis of many various diseases of the GI tract [Coelho *et al.*, 1987; Drumm *et al.*, 1987; Fischer, 1992; Gagliardi *et al.*, 1998].

Microbiological balances have been shown to form between various members of the microflora in the PMEU (Portable Microbe Enrichment Unit) [Hakalehto *et al.*, 2008; Hakalehto *et al.*, 2010]. This method gives space for more strains of *Enterobacteriaceae*, for instance [Pesola *et al.*, 2009]. Utilizing the PMEU method, when the development of the health condition of an individual child was followed for more than two years, starting from birth, indicated the importance of close microbiological examinations for recording the advancement of diseases and the phase of the recovery [Pesola and Hakalehto, 2011]. The numbers of isolated enterobacterial species at different ages is shown in Figure 1. The use of PMEU enrichment was very important in the monitoring of enterobacterial microbiota because 72% of the isolates were detected only after enrichment and some species, like *Klebsiella pneumoniae* and *Proteus mirabilis*, were not at all detectable without the PMEU enrichment (Figure 2) [Pesola and Hakalehto, 2011].

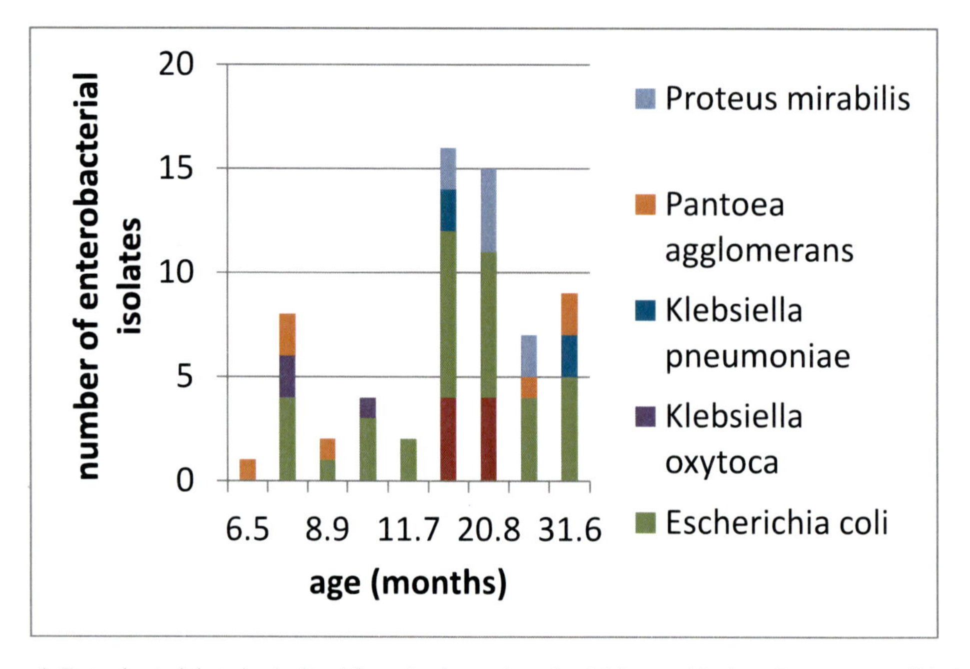


Figure 1. Enterobacterial strains isolated from fecal samples of a child treated by broad-spectrum antibiotics right after birth because of a neonatal septic infection. No enterobacterial strain was detected before the age of 6,5 months [Pesola and Hakalehto, 2011].

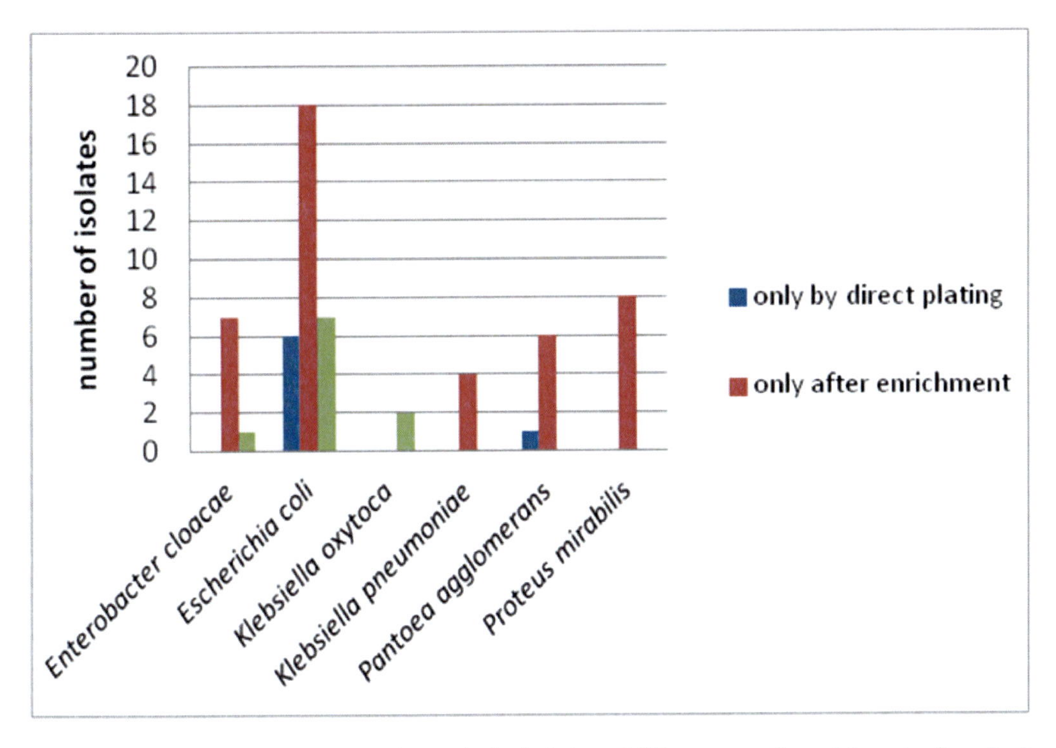


Figure 2. Importance of the PMEU enrichment in the isolation of different enterobacterial species from fecal samples of an child followed after treatment of neonatal sepsis [Pesola and Hakalehto, 2011].

In the pediatric clinical practice the most often performed microbiological examinations from feces are immunological and PCR testing of viruses that are taken when the etiology of viral gastroenteritis is studied. When travel related gastroenteritis is suspected, *Campylobacter, Salmonella, Serratia* and *Yersinia* are studied from the fecal samples [Bottieau *et al.*, 2011], but also toxin-producing *E. coli* strains [Okamoto *et al.*, 1993] and other pathogens are occasionally detected [Serichantalergs *et al.*, 2007; Woo *et al.*, 2005]. Other bacteriological tests performed from clinical fecal samples are the detection of *Clostridium difficile* and its toxins and the examination on so called "dominating flora of the feces" [Kühn *et al.*, 1991; Simojoki, 2011].

Surveillance stool cultures have proven especially useful during periods of severe neutropenia in bone marrow transplant patients [Wells *et al.*, 1987]. In the research of neonatal blood samples, the septicemia causing bacteria were isolated from the patients' blood after enhanced detection by the gas analysis in the PMEU Scentrion® [Hakalehto *et al.*, 2009]. This approach has been validated in the Kuopio University Hospital at the Pediatric Hematology and Oncology Ward and at the Neonatal Intensive Care Unit [Pesola et al. 2011].

The examination of the dominating flora is occasionally applied during treatment of severely immunocompromised patients, like with patients treated because of an allogenous bone marrow transplant [Simojoki, 2011]. Even in this group of patients it is challenging to interpret the results. Clinician would like to have some knowledge about the microbiome that the patients carry in their intestines and about the antibiotic resistance profiles of the microbes. However, the clinical microbiology laboratory personnel do not know what they should look for and how they should answer the clinicians in a proper way. Suitable methods for monitoring the microbiota have not yet been available. From the clinician's point of view it has been difficult to determine what to do with the results of the fecal microbiota analysis in normal hospital practice because the results show only a few strains in a rather random fashion.

The authors would advocate a wider search with respect to the pathogenic organisms, and also for the interactions between the members of the microbial communities in the body.

The correlation of the examination of dominating flora from samples taken from oral cavity and feces on one hand and blood culture analysis on the other hand has been studied in patients being treated because of allogenous hematological stem cell transplantation. With routine methods in 31% of blood culture positive cases the microbe detected from blood culture was earlier isolated also from dominating flora samples [Simojoki, 2011]. This percentage could have been more than doubled by using a PMEU (Portable Microbe Enrichment Unit; Finnoflag Oy and Samplion Oy, Kuopio, Finland) cultivation method [Pesola *et al.*, 2009].

The PMEU method has been useful in the microbiological mapping of the intestinal microbiota. More strains have been isolated by the PMEU method than by conventional culture methods [Pesola *et al.*, 2009; Pesola and Hakalehto, 2011]. The PMEU method has also been used in the study of microbial strains from esofago-gastro-duodenoscopy samples [Hakalehto *et al.*, 2010].

## Colonization of the Intestines after Birth and Development of the Intestinal Microbiota during the Growth

Intestinal colonization begins immediately after the fetal membranes are pierced [Brook *et al.*, 1979] when the baby is exposed to vaginal and perianal microbes of the mother. In a couple of days after vaginal delivery, intestinal bacteria similar to those belonging to the mother's vaginal and fecal microbiota can be found in the fecal samples of the baby [Long and Swenson, 1977; Rotimi and Duerden, 1981].

The environment is an important source of colonizing microbes when the infant is born by Caesarean section or nursed apart from the mother, for example in a neonatal intensive care unit. In this kind of a setting also the cross-spreading of strains between babies by nurse's hands has been documented [Lennox-King *et al.*, 1976]. When infants were nursed in a neonatal intensive care unit in Sweden, vertical transmission of gram-negative bacteria from mother to infant occurred in 12% of vaginally delivered infants and in 0% of 18 infants delivered by Caesarean section [Fryklund *et al.*, 1992]. Also in this study the principal source of the colonizing strains was the fecal flora of other infants.

Delayed acquisition of the intestinal flora is seen, if the baby is born by Caesarean section [Bennet and Nord, 1987; Long and Swenson, 1977] and the formation of the primary gut microbiota may be disturbed. Particularly the acquisition of *Bacteroides fragilis* may be delayed for up to 6 months after the birth [Grönlund *et al.*, 1999]. In the case of enteric coliforms, the antibiotic treatments were documented to delay the colonization for about 6 months [Pesola and Hakalehto, 2011].

There is also evidence that there could be some genetic regulation of the composition of the intestinal microbiota [Toivanen *et al.*, 2001; Van de Merwe *et al.*, 1983; Zoetendal *et al.*, 2002]. De Palma *et al.* have also shown that some interplay exists between human leukocyte antigen genes and the microbial colonization of the newborn intestine [De Palma *et al.*, 2010]. The immunological factors are most likely involved in the determination of the microflora.

## Microbiota in Different Parts of the Gastrointestinal Tract

The fecal bacterial counts of normal individuals average $10^{10}$ - $10^{12}$ cfu / g representing 400-500 different species [Nord and Kager, 1984]. Biochemical and physiological factors have influence on the composition of the microbiota in different parts of the gastrointestinal tract. Diet, gastric acid, exocrine enzymes secreted from salivary glands, pancreas and biliary tract and the velocity of the passage of the digesta play important roles in this process. For example, the bile secretion of a neonate is induced as soon as the umbilical supply is interrupted [Beath, 2003].

The numbers of bacteria cultured from different parts of gastrointestinal tract and the relative proportion of anaerobic and aerobic bacteria investigated by traditional culture methods [Hill, 1995] are presented in tables 1a and 1b.

*Oro-pharyngeal area:* Different parts of the mouth (tooth surfaces, gingival crevices, the surfaces of the tongue, palatinum, cheek surface and pharynx) have their own characteristic microbiota [Gibbons and Houte, 1975; Könönen, 2000]. The oral microbiota is dominated by gram-positive rods and cocci [Hill, 1995]. The microbiota of the oro-pharyngeal area is discussed in Chapter 3.

**Table 1a. Numbers of bacteria in different parts of the human gut studied by traditional culture methods [Hill, 1995] (cfu, colony forming unit)**

| Part of the gastrointestinal tract | cfu / g of luminal content |
|---|---|
| mouth | $10^7 - 10^8$ |
| stomach | $< 10^3$ |
| jejunum | $< 10^3$ |
| terminal ileum | $10^5$ |
| colon | $10^{11}$ |

**Table 1b. Relative proportions of anaerobic and aerobic bacteria at various parts of the human gut studied by traditional culture methods [Hill, 1995]**

| Part of the gastrointestinal tract | counts of anaerobic : aerobic bacteria |
|---|---|
| mouth | 10-100 : 1 |
| stomach | <1 : 1 |
| jejunum | 1-10: 1 |
| terminal ileum | 1-100 : 1 |
| cecum | 100-10000 : 1 |
| colon | 100-1000 : 1 |
| colonic mucosa | 1-10 : 1 |

*Esophagus and ventricle:* The normal resting pH of the gastric juice is below 3 resisting microbial growth, but a hypochlorhydric stomach supports the bacterial growth [Hill, 1995; Väkeväinen *et al.*, 2000]. *Helicobacter pylori* is frequently isolated from samples taken from gastric or duodenal ulcerations and it is regarded as a co-carcinogen in the pathogenesis of gastric carcinoma by damaging the mucosa [Misra *et al.*, 2007].

*Duodenum and jejunum:* Lactobacilli and streptococci are the principal bacterial genera isolated from the upper small intestine [Berg, 1996]. On the other hand, in a wide survey of bile tract isolates in an Austrian hospital, a domination of coliformic bacteria and enterococci in the samples was detected [Hakalehto *et al.*, 2010]. This could indicate the permanent strain on the microbiota caused by the bile substances that would eventually select the most bile resistant strains in the duodenum. The bacterial colonization of these mucosal areas is reduced, besides by bactericidal biliary and pancreatic secretions, also by extensive fluid secretion from the mucosa and by the short small bowel transit time [Holzapfel *et al.*, 1998]. On the other hand, the secretions intensely modulate the microbiota of the duodenum and jejunum leading to the selection of bile and pancreatic fluid resistant bacteria.

*Ileum:* Ileum is colonized both by the microbes transiting from the small bowel and as a result of reflux from the colon through the ileocecal junction [Nord and Kager, 1984]. The microbiota of the ileum is assumed to be similar to that of the cecum [Hill, 1995]. However, it is very difficult to get reliable samples from the terminal ileum. On the other hand, this region gets inocula also from the upper GI tract, originating from the ventricle and duodenum, which could influence the microbiota composition. These bacteria participate in the bile circulation [Garbutt *et al.*, 1970]. Additionally, there are microbial organisms attached to the intestinal epithelial tissue and within the crypts of the intestinal mucosa [Onderdonk, 1999]. The cells of the crypts may be associated with the food consumption of the host together with the microbiota [Hakalehto *et al.*, 2008].

*Large intestines:* The largest number of bacteria is found in the large intestines where bacteria comprise approximately 55% of the fecal mass [Stephen and Cummings, 1980]. Especially the cecum is densely occupied by mostly anaerobic strains [Hill, 1995]. The anaerobic bacteria outnumber the aerobic ones within the large bowel (Table 1b.). The most dominant bacterial genus in this area is *Bacteroides* that accounts for up to 30% of all isolates [Nord and Kager, 1984]. Other culturally dominant anaerobic genera of the colon are *Peptostreptococcus*, *Eubacterium*, *Bifidobacterium*, *Clostridium* and *Fusobacterium*. The proportion of facultatively anaerobic enterobacterial species, such as *Escherichia coli*, is less than 0.1% of the total culturable population of the colon [Onderdonk, 1999]. However, they form the most widely used indication of fecal contamination due to their presence in almost all cases of such contamination together with another duodenal group, namely enterococci. Also other indicator species are studied in order to find more tools for hygiene analysis. For example, the bacteriophages of *Bacteroides* sp. have proven to be a promising means for biological source-tracking and water microbiology [Gomez-Donate *et al.*, 2011; Wicki *et al.*, 2011].

*Mucosal-associated flora (MAF):* The nutrient-rich mucus covering the mucosa of the intestines provides an excellent niche for bacteria forming a MAF [Midtvedt, 1986; Savage, 1970]. Within the MAF the aerobe : anaerobe ratio is approximately 1:1 [Hill, 1995; Marks *et al.*, 1979; Peach *et al.*, 1978] (Table 1b.). Here the *Enterobacteriaceae* family is well represented [Marks *et al.*, 1979; Peach *et al.*, 1978]. Most of the anaerobes belong to the genus *Bacteroides* [Poxton *et al.*, 1997]. The higher proportion of aerobes within the MAF has been explained by the diffusion of oxygen from the capillary vessels through the mucosal membrane [Hill, 1995].

## Nutrition and Microbiota

The early diet of an infant has a great impact on the development of the intestinal microbiota. In breast milk -fed infants the fecal microbiota consists mostly of *Bifidobacterium* species [Benno *et al.*, 1984; Favier *et al.*, 2002; Harmsen *et al.*, 2000; Kleessen *et al.*, 1995]. Formula-fed infants have clearly more enterococci and clostridia than their counterparts fed by breast milk [Kleessen *et al.*, 1995]. After weaning and starting with other foods, the species belonging to the family *Enterobacteriaceae* appear and the microbiota becomes more complex [Pesola *et al.*, 2009; Pesola and Hakalehto, 2011].

Later, after the first years of life when intestinal microbiota has stabilized, the dietary intake does not cause any significant changes in the microbiota [Rautio, 2002]. If some changes happen they are usually reversible.

Different gases, like hydrogen, methane and carbon dioxide, are produced by the microbes of the large intestines. The amounts and proportions of the gases are affected by the diet. The consumption of beans or peas rich in non-absorbable carbohydrates leads to excessive production of gases in the large intestines. This happens also in the case of lactose intolerance when excessive lactose is consumed by the colonic bacteria. The strong odor of the intestinal gas is mainly caused by hydrogen sulfide. The excessive intestinal gases may lead to colic-like abdominal pains resulting from stretching of the gut wall.

## Bile and Intestinal Microbiota

Bile secreted through the biliary tract to the duodenum has antibacterial properties. Only some bile resistant bacteria can survive within its presence. Hence the bile is an important agent causing the selection of microbial species of the intestinal microbiota [Midtvedt and Norman, 1967]. From the duodenum, bile-resistant microbes migrate to lower parts of the intestines [Hakalehto *et al.*, 2010].

Hygiene-indicator bacteria that are used as a sign of fecal contamination in water safety monitoring, are typically bile-resistant bacteria originating from the duodenal area [Edberg *et al.*, 2000; Heitto *et al.*, 2009; Midtvedt and Norman, 1967].

## Antibiotic Treatments and the Intestinal Microbiota

Neonatal sepsis is a relatively common reason for admission of a newborn baby to the neonatal intensive care unit. When the disease is suspected, an empiric broad-scale antibiotic treatment is started right after the collection of diagnostic samples. The duration of the antibiotic treatment depends on the overall condition of the baby and on the results of the microbiological tests, *e.g.* blood cultures. Even though the antibiotic treatment of a neonate may last only a few days, it may have a severe impact on the development of intestinal microbiota. It has been shown that the appearance of Gram – negative rods, for example, may be delayed for six months because of a neonatal antibiotic treatment [Pesola and Hakalehto, 2011]. In a murine model the neonatal antibiotic treatment also altered the developmental gene expression of the gastrointestinal tract and the development of gut barrier function [Schumann *et al.*, 2005].

### Mutual Relationship between Man and His Microbiota

The microbes of the colon produce several vitamin B group vitamins and vitamin K. They also form some toxic compounds that are normally detoxified by the intestinal mucosa and liver. Also intestinal gases originate partly from bacterial metabolic activity. There is a certain symbiotic relationship between a man and his intestinal microbiota. The normal flora prevents the pathogenic microbes from intruding through the intestinal wall into the body. If the normal microbiota is disturbed, some symptoms and disorders may appear. Disturbance of the flora may be seen during traveling to different kinds of microbial environments or after antibiotic treatment. This may lead to vomiting, diarrhea, constipation, stomachache, fatigue or swallowed belly, for example, depending on the situation.

## Microbiological Factors Related to Immunocompromised Children

There are several factors causing immunosuppression and increasing the proneness to infections in children.

**Table 1. Factors causing or related to immunosuppression**

- prematurity
- neonatal period
- infections
    * prolonged infections
    * parvovirus
    * AIDS (acquired immune deficiency syndrome)
- specific immunodeficiency syndromes
    * chronic neutropenia
    * SCID (severe combined immunodeficiency)
- malnutrition
    * unusual diets
    * maldigestion
    * starvation
    * anorexia
- leukemia and other malignancies
- chemotherapy

## Immunosuppression and Infections – How to Keep the Weakest Alive?

Immunosuppression makes it easier for the harmful microbes to overcome the defense mechanisms of the body. This increases the risk of severe morbidity and mortality related to different infections. This is common for all the immunocompromised patients. However, there are certain characteristics that are unique in different clinical situations.

### Prematurity and Neonatal Period

The basis of the intestinal microbiota is founded during the birth. During their first days, children are vulnerable because of the immaturity of both the immunological system and microbiota. The pH of the ventricle of babies is higher keeping the colonization window open for the first bacteria that are ready to settle down into the native intestines of the baby. Unfortunately, during this process, also pathogenic microbes can enter the intestines more easily. The babies treated at the neonatal intensive care units are also prone to nosocomial infections and foreign body related infections. The longer they stay at the hospital the longer is the risk of developing infections caused by some antibiotic resistant bacteria.

### Infections

Prolonged infection periods lead to a consumption of the compounds of the immune system making the patient even more prone to get new infections. The treatment of infections by antibiotics may unfortunately increase the risk of certain infections because of the changes in the normal microbiota. Viral infections like adenovirus may cause temporary aplasia of the bone marrow leading to cytopenias lasting from weeks to months. In respect to bacterial infections, the most important cytopenia is neutropenia (the low number of neutrophils).

AIDS (acquired immune deficiency syndrome) caused by HIV (human immunodeficiency virus) leads to severe vulnerability to infections because the virus attacks specifically the body's defensive mechanisms.

### Nutritional Abnormalities

Unusual diets, maldigestion, starvation and anorexia may lead to the deterioration of the general condition and severe immunosuppression because of the depletion of the body's nutrition. In certain kidney diseases, like nephrotic syndrome and nephrosis, the permeability of the glomerular capillaries is increased leading to the loss of anti-infective proteins.

### Leukemia and Other Malignancies

The most common leukemia in childhood is acute lymphatic leukemia with an incidence of 4% in Nordic population, while acute myeloid leukemia is more rarely seen (incidence 0,7-

0,8%) [Heyman et al., 2011]. In case of acute leukemia the rapidly growing leukemic blasts take over the bone marrow leaving no space for healthy haemopoietic cells. In addition to the bone marrow leukemic blasts are also frequently seen in circulating blood at the time of leukemia diagnosis. Other organs where infiltration of leukemic blasts may be seen are liver, spleen, kidneys, lymph nodes, intestinal wall, testicles and the central nervous system.

Leukemia itself causes a severe risk for life-threatening infections because of the leucopenia resulting from the bone marrow infiltration of the blasts.

Oncologic treatments cause damage to all dividing cells of the body. That is why the bone marrow is especially sensitive to the chemotherapy. Cytopenias – anemia, thrombocytopenia and leucopenia – are frequently seen. Also mucous membranes are often affected as a side effect of cytostatic drugs.

In the treatment of pediatric malignant diseases roughly half of the work is diagnostics of the malignancy itself but the other half goes for the treatment of infections. Neutropenia means the depletion of specific leucocytes mainly responsible for the protection against bacterial infections, namely neutrophilic granulocytes. Neutropenia may be related to acute leukemia or chemotherapy of the malignancies. During neutropenia, the risk for severe infections is very high.

When neutropenic sepsis is suspected, blood culture samples should be immediately taken and empiric antibiotic treatment started. Blood culture samples are taken repeatedly during prolonged infections. Blood culture analysis is essential because no clinical or laboratory parameters have been found to differentiate bacterial infections from non-bacterial ones [Riikonen *et al.*, 1993]. The speed of the analysis of the blood cultures is of great importance because of the rapid course of neutropenic sepsis. After the detection of fever, remarkable mortality is seen within hours if the onset of wide-spectrum antibiotics is delayed [Castagnola and Faraci, 2009].

In the treatment of neutropenic sepsis it is of great importance to detect the causative agents as quickly as possible [Hakalehto, 2006]. Because the bacterial concentration in the circulating blood is low, the samples should be enriched in an incubator before plate culture. There are a few automated incubators of different manufacturers equipped with a built-in monitoring of the microbial growth in the blood culture bottles. 16S rDNA PCR method has also been used in the monitoring of pathogens from blood samples. The specificity of the method can be increased by preanalytical removal of human DNA from the sample [Handschur *et al.*, 2009]. Matrix-assisted laser desorption/ionization time-of-flight mass spectrometry (MALDI-TOF MS) offers rapid detection of bacteria from growing colonies but prior enrichment is recommended in order to get more reliable results [Kroumova *et al.*, 2011]. Lately also the PMEU (Portable Microbe Enrichment Unit, Finnoflag Oy and Samplion Oy, Finland) has been used in the examination of blood culture samples offering optimal growth conditions and an ultrasensitive detection method for the microbial growth [Hakalehto *et al.*, 2009; Hakalehto *et al.*, 2010].

Because of the high risk of invasive bacterial and fungal infections, wide spectrum antibiotics are used very frequently within this patient group posing a potential risk for the development of the resistance towards the antimicrobial agents [Hakalehto, 2011].

The infectious complications are most difficult when the duration of neutropenia is prolonged. This is the case during the chemotherapy block treatments of high risk acute lymphatic leukemia (ALL), AML treatment and especially after autologous and allogenous haematopoietic stem cell transplantations. The treatment of patients after allogenous transplantation is by far the most challenging because of the risk of graft versus host disease resulting from the mismatch between the donor and the recipient. The severity of the graft vs. host disease of the gut may also be related to intestinal microbiota.

An increased risk of infections during oncological treatments is also related to foreign bodies like peripheral and central venous catheters. A biofilm is formed on the surfaces of these foreign bodies. Because deep venous catheters are important sources of invasive infections, the patients are taught to bring their child to the hematology/oncology ward within six hours after the appearance of the fever. After blood culture samples, intravenous wide-spectrum antibiotics are immediately introduced at the ward.

Nosocomial infections are frequently faced complications in hematological and oncological patients who have several long-lasting treatments in the hospital. Every means to minimize the risk of infections should be used. Laminar chambers or clean room facilities are used by the pharmacological industry but they are also needed in hospitals in preparing medicines for use. The microbiological monitoring is an important part of the work in the clean room facilities [Hakalehto, 2010].

## Antibiotic Resistance – A Growing Concern

The antibiotic susceptibility monitoring of pathogens is of growing importance because of increasing tendency towards bacterial resistance development [Ariffin *et al.*, 2000; Ayalew *et al.*, 2003; Siu *et al.*, 1999].

The antibiotic resistance pressure is highest at hospital wards using long courses of broad-spectrum antibiotics. This goes especially for neonatological, hematological, oncological and plastic surgery wards.

For example, the increasing intensity of the chemotherapy of malignant diseases is inevitably related to more prolonged neutropenias and more difficult infections [Crokaert, 2000]. The time for antibiotic susceptibility monitoring should be reduced to minimum. Otherwise the proper treatment would be delayed and the outcome of the patients endangered.

## Pathogenesis of Infections in Immunocompromised Children

Most infections of neutropenic patients originate from their own microbiota through mucous membranes of the oral cavity or other parts of the alimentary tract or from the biofilms formed on the foreign bodies, like central venous catheters.

## Invasion of Microbes through the Intestinal Mucosa

Intestinal mucosa, the microbiota and immune system normally form a barrier that protects from the invasion or intrusion of pathogenic microbes. *E.g.* macrophages are normally able to engulf microbes by phagocytosis. Then macrophages can degrade the engulfed microbes, cells and tissue debris by enzymes released from their endosomes. However, bacteria like *Salmonella* and *Yersinia* have also intracellular phases during their invasion. They can live and multiple inside macrophages that have engulfed them by phagocytosis.

When some part of the preventive measures is declined or totally lacking, the invasive pathogens may find their way into the blood stream. The pathogenic microbes have specific structures for the movement towards the intestinal wall and adhesion to it [Hakalehto et al. 2000; Hakalehto *et al.*, 2007].

## Microbiota and Pediatric Morbidity

The disturbance of the development of the intestinal microbiota during infancy may result *e.g.* from antibiotic treatment [Schumann *et al.*, 2005], enteral infections, dietary problems or intestinal food allergy. Imbalanced intestinal microbiota increases the proneness to infections but it might also be associated with the risk of developing autoimmune diseases [Lodes *et al.*, 2004].

The PMEU method has proven useful in the microbiological mapping of the intestinal microbiota. It has also been developed for the blood culture analysis. It offers also the possibility to compare microbial strains isolated from blood culture samples and intestinal cultures.

*Clostridium difficile*, a spore-forming Gram – positive rod, is a common member of intestinal microbiota being usually harmless to a healthy host. As a result of broad-spectrum antibiotic treatment or chemotherapy some strains capable of producing toxins may, however, cause severe bloody diarrhea and even life-threatening pseudomembraneous colitis. The bacterium cannot normally be eradicated by antibiotics although the symptoms can be cured by *e.g.* metronidazole or vancomycin. Because the spores of the bacterium are very resistant to the environmental stress and also to ethanol-based disinfectants, they are easily transmitted via hands. Both mechanical washing of hands and the use of disinfectants is recommended after contact with a patient with *C. difficile.* Also the sterilization of the equipment used in the diagnostic examinations, like colonoscopes need to be handled with special attention [Hell *et al.*, 2010].

## Prevention of the Diseases and Microbiota

During recent years probiotics and prebiotics have been extensively studied. These products can be used to modify the intestinal flora temporarily, but normally no permanent changes can be achieved.

Diet has a major impact on the composition of the microbiota. If severe imbalance of the microbiota is seen, *e.g.* microbial overgrowth syndrome, also antibiotics can be used for the treatment [Ford *et al.*, 2009; Pimentel, 2009].

There is a growing interest that focuses on the intestinal microbiota. This is related to the development of the intestinal immunity, production of some vitamins, immunological defense of the intestines and several pathogenic processes. The careful investigation of the composition of the microbiota would be important in order to find out indicator microbes that might be involved in certain pathogenic processes. But it is still a challenge to find the best ways in order to make conclusions on the microbiota.

The intestinal bacteria interact with each other but also with the host. It is a demanding task to investigate these interactions. Especially the study of the flora of the small intestines is rather complicated. However, the interactions of intestinal microbes can be ideally simulated in the PMEU [Hakalehto, 2011]. New interesting interactions have been found using this approach. For example, in mixed culture *E. coli* and *Klebsiella* have been shown to have a dualistic balance being beneficial for both of them and also for the host [Hakalehto *et al.*, 2008]. In the PMEU, this kind of symbiosis has been shown to form the basis for building up balanced microbial communities [Hakalehto *et al.*, 2010].

Intensive research has focused on the intestinal microbiota and its impact on the human welfare. There are still many open questions: Which are the most important representatives of the intestinal microbiota and how to find them? How could the intestinal microbiota be safely modulated? Further studies are warranted in order to answer these questions.

# Conclusion

The severely threatened first days of life necessitate special care. No risks can be taken, whatsoever, with respect to microbial infections. At the same time the normal flora should be developing in as balanced fashion as possible. Nutrition and developing immune system set the framework for the growth and survival. However, any alarming signs of infections, for example in the blood test results, deserve extreme measures in case of seriously diseased pediatric patients. When an infection is suspected it is not always possible to get any idea on the causative reasons, or on the pathogenic agent. The results from the chosen treatment have to be continuously monitored and recorded. The PMEU system has proven out to be an extremely fast tool for fast and efficient follow up of infections in the hospital wards. It is often important to find out also the hiding anaerobic agents from patient samples, thus giving the doctor a possibility to direct the effective treatments on them. Also in monitoring blood and blood products this technology could be useful.

# References


Ariffin, H; Navaratnam, P; Mohamed, M; Arasu, A; Abdullah, WA; Lee, CL; Peng, LH. Ceftazidime-resistant *Klebsiella pneumoniae* bloodstream infection in children with febrile neutropenia. *Int. J. Infect. Dis.,* 2000; 4, 21-25.

Ayalew, K; Nambiar, S; Yasinskaya, Y; Jantausch, BA. Carbapenems in pediatrics. *Ther. Drug. Monit.,* 2003; 25, 593-599.

Beath, SV. Hepatic function and physiology in the newborn. *Semin. Neonatol.,* 2003; 8, 337-346.

Bennet, R; Nord, CE. Development of the faecal anaerobic microflora after caesarean section and treatment with antibiotics in newborn infants. *Infection,* 1987; 15, 332-336.

Benno, Y; Sawada, K; Mitsuoka, T. The intestinal microflora of infants: composition of fecal flora in breast-fed and bottle-fed infants. *Microbiol. Immunol.*, 1984; 28, 975-986.

Berg, RD. The indigenous gastrointestinal microflora. *Trends Microbiol.,* 1996; 4, 430-435.

Bottieau, E; Clerinx, J; Vlieghe, E; Van Esbroeck, M; Jacobs, J; Van Gompel, A; Van Den Ende, J. Epidemiology and outcome of *Shigella, Salmonella* and *Campylobacter* infections in travellers returning from the tropics with fever and diarrhoea. *Acta Clin. Belg.,* 2011; 66, 191-195.

Brook, I; Barrett, CT; Brinkman, CR,3rd; Martin, WJ; Finegold, SM. Aerobic and anaerobic bacterial flora of the maternal cervix and newborn gastric fluid and conjunctiva: a prospective study. *Pediatrics,* 1979; 63, 451-455.

Castagnola, E; Faraci, M. Management of bacteremia in patients undergoing hematopoietic stem cell transplantation. *Expert Rev. Anti. Infect. Ther.,* 2009; 7, 607-621.

Coelho, LG; Das, SS; Karim, QN; Walker, MM; Queiroz, DM; Mendes, EN; Lima Junior, GF; de Oliveira, CA; Baron, JH; Castro Lde, P. *Campylobacter pyloridis* in the upper gastrointestinal tract: a Brazilian study. *Arq. Gastroenterol.,* 1987; 24, 5-9.

Crokaert, F. Febrile neutropenia in children. *Int. J. Antimicrob. Agents,* 2000; 16, 173-176.

De Palma, G; Capilla, A; Nadal, I; Nova, E; Pozo, T; Varea, V; Polanco, I; Castillejo, G; Lopez, A; Garrote, JA; Calvo, C; Garcia-Novo, MD; Cilleruelo, ML; Ribes-Koninckx, C; Palau, F; Sanz, Y. Interplay between human leukocyte antigen genes and the microbial colonization process of the newborn intestine. *Curr. Issues Mol. Biol.,* 2010; 12, 1-10.

Drumm, B; O'Brien, A; Cutz, E; Sherman, P. *Campylobacter pyloridis*-associated primary gastritis in children. *Pediatrics,* 1987; 80, 192-195.

Edberg, SC; Rice, EW; Karlin, RJ; Allen, MJ. *Escherichia coli*: the best biological drinking water indicator for public health protection. *Symp. Ser. Soc. Appl. Microbiol.,* 2000; 29, 106S-116S.

Favier, CF; Vaughan, EE; De Vos, WM; Akkermans, AD. Molecular monitoring of succession of bacterial communities in human neonates. *Appl. Environ. Microbiol.,* 2002; 68, 219-226.

Fischer, R. '*Gastrospirillum hominis*', another gastric spiral bacterium. *Dig. Dis.,* 1992; 10, 144-152.

Ford, AC; Spiegel, BM; Talley, NJ; Moayyedi, P. Small intestinal bacterial overgrowth in irritable bowel syndrome: systematic review and meta-analysis. *Clin. Gastroenterol. Hepatol.,* 2009; 7, 1279-1286.

Fryklund, B; Tullus, K; Berglund, B; Burman, LG. Importance of the environment and the faecal flora of infants, nursing staff and parents as sources of gram-negative bacteria colonizing newborns in three neonatal wards. *Infection,* 1992; 20, 253-257.

Gagliardi, D; Makihara, S; Corsi, PR; Viana Ade, T; Wiczer, MV; Nakakubo, S; Mimica, LM. Microbial flora of the normal esophagus. *Dis. Esophagus,* 1998; 11, 248-250.

Garbutt, JT; Wilkins, RM; Lack, L; Tyor, MP. Bacterial modification of taurocholate during enterohepatic recirculation in normal man and patients with small intestinal disease. *Gastroenterology,* 1970; 59, 553-566.

Gibbons, RJ; Houte, JV. Bacterial adherence in oral microbial ecology. *Annu. Rev. Microbiol.,* 1975; 29, 19-44.

Gomez-Donate, M; Payan, A; Cortes, I; Blanch, AR; Lucena, F; Jofre, J; Muniesa, M. Isolation of bacteriophage host strains of *Bacteroides* species suitable for tracking sources of animal faecal pollution in water. *Environ. Microbiol.,* 2011; 13, 1622-1631.

Grönlund, MM; Lehtonen, OP; Eerola, E; Kero, P. Fecal microflora in healthy infants born by different methods of delivery: permanent changes in intestinal flora after cesarean delivery. *J. Pediatr. Gastroenterol. Nutr.*, 1999; 28, 19-25.

Hakalehto, E. Hygiene monitoring with the Portable Microbe Enrichment Unit (PMEU). *41st R3 -Nordic Symposium. Cleanroom technology, contamination control and cleaning. VTT Publications 266.* Espoo, Finland: VTT (State Research Centre of Finland); 2010.

Hakalehto, E. Semmelweis' present day follow-up: Updating bacterial sampling and enrichment in clinical hygiene. *Pathophysiology,* 2006; 13, 257-267.

Hakalehto, E. Simulation of enhanced growth and metabolism of intestinal *Escherichia coli* in the Portable Microbe Enrichment Unit (PMEU). In: Rogers MC, Peterson ND, editors. *E. coli infections: causes, treatment and prevention.* New York, USA: Nova Publishers; 2011.

Hakalehto, E; Hell, M; Bernhofer, C; Heitto, A; Pesola, J; Humppi, T; Paakkanen, H. Growth and gaseous emissions of pure and mixed small intestinal bacterial cultures: Effects of bile and vancomycin. *Pathophysiology,* 2010; 17, 45-53.

Hakalehto, E; Hujakka, H; Airaksinen, S; Ratilainen, J; Närvänen, A. Growth-phase limited expression and rapid detection of *Salmonella* type 1 fimbriae. In: Hakalehto E. *Characterization of Pectinatus cerevisiiphilus and P. frisingiensis surface components. Use of synthetic peptides in the detection fo some Gram-negative bacteria.* Kuopio, Finland: Kuopio University Publications C. Natural and Environmental Sciences 112; 2000. Doctoral dissertation.

Hakalehto, E; Humppi, T; Paakkanen, H. Dualistic acidic and neutral glucose fermentation balance in small intestine: Simulation *in vitro. Pathophysiology,* 2008; 15, 211-220.

Hakalehto, E; Pesola, J; Heitto, A; Bhanj Deo, B; Rissanen, K; Sankilampi, U; Humppi, T; Paakkanen, H. Fast detection of bacterial growth by using Portable Microbe Enrichment Unit (PMEU) and ChemPro100i((R)) gas sensor. *Pathophysiology,* 2009; 16, 57-62.

Hakalehto, E; Pesola, J; Heitto, L; Närvänen, A; Heitto, A. Aerobic and anaerobic growth modes and expression of type 1 fimbriae in *Salmonella. Pathophysiology,* 2007; 14, 61-69.

Handschur, M; Karlic, H; Hertel, C; Pfeilstocker, M; Haslberger, AG. Preanalytic removal of human DNA eliminates false signals in general 16S rDNA PCR monitoring of bacterial pathogens in blood. *Comp. Immunol. Microbiol. Infect. Dis.*, 2009; 32, 207-219.

Harmsen, HJ; Wildeboer-Veloo, AC; Raangs, GC; Wagendorp, AA; Klijn, N; Bindels, JG; Welling, GW. Analysis of intestinal flora development in breast-fed and formula-fed infants by using molecular identification and detection methods. *J. Pediatr. Gastroenterol. Nutr.*, 2000; 30, 61-67.

Heitto, L; Heitto, A; Hakalehto, E. Tracing wastewaters with faecal enterococci. (Poster). *Second European Large Lakes Symposium.* Norrtälje, Sweden; 2009.

Hell, M; Bernhofer, C; Huhulescu, S; Indra, A; Allerberger, F; Maass, M; Hakalehto, E. How safe is colonoscope-reprocessing regarding *Clostridium difficile* spores? *J. Hosp.Inf.*, 2010; 76, 21-22.

Heyman, M; Wesenberg, F; Kock Lie, H; Frandsen, T; Lähteenmäki, P; de Verdier, B., editors. Childhood cancer in the Nordic countries. Report on epidemiologic and therapeutic results from registries and working groups. Turku, Finland: Nordic Society of Pediatric Haematology and Oncology; 2011.

Hill, MJ. The normal gut bacterial flora. In: Hill MJ, editor. *Role of the gut bacteria in human toxicology and pharmacology*. London, United Kingdom: Taylor and Francis Ltd; 1995; 3-17.

Holzapfel, WH; Haberer, P; Snel, J; Schillinger, U; Huis in't Veld, JH. Overview of gut flora and probiotics. *Int. J. Food Microbiol.,* 1998; 41, 85-101.

Kleessen, B; Bunke, H; Tovar, K; Noack, J; Sawatzki, G. Influence of two infant formulas and human milk on the development of the faecal flora in newborn infants. *Acta Paediatr.,* 1995; 84, 1347-1356.

Könönen, E. Development of oral bacterial flora in young children. *Ann. Med.,* 2000; 32, 107-112.

Kroumova, V; Gobbato, E; Basso, E; Mucedola, L; Giani, T; Fortina, G. Direct identification of bacteria in blood culture by matrix-assisted laser desorption/ionization time-of-flight mass spectrometry: a new methodological approach. *Rapid Commun. Mass. Spectrom.,* 2011; 25, 2247-2249.

Kühn, I; Allestam, G; Stenström, TA; Möllby, R. Biochemical fingerprinting of water coliform bacteria, a new method for measuring phenotypic diversity and for comparing different bacterial populations. *Appl. Environ. Microbiol.,* 1991; 57, 3171-3177.

Lennox-King, SM; O'Farrell, SM; Bettelheim, KA; Shooter, RA. *Escherichia coli* isolated from babies delivered by caesarean section and their environment. *Infection,* 1976; 4, 139-145.

Lodes, MJ; Cong, Y; Elson, CO; Mohamath, R; Landers, CJ; Targan, SR; Fort, M; Hershberg, RM. Bacterial flagellin is a dominant antigen in Crohn disease. *J. Clin. Invest.,* 2004; 113, 1296-1306.

Long, SS; Swenson, RM. Development of anaerobic fecal flora in healthy newborn infants. *J. Pediatr.,* 1977; 91, 298-301.

Marks, CG; Hawley, PR; Peach, SL; Drasar, BS; Hill, MJ. The effects of phthalylsulphathiazole on the bacteria of the colonic mucosa and intestinal contents as revealed by the examination of surgical samples. *Scand. J. Gastroenterol.,* 1979; 14, 891-896.

Midtvedt, T. Intestinal microflora-associated characteristics. *Microecol. Ther.,* 1986; 16, 121-130.

Midtvedt, T; Norman, A. Bile acid transformations by microbial strains belonging to genera found in intestinal contents. *Acta Pathol. Microbiol. Scand,* 1967; 71, 629-638.

Misra, V; Misra, SP; Singh, MK; Singh, PA; Dwivedi, M. Prevalence of *H. pylori* in patients with gastric cancer. *Indian J. Pathol. Microbiol.,* 2007; 50, 702-707.

Nord, CE; Kager, L. The normal flora of the gastrointestinal tract. *Neth. J. Med.,* 1984; 27, 249-252.

Okamoto, K; Fujii, Y; Akashi, N; Hitotsubashi, S; Kurazono, H; Karasawa, T; Takeda, Y. Identification and characterization of heat-stable enterotoxin II-producing *Escherichia coli* from patients with diarrhea. *Microbiol. Immunol.,* 1993; 37, 411-414.

Onderdonk, AB. The intestinal microflora and intra-abdominal sepsis In: Tannock GW, editor. *Medical importance of the normal microflora.* United Kingdom: Kluwer Academic Publishers; 1999; 164-176.

Peach, S; Lock, MR; Katz, D; Todd, IP; Tabaqchali, S. Mucosal-associated bacterial flora of the intestine in patients with Crohn's disease and in a control group. *Gut,* 1978; 19, 1034-1042.

Pesola, J; Hakalehto, E. Enterobacterial microflora in infancy - a case study with enhanced enrichment. *Indian J. Pediatr.,* 2011; 78, 562-568.

Pesola, J; Heitto, A; Myöhänen, P; Laitiomäki, E; Sankilampi, U; Paakkanen, H; Riikonen, P; Hakalehto, E. Enhanced bacterial enrichment in the microbial diagnostics of pediatric neutropenic sepsis. (Poster abstract). In: Lähteenmäki P, Grönroos M, Salmi T, editors. *NOPHO 29th Annual Meeting.* Turku, Finland: NOPHO; 2011; 59.

Pesola, J; Vaarala, O; Heitto, A; Hakalehto, E. Use of portable enrichment unit in rapid characterization of infantile intestinal enterobacterial microbiota. *Microb. Ecol. Health Dis.,* 2009; 21, 203-210.

Pimentel, M. Review of rifaximin as treatment for SIBO and IBS. *Expert Opin. Investig. Drugs,* 2009; 18, 349-358.

Poxton, IR; Brown, R; Sawyerr, A; Ferguson, A. Mucosa-associated bacterial flora of the human colon. *J. Med. Microbiol.,* 1997; 46, 85-91.

Rautio, M. Indigenous intestinal microbiota and disease – different approaches to characterize the composition and products of the human microbiota. Helsinki, Finland: Publications of the National Public Health Institute, A5/2002; 2002.

Riikonen, P; Jalanko, H; Hovi, L; Saarinen, UM. Fever and neutropenia in children with cancer: diagnostic parameters at presentation. *Acta Paediatr.,* 1993; 82, 271-275.

Rotimi, VO; Duerden, BI. The development of the bacterial flora in normal neonates. *J. Med. Microbiol.,* 1981; 14, 51-62.

Savage, DC. Associations of indigenous microorganisms with gastrointestinal mucosal epithelia. *Am. J. Clin. Nutr.,* 1970; 23, 1495-1501.

Schumann, A; Nutten, S; Donnicola, D; Comelli, EM; Mansourian, R; Cherbut, C; Corthesy-Theulaz, I; Garcia-Rodenas, C. Neonatal antibiotic treatment alters gastrointestinal tract developmental gene expression and intestinal barrier transcriptome. *Physiol. Genomics,* 2005; 23, 235-245.

Serichantalergs, O; Bhuiyan, NA; Nair, GB; Chivaratanond, O; Srijan, A; Bodhidatta, L; Anuras, S; Mason, CJ. The dominance of pandemic serovars of *Vibrio parahaemolyticus* in expatriates and sporadic cases of diarrhoea in Thailand, and a new emergent serovar (O3 : K46) with pandemic traits. *J. Med. Microbiol.,* 2007; 56, 608-613.

Simojoki, S-T. Kantasolusiirron saaneiden lapsipotilaiden kolonisaatioviljelyt. (in Finnish). Helsinki, Finland: University of Helsinki, Faculty of Medicine; 2011.

Siu, LK; Lu, PL; Hsueh, PR; Lin, FM; Chang, SC; Luh, KT; Ho, M; Lee, CY. Bacteremia due to extended-spectrum beta-lactamase-producing *Escherichia coli* and *Klebsiella pneumoniae* in a pediatric oncology ward: clinical features and identification of different plasmids carrying both SHV-5 and TEM-1 genes. *J. Clin. Microbiol.,* 1999; 37, 4020-4027.

Stephen, AM; Cummings, JH. The microbial contribution to human faecal mass. *J. Med. Microbiol.,* 1980; 13, 45-56.

Toivanen, P; Vaahtovuo, J; Eerola, E. Influence of major histocompatibility complex on bacterial composition of fecal flora. *Infect. Immun.,* 2001; 69, 2372-2377.

Väkeväinen, S; Tillonen, J; Salaspuro, M; Jousimies-Somer, H; Nuutinen, H; Färkkilä, M. Hypochlorhydria induced by a proton pump inhibitor leads to intragastric microbial production of acetaldehyde from ethanol. *Aliment. Pharmacol. Ther.,* 2000; 14, 1511-1518.

Van de Merwe, JP; Stegeman, JH; Hazenberg, MP. The resident faecal flora is determined by genetic characteristics of the host. Implications for Crohn's disease? *Antonie Van Leeuwenhoek,* 1983; 49, 119-124.

Wells, CL; Ferrieri, P; Weisdorf, DJ; Rhame, FS. The importance of surveillance stool cultures during periods of severe neutropenia. *Infect. Control.,* 1987; 8, 317-319.

Wicki, M; Auckenthaler, A; Felleisen, R; Tanner, M; Baumgartner, A. Novel *Bacteroides* host strains for detection of human- and animal-specific bacteriophages in water. *J. Water Health,* 2011; 9, 159-168.

Woo, PC; Lau, SK; Teng, JL; Yuen, KY. Current status and future directions for *Laribacter hongkongensis*, a novel bacterium associated with gastroenteritis and traveller's diarrhoea. *Curr. Opin. Infect. Dis.,* 2005; 18, 413-419.

Zoetendal, EG; von Wright, A; Vilpponen-Salmela, T; Ben-Amor, K; Akkermans, AD; de Vos, WM. Mucosa-associated bacteria in the human gastrointestinal tract are uniformly distributed along the colon and differ from the community recovered from feces. *Appl. Environ. Microbiol.,* 2002; 68, 3401-3407.

In: Alimentary Microbiome: A PMEU Approach
Editor: Elias Hakalehto
ISBN: 978-1-61942-692-4
© 2012 Nova Science Publishers, Inc.

Chapter IX

# Fecal Microbiological Analysis in the Health Monitoring

*Elias Hakalehto*[1], *Anneli Heitto*[2], *and Jouni Pesola*[3]
[1]Department of Biosciences, University of Eastern Finland, FI, Kuopio, Finland
[2]Finnoflag Oy, FI-70101 Kuopio, Finland
[3]Clinic of Children and Adolescents, Kuopio University Hospital, FI, Kuopio, Finland

## Abstract

Numerous microbial parameters are followed up by fecal analysis. It is noteworthy that the results from these tests always more or less demonstrate, at least quantitatively, the condition of the microbiota particularly in the colon. In the more distal part of the intestines the human self-defense mechanisms play a lesser role than on the mucosal membranes of the upper parts of the alimentary tract. Therefore, any results from fecal tests have to be considered taking into account this point of view. The evaluation of the alimentary microflora on the basis of fecal analysis only, is usually giving an inadequate picture of the entire microbiome. However, it could provide us important clues on the condition and recovery of the patient, as well as on the pathogenic processes.

While fecal microbiota is also reflecting the general health status of a patient, it gives information also on his or her nutritional status. The gut bacteria provide the host body system with some important vitamins. In metabolic sense, studying the fecal microbes serves as an information source on the outcome of joint activities of the host organs and the microbial inhabitants of the body. The fecal microbes live in the waste outlet of the host body system. Microbes set up communities on the colon, taking part in building up suitable conditions for themselves also in such conditions as diverticulosis, Crohn's disease, and many other pathological or physiological conditions. In most cases the effect on the host is more a result of the combinations of various microbial activities and interactions, than an outcome of the action of a single micro-organism. Essentially, the diseased status of a human individual can be considered as a "misorientation" of the intestinal ecosystem, which could lead to a "negatively balanced" gut ecosystem. These adverse effects on our health could be followed by monitoring the fecal microflora, and sometimes also by simulating its behavior in the PMEU (Portable Microbe Enrichment Unit).

In order to understand better the situation in the intestines, more tests should be developed taking into account the microbial interactions in this area. In the current survey we briefly sum up the present microbiological testing taking place in the Finnish hospitals as well as comment on the need of spreading up the analysis, as well as widening its scope. This requires a more alerted attitude toward the follow up of the microbiological condition of the patient. Because changes and reactions on the microflora level take place swiftly, the monitoring needs to be set up accordingly. The results from the analysis should be available for the physician responsible for the treatments preferably in real time (as fast as they are getting produced). The hygiene control at large is based on monitoring specific hygiene indicator species. These indicators are usually of fecal origin. In this chapter, the importance of developing new methods for fecal microbiological studies is demonstrated and further discussed.

## Introduction

Over 400 different microbial species, *i.e.* bacteria, viruses, yeasts and parasites can be detected from stool samples. Also multiple non-microbiological examinations are available for the fecal samples. These include the analysis of occult blood, bile, calprotectin, different intestinal enzymes, markers of the colon cancer and also microbial metabolites. Also the use and degradation of dietary components can be studied.

The origins of the bacteria of the large intestines are in cecum, where a lot of microbiological activities take place. The type of the microbiota of this region has been shown to correlate with the obesity, in which case the butyrate bacteria seem to dominate [Ley *et al.*, 2006; Turnbaugh *et al.*, 2006]. This was shown to lead to increased ratio of the *Firmicutes* bacterial phylum compared with the *Bacteroidetes*. On the other hand, the cecum contains only 20% of the last remaining nutrients of the human diet. Butyric acid bacteria are the most prevailing groups of organisms in this area. They are obligate anaerobes which belong mostly to genus *Clostridium*. They have been shown to benefit from the $CO_2$ produced by other bacterial species such as lactobacilli [Hakalehto and Hänninen, 2012]. This kind of situation is also occurring in the silage production, where too high a number of clostridia, or a production of butyrate is a cause for spoilage of the product [Borreani and Tabacco, 2008]. The $CO_2$ has been shown to contribute to the onset of bacterial growth in general [Walker, 1932].

The current approach in the fecal diagnostics is the elimination of specific problems or pathogens as causative agents responsible for the patient's disease and condition. However, alternatively it could be possible to look at the issue "with an open mind" in order to observe any major microbial strains in the specimens. Finding these dominating species, whatever categories they belong, would then give as complete basis as possible for investigating the pathogenesis. Then the potential reasons and routes for contagion or contamination could be deduced. If this analysis is carried out with the PMEU technology, it gives more viable strains from the cultivations [Pesola *et al.*, 2009; Pesola and Hakalehto, 2011]. In the PMEU it is also possible to investigate the development of any mixed culture of the isolates, that could potentially give further clues regarding the origins and development of the particular disease.

Nevertheless, it could well be stated that most microbial infections are multifactorial [Giannella, 1981; Thompson and Bizzarro, 2008]. The infections are also directly dependent on the overall health condition, nutritional status, and regulatory functions of the patient [Kelly, 2010]. Thus it is not possible to separate any individual strain from the entity of multiple parameters behind the illnesses. In the PMEU it is possible to continue the study of the fecal samples in the laboratory by dividing the mixed cultures into several sects, for example, and by performing further research on them, and on their interactions.

In the hygienic analysis of the hospital wards the monocultures or decreased amounts of strains in the samples have been shown to represent a deteriorated hygienic condition [Kühn *et al.*, 1993]. In order to effectively monitor the potential dissemination of hazardous microbes, such as pathogens or antibiotic resistant strains, one needs comparative studies between the patient specimen and the environmental samples. One of the major groups to be followed up is the wide spectrum of fecal indicator strains. Their screening up has been intensified by using the PMEU Scentrion® techniques in the hospital wards, and also in the control of clean water hygiene [Hakalehto *et al.*, 2009; Hakalehto *et al.*, 2011; Pesola *et al.*, 2011].

## Fecal Microbiological Analysis in Clinical Use

The amount of fecal routine analysis in the clinical laboratories is relatively low compared with *e.g.* the volume of blood or urine microbiology in these labs. We have roughly estimated, for example, that the amount of fecal testing quantitatively corresponds to 1/7 of the urine cultivations and 1/3 of the blood cultures with respect to the number to the number of tests in Finland. However, the proper analysis of the individual fecal flora of a wider range of patients could improve the general level of hospital care by giving an idea on the patient condition, and on the phase of the recovery process, also in case of other diseases than those dealt with in the gastrointestinal wards. It is also noteworthy that septic infections often originate from intestinal microbes.

Surveys on microflora composition are amounting less than 2% of the entire number of fecal analysis in Finland. For example, they equal more or less with the number of analysis of one individual species *Clostridium difficile*, whereas fecal salmonellas are analyzed approximately 30 times more often. These tests for *Salmonella* sp., in turn, are almost equaled by the breath tests for *Helicobacter pylori*. The latter testing was increasing with 50% from 2009 to 2010. These figures related to the microbiological testing frequencies give ideas about the appreciation of the different diagnostic procedures in practical clinical care. In order to better understand the contribution of human microbiota to the human health, the microbiological tests should be developed with a scope for more frequent usages. Also, their correlations with host physiological conditions need to be understood and evaluated better.

Supposed that the Finnish case reflects the situation worldwide, we suggest that the importance of fecal microbiology, and also that of microbiology from endoscopic and colonoscopic samples, should be re-estimated and revalidated. If the methodology would be further developed, it could make sense to estimate balances in the microflora composition. In 2011 it was suggested in a wide international study that the intestinal microflora of human individuals could be divided into three categories, and consequently people into three enterotypes, by the molecular biology analysis [Wu *et al.*, 2011]. This approach of overall, and also more precise, characterization of the flora could give guidelines for the clinical and nutritional treatments. On the basis of microbiological background analysis, important deductions could be made on the patient condition. Even with the existing medical microbiology methods, the causative agent for septic infections in small children were found in 31% of the cases amongst the normal flora causing often remarkable overgrowth in the gastrointestinal tract [Simojoki, 2011]. With improved methodology, this proposition of cases of overall infections with roots in the intestinal flora could considerably increase. For example, altogether 7/21 (33%) and 24/46 (52%) enterobacterial isolates from the fecal samples of eight small children at the ages of 3 and 12 months, respectively, were not detected by plate culture, but their isolation was possible after the PMEU pre-enrichment of the samples [Pesola *et al.*, 2009]. If the numbers of isolates from the fecal samples could be increased, their chances for distribution into the blood stream could be better estimated.

In Finland there is a set of fecal bacteriological tests available for the clinical use (Eastern Finland Laboratory Centre Joint Authority Enterprise,ISLAB, Finland). When traveller's diarrhea is suspected, *Campylobacter* sp., *Salmonella* sp., *Shigella* sp. and *Yersinia* sp. are detected from the fecal samples by bacterial culture. Also some other enteric pathogens, like enterohemorrhagic *E. coli* (EHEC) and *Vibrio cholera,* can be detected. These pathogens are very rare findings among Finnish population. In cases of antibiotic diarrhea, *Bacillus cereus, Clostridium difficile* and yeasts can be studied. When an epidemic gastroenteritis is investigated, *Aeromonas* sp., *Bacillus cereus, Campylobacter* sp., *Clostridium perfringens, Plesiomonas* sp., *Salmonella* sp., *Shigella* sp., *Staphylococcus aureus* and *Yersinia* sp. are usually suspected and searched for. When *B. cereus, Cl. difficile, Cl. perfringens* or *S. aureus* cells are detected, also their toxins can usually be found in the specimen.

Nowadays the general analysis of fecal microbiota is very seldom used in the clinical practise. A test called dominating fecal flora is available in Finnish clinical microbiological laboratories and it has been used in the monitoring of the intestinal microbiota of severely immunocompromised patients, like recipients of an allogenous bone marrow transplant. By this test a couple of dominating fecal strains are identified and their antibiotic susceptibilities studied.

Even in the case of the most immunocompromised patients the analysis of fecal microbiota is problematic because the clinical microbiology laboratory personnel do not know what the clinicians expect to be tested from the samples.

A polymerase chain reaction denaturing gradient gel electrophoresis (PCR-DGGE) fingerprinting has been used in conjunction with fluorescent *in situ* hybridization analysis with specific bacterial oligonucleotide probes in order to quantify the fecal bacteria [van Vliet *et al.*, 2009]. It was shown that during chemotherapy remarkable changes are seen in the intestinal microbiota.

Methods suitable for routine use in the evaluation of the microbiome have not been available in the clinical microbiological laboratories yet. Normally the fecal microbiota is studied by cultivation of the samples on couple of agar plates and only the most dominating strains are isolated, identified and their antimicrobial resistances measured.

On the other hand, the clinicians do not know what to do with the results of the fecal microbiota analysis because the results show only few strains in a very random fashion. When so called colonization samples have been taken from mouth or feces from patients being treated because of the allogenic hematological stem cell transplantation, in 31% of cases the microbe detected from blood culture had been earlier isolated from standard colonization samples taken either from throat or feces [Simojoki, 2011]. This amount of bacterial isolations from the feces could be more than doubled by using the PMEU [Pesola and Hakalehto, 2011]. The antibiotic susceptibilities of any components of the patient microflora can also be effectively studied in the PMEU [Hakalehto, 2012]. The remote control uses of the PMEU devices are presented in Chapter 11 of this book. This information transfer technology could be taken into service in case of urgenntly analyzed samples, or when the results are needed to be widely distributed for epidemiological or other reasons.

The fecal colonization samples have been studied using [Simojoki, 2011]:

- different agar plates intended for aerobic and anaerobic bacteria
- Sabouraud glucose (SG) plate for fungi,
- CHROMagar Candida plate for yeasts,
- Cycloserine Cefoxitin Fructose Agar (CCFA) for *Clostridium difficile*
- Müller Hinton Agar plate for the analysis of the antibiotic resistance of bacteria

# Clinical Infection Related to Fecal Microbiota

A large proportion of urinary infections are caused by fecal contaminants like *Escherichia coli*, *Klebsiella* sp. or enterococci. The studies on intestines and urinary tract colonizing *E. coli* have been documented earlier [Hakalehto. 2011]. Blood, urinary and liquor cultures are important in the work-up of severe septic infections. The life-threatening infections are, fortunately, rare with individuals possessing normal immunity, but they belong to everyday life at the hospital wards where hematological and oncological patients are treated, for instance. The problems of dealing with the infections of these weakened patients are reviewed in the Chapter 8 of this book.

In practice a majority of the microbial cultivation tests performed during septic infections do not indicate the presence of any bacterium although the clinical status of the patient would suggest an invasive bacterial infection. The site of a microbe to enter the urinary tract, blood circulation or even central nervous system is often the alimentary tract [Sherman, 2010; Ubeda *et al.*, 2010].

## Importance of the Study of Microbiota for Individual Health

A multitude of clinical situations can disturb the anti-infective ability of the body and the balance of the microbiota in the intestines. In the first place this may cause dyspepsia, colic, changes in the consistence of the feces, diarrhea or obstipation, malodour of the feces, problems resulting from extensive intestinal gas production, vomiting, etc. – All these symptoms are typical for the irritable bowel syndrome (IBS) that is characterized by some harmful changes in the behavior of the intestines. Under suitable circumstances certain bacterial strains tooled with pathogenic potential may cause more severe infections like appendicitis, diverticulitis or even peritonitis. Furthermore, the imbalance of the intestinal microbiota increases the risk for bacterial intrusion through the mucosa of the gut into blood circulation. The risk of bacterial translocation through the gut wall is high in premature babies [Mshvildadze and Neu, 2010; Sherman, 2010] and in cancer patients during intensive chemo- or radiotherapy [Ubeda *et al.*, 2010].

When some of the above-mentioned symptoms take place, the changes of the intestinal microbiota could be measured as reflecting the health status of an individual. This approach could also offer new preventive options and guidance for the treatment of the patients with these problems. In comparison with the current protocols used for clinical microbiology monitoring in most hospital wards, a wider approach toward the samples of individual patients could be justified. This broader scope including a survey on the microbiological situation as a whole gives grounds for more reliable estimates on the development of the patient health status especially in chronic or long-lasting infections or general illnesses. The PMEU unit effectively promotes the picking up of the active strains from the samples. A graph regarding a comparative study between the PMEU method and the direct plate count is presented in the Chapter 2 of this book, and further reviews could be found in other articles [Hakalehto *et al.*, 2009; Pesola *et al.*, 2009; Pesola *et al.*, 2011]. The general views on hospital microbiology are also discussed elsewhere [Hakalehto, 2006; Hakalehto, 2012].

## Other Investigations That Can Be Performed from Stool Samples

The clinical use of stool samples is limited to few indications. However, it would be possible to perform a much wider range of investigations from stool samples. For example, a test kit called Comprehensive Digestive Stool Analysis 2.0™ (CDSA 2.0™; Genova Diagnostics®, Asheville, NC, USA) that includes evaluation of digestion (*e.g.* primary bile acids, β-glucuronidase and pancreatic elastase), absorption, intestinal microbes (bacteria, yeasts and parasites), and also detection of parasites from stool samples. The test can be used for evaluating different chronic GI problems, acute bowel pattern changes, and also many systemic diseases. A test panel for the analysis of the antibiotic resistance of detected pathogenic flora is also available. This product offers a view into the stool testing with both microbiological and physiological aspects being included in the repertoire.

The microbiological tests include aerobic and anaerobic bacterial culture, yeast culture, examination of parasites (*e.g. Entamoeba histolytica*) by using concentrate preparates and microscopy, and immunological diagnostics of *Cryptosporidium* and *Giardia lamblia*.

The analysis of pancreatic elastase can be used to distinguish gastric and pancreatic maldigestion. Decreased exocrine pancreatic function may be related to gallstones, osteoporosis, diabetes, and other autoimmunities. Exocrine pancreatic insufficiency may play a role also in post-prandial bloating, pain or nausea, loose or watery stools, undigested food in the stool, hypochlorhydria, food intolerances and gastroesophageal reflux syndrome.

CDSA 2.0™ can also be used for the measurement of metabolic byproducts of intestinal bacteria, like secondary bile acids, deoxycholic acid and lithocholic acid. The test also includes the examination of unabsorbed nutrients, like beneficial and putrefactive small chain fatty acids (SCFA's), fecal n-butyrate and pH. Abnormal levels of SCFA's may indicate alterations in the gut microbiota, insufficient dietary fiber, altered transit time and small bowel bacterial overgrowth (SIBO). Metabolic markers also help to identify imbalances that are associated with increased toxic burden within the colon that is related to a long-term risk of colon and breast cancers.

The intestinal immunology status is monitored in CDSA 2.0™ by analyzing the activation and degranulation of eosinophils (eosinophil protein X, EPX) and neutrophils (calprotectin). EPX is related to intestinal inflammation and tissue damage, and can be elevated in food allergies, celiac disease, helminthic infections and inflammatory bowel disease (IBD). Calprotectin is an inflammation specific marker that can be elevated during chronic diarrhea related to allergies, infection, post infectious irritable bowel syndrome (IBS), IBD, non-steroidal anti-inflammatory drug (NSAID) enteropathy, protein-sensitive enteropathy and cancer.

In a wider sense, studying the serum immunoglobulin levels of an individual against various intestinal or other microbes could produce useful information on the microbial challenges confronted by the patient. Such methods as the immunoblot analysis could give useful infosmation about the pathogenic strains, their epitopes, and markers in the infections or contaminations (Hakalehto and Finne, 1990).

The microbes isolated from the samples in the test of Genova Diagnostics® can be divided to different entities. Non-pathogens are regarded as normal participants of the intestinal microbiota, whereas pathogens are well-recognized microbes able to cause symptomatic diseases that are considered significant regardless of their quantity. When fungi are detected, they are normally participants of the colonic microbiota. However, when a disruption of the mucosal lining takes place, the fungi may become potential pathogens causing local or even systemic infection. For example, clinical yeasts can form hyphal structures into the tissues, and even within the host cells (Dr. Kaarlo Jaakkola, oral communication).

Antibiotic susceptibility monitoring of the representants of the intestinal microbiota gives information on the pool of antibiotic resistance that prevails in the intestinal microbiota. Especially in immunocompromised patients this information is of great importance, because their septic infections may originate from intestinal area [Ubeda *et al.*, 2010]. Then the panel of antibiotic susceptibilities can be used in the selection of a proper antibiotic treatment.

There may be multiple consequencies as a result of an imbalanced gastrointestinal health. The interpretation of the situation is often complex and multifactorial. The IBS may be a result of maldigestion, malabsorption, dysbiosis and/or inflammation. Maldigestion may cause several symptoms, like gas production, bloating, abdominal pain, diarrhea or constipation.

Chronic maldigestion may lead to bacterial and/or fungal overgrowth and alterations in gut permeability. Toxins of microbes or parts of the microbial cells may enter blood circulation and cause severe inflammatory diseases like reactive arthritis [Laugerette *et al.*, 2011].

Malabsorption may lead to deficiencies of nutrients. This can result in long term health complications such as anemia, impaired metabolism and other disesases.

Chronic dysbiosis can lower the levels of beneficial SCFA's and alter metabolic activity of the microbes. This increases the risk of carcinogenesis, hormonal imbalances and gastrointestinal inflammation. Dysbiosis makes the intestinal ecosystem also more vulnerable when pathogenic microbes enter the intestines. This may lead to diarrhea, vomiting, mucosal inflammation, increased intestinal permeability, toxin production and auto-immune disorders. The dysbiotic conditions are evaluated more precisely in the Chapter 7 of this book. From the diagnostic point of view, it is of crucial importance to recognize any dysbiosis, or overgrowth, as early as possible. Especially the intervention into the small bowel, gall bladder, or pancreas is rarely carried out for analytical purposes. This underlines the importance of developing better analysis tools for fecal monitoring. Such physiological functions as the collection of measurement data on the fatty acids could be combined with the results from the microbiological studies.

## Conclusion

On the basis on our clinical and research experience, we strongly recommend increasing the use of the fecal microbiological analysis for the follow-up of various patients in the hospital wards, as well as in long-term monitoring of health. Alimentary microbiome reflects the health status of the body. The indirect information from the fecal analysis could be coupled with results from standard blood tests, and also with stool surveys monitoring host enzymes, and levels of food degradation. Especially groups of chronic or severely ill patients, and those suffering from intestinal infections or disorders, could be monitored in intervals in order to get information on the development of their intestinal conditions and their microflora. It is possible to use the fecal analysis also for screening up the effects of probiotic treatments. In future this kind of holistic approach to the individual health problems is supposed to be more commonly used for health monitoring. Such devices as the PMEU could also be taken out of the laboratories and exploited as point-of-care detection and surveillance methods. The information from these studies could be followed up by the doctors in real-time by the mobile or internal hospital network connections.

## References

Borreani, G; Tabacco, E. Low permeability to oxygen of a new barrier film prevents butyric acid bacteria spore formation in farm corn silage. *J. Dairy Sci.,* 2008; 91, 4272-4281.

Giannella, RA. Pathogenesis of acute bacterial diarrheal disorders. *Annu. Rev. Med.,* 1981; 32, 341-357.

Hakalehto, E. Antibiotic resistant traits of facultative *Enterobacter cloacae* strain studied with the PMEU (Portable Microbe Enrichment Unit). In Press. In: Méndez-Vilas A, editor. *Science against microbial pathogens: communicating current research and technological advances. Microbiology book series Nr. 3.* Badajoz, Spain: Formatex Research Center; 2012.

Hakalehto, E. Semmelweis' present day follow-up: Updating bacterial sampling and enrichment in clinical hygiene. *Pathophysiology,* 2006; 13, 257-267.

Hakalehto, E. Simulation of enhanced growth and metabolism of intestinal *Escherichia coli* in the Portable Microbe Enrichment Unit (PMEU). In: Rogers MC, Peterson ND, editors. *E. coli infections: causes, treatment and prevention.* New York, USA: Nova Publishers; 2011.

Hakalehto, E; Finne, J. Identification by immunoblot analysis of major antigenic determinants of the anaerobic beer spoilage bacterium genus Pectinatus. *FEMS Microbiol. Let.,t* 1990; 67, 307-312

Hakalehto, E; Hänninen, O. Lactobacillic $CO_2$ signal initiate growth of butyric acid bacteria in mixed PMEU cultures. Manuscript in preparation. 2012.

Hakalehto, E; Heitto, A; Heitto, L; Pesola, J; Hänninen, O. Bile resistant duodenal strains comprise the backbone of gut flora and serve as hygienic indicators. Poster. *19th International Conference on Environmental Indicators. September 11-14, 2011.* Haifa, Israel; 2011.

Hakalehto, E; Pesola, J; Heitto, A; Bhanj Deo, B; Rissanen, K; Sankilampi, U; Humppi, T; Paakkanen, H. Fast detection of bacterial growth by using Portable Microbe Enrichment Unit (PMEU) and ChemPro100i((R)) gas sensor. *Pathophysiology,* 2009; 16, 57-62.

Kelly, P. Nutrition, intestinal defence and the microbiome. *Proc. Nutr. Soc.,* 2010; 69, 261-268.

Kühn, I; Ayling-Smith, B; Tullus, K; Burman, LG. The use of colonization rate and epidemic index as tools to illustrate the epidemiology of faecal *Enterobacteriaceae* strains in Swedish neonatal wards. *J. Hosp. Infect.,* 1993; 23, 287-297.

Laugerette, F; Vors, C; Peretti, N; Michalski, MC. Complex links between dietary lipids, endogenous endotoxins and metabolic inflammation. *Biochimie,* 2011; 93, 39-45.

Ley, RE; Turnbaugh, PJ; Klein, S; Gordon, JI. Microbial ecology: human gut microbes associated with obesity. *Nature,* 2006; 444, 1022-1023.

Mshvildadze, M; Neu, J. The infant intestinal microbiome: friend or foe? *Early Hum. Dev.,* 2010; 86 (Suppl 1), 67-71.

Pesola, J; Hakalehto, E. Enterobacterial microflora in infancy - a case study with enhanced enrichment. *Indian J. Pediatr.,* 2011; 78, 562-568.

Pesola, J; Heitto, A; Myöhänen, P; Laitiomäki, E; Sankilampi, U; Paakkanen, H; Riikonen, P; Hakalehto, E. Enhanced bacterial enrichment in the microbial diagnostics of pediatric neutropenic sepsis. (Poster abstract). In: Lähteenmäki P, Grönroos M, Salmi T, editors. *NOPHO 29th Annual Meeting.* Turku, Finland: NOPHO; 2011; 59.

Pesola, J; Vaarala, O; Heitto, A; Hakalehto, E. Use of portable enrichment unit in rapid characterization of infantile intestinal enterobacterial microbiota. *Microb. Ecol. Health Dis.,* 2009; 21, 203-210.

Sherman, MP. New concepts of microbial translocation in the neonatal intestine: mechanisms and prevention. *Clin. Perinatol.,* 2010; 37, 565-579.

Simojoki, S-T. Kantasolusiirron saaneiden lapsipotilaiden kolonisaatioviljelyt (in Finnish). Helsinki, Finland: University of Helsinki, Faculty of Medicine; 2011.

Thompson, AM; Bizzarro, MJ. Necrotizing enterocolitis in newborns: pathogenesis, prevention and management. *Drugs,* 2008; 68, 1227-1238.

Turnbaugh, PJ; Ley, RE; Mahowald, MA; Magrini, V; Mardis, ER; Gordon, JI. An obesity-associated gut microbiome with increased capacity for energy harvest. *Nature,* 2006; 444, 1027-1031.

Ubeda, C; Taur, Y; Jenq, RR; Equinda, MJ; Son, T; Samstein, M; Viale, A; Socci, ND; van den Brink, MR; Kamboj, M; Pamer, EG. Vancomycin-resistant *Enterococcus* domination of intestinal microbiota is enabled by antibiotic treatment in mice and precedes bloodstream invasion in humans. *J. Clin. Invest.,* 2010; 120, 4332-4341.

van Vliet, MJ; Tissing, WJ; Dun, CA; Meessen, NE; Kamps, WA; de Bont, ES; Harmsen, HJ. Chemotherapy treatment in pediatric patients with acute myeloid leukemia receiving antimicrobial prophylaxis leads to a relative increase of colonization with potentially pathogenic bacteria in the gut. *Clin. Infect. Dis.*, 2009; 49, 262-270.

Walker, HH. Carbon Dioxide as a Factor Affecting Lag in Bacterial Growth. *Science,* 1932; 76, 602-604.

Wu, GD; Chen, J; Hoffmann, C; Bittinger, K; Chen, YY; Keilbaugh, SA; Bewtra, M; Knights, D; Walters, WA; Knight, R; Sinha, R; Gilroy, E; Gupta, K; Baldassano, R; Nessel, L; Li, H; Bushman, FD; Lewis, JD. Linking long-term dietary patterns with gut microbial enterotypes. *Science,* 2011; 334, 105-108.

In: Alimentary Microbiome: A PMEU Approach
Editor: Elias Hakalehto
ISBN: 978-1-61942-692-4
© 2012 Nova Science Publishers, Inc.

*Chapter X*

# Monitoring *Clostridium difficile*

***Markus Hell***[1] ***and Elias Hakalehto***[2]
[1]Department of Hospital Epidemiology and Infection Control,
University Hospital Salzburg, Salzburg, Austria
[2]Department of Biosciences, University of Eastern Finland,
FI, Finland

## Abstract

The excessive use of antibiotics for safeguarding our health against infections or communicable microbial diseases has, quite paradoxically, lead to the emerging of new groups of harmful microbes such as antibiotic resistant strains, yeasts, or *Clostridium difficile*. The removal of normal microflora has given room for these tedious microbes, which conquer space from the balanced symbiotic microbiota. Moreover, they could intrude the body system with their toxins causing life-threatening illnesses. One reason for their strong abilities to produce diseases could be deriving from their origin. Members of *Clostridium* sp. are typically soil bacteria. Another reason of course is their capability of sporulation. Their spores can persist in the environment for decades and they resist any antibiotics in the human gut and disinfectants in the environment. Therefore preventive - and new therapeutic strategies have to be developed and evaluated to minimize frequency and severity of *Clostridium difficile* Infection (CDI).

## Introduction

As discussed elsewhere in this book in Chapter 6, the antibiosis is a common phenomenon in soil where the microbes have to compete against each other for the relatively scarce nutrients. On the contrary, in the human body, or in the alimentary tract, most strains strive for cooperation. Any monocultures would be detrimental for the host as living organism, and thus counterproductive for the common members of the normal flora.

We have been using the PMEU technologies for mapping the microflora in different samples, and for simulating their joint influences and molecular communication [1a,1b]. An example related to the bacterial interactions is the growth promoting signal effect of lactobacilli on the clostridial strains in the PMEU which is mediated by the $CO_2$ in the intestinal environment [2]. This effect is provoking the onset of growth of the anaerobic culture of the *Clostridium* sp. In the intestinal environment, this activity of the mixed microflora could thus keep these strains in the active state which, in turn, could prevent them from colonizing the gut walls, or producing endospores.

In case of the CDI (*C. difficile* infections) the causative agents (strains) are mostly monophasic [3]. This is an indication of the flourishing of one pathogenic strain in the background of the illness. This behavior of the clostridial strains is a consequence of destroyed BIB (Bacterial Intestinal Balance). The routes of the infection need to be further studied on in particular cases of CDI. It needs to be investigated also the proportion of healthy carriers among the population and CDI exposed healthy health care workers. In both groups carrying CD seems to be very low at the moment [4] but further investigation at a larger scale is needed.

This emerging pathogen has caused severe threats for the healthcare system. It is typically belonging to the hospital acquired infections, whether originating from the wards or from the patient self. Therefore, this new widely spread pathogen should be mapped amongst the population [5]. Since 2005 The European institutions have made similar experience like the north American ones with several hospital outbreaks related to new, probably selected hypervirulent strains such as the NAP 1/ PCR Ribotype 027 related with higher morbidity and mortality among hospitalized patients [6]. But more over CDI has become also a community problem nowadays in some cases related to the food chain esp. pork and ground meat [7]. So epidemiology has taken a dynamic global development and made this new strains successful pathogens.

## *Clostridium* Infections

There are several species of the genus *Clostridium* causing diseases in man. These bacteria are obligate anaerobic spore-formers. The endospores facilitate their survival in soil, from where they mostly spread into humans.

Infections by clostridia always mean infection via toxins and therefore one could also talk about "intoxication" related to these organisms. The classical clostridial infections such as gas gangrene by *C. perfringens*, botulism by *C. botulinum* or tetanus by *C. tetani* are very dangerous and still associated with a high fatality in clinical outcome. However, they are very rare at least in the developed countries where vaccination coverage is good (*e.g. C. tetani*).

All these clostridia come from soil, and are very dangerous. Thus the microbial ecosystems where they normally operate are totally differing from the alimentary microbiome (see also Chapter 6). They should not belong to the normal microbiota of man, at least not to the permanent/resident flora. Beside the most common species *Clostridium difficile* within the genus *Clostridium* there are also emerging species in health care setting. *Clostridium sordellii* may cause myonecrosis and severe shock syndrome among previously healthy persons which has been described recently in a small number of *C. sordellii* cases, but most often it is associated with gynecologic infections in women and infection of the umbilical stump in newborns [8].

## *Clostridium difficile*: Epidemiology

The most common type of infection is *C. difficile* infection (CDI), which emerged and re-emerged the last ten years globally. Some experts are even talking about a *C. difficile* pandemic.

This phenomenon is not only due to antibiotic-overuse during the last decades in the health care setting but also food related, esp. linked to meat (ground meat). So CDI has also become a community associated infection and not only a hospital-acquired infection any longer. But these cases are still rare compared to the antibiotic-related cases at least in Europe. It is estimated that in North America there are 25, 000 deaths each year linked to *Clostridium difficile* infections (CDI). The economic burden of CDI is estimated to be $10B per year with the *C. difficile* control sector market being in the order of $7B and growing. Traditionally *C. difficile* was considered to be exclusively linked to hospital-acquired infections primarily affecting the elderly or those with weakened immune system treated by antibiotics. Here, the antibiotics disrupt the gastrointestinal tract microflora thereby enabling the drug resistant *C. difficile* to proliferate and produce toxins leading to potentially fatal diarrhea. Over the last 3 years there has been an increasing occurrence of Community Acquired *C. difficile* infections and are thought to account for over 40% of all CDI reported. The case definition of Community Acquired CDI is patients who have no recent contact with health care facilities or taking antibiotics. There are many aspects of Community Acquired CDI that remain unknown especially with relation to sources of the pathogen. There is evidence for foodborne, environmental and zoonotic transfer of *C. difficile* [9]. In the hospital setting CDI is occurring only when the patient's normal flora and its balance has been destroyed by excessive antibiotic treatments. Patients who have other illnesses or conditions requiring prolonged use of antibiotics, and the elderly, are at greater risk of acquiring this disease. The bacteria are found in the feces. People can become infected if they touch items or surfaces that are contaminated with feces and then touch their mouth or mucous membranes. Healthcare workers can spread the bacteria to patients or contaminate surfaces through hand contact.

Clinical symptoms of CDI range from a single episode of diarrhea, to watery and bloody diarrhea, to pseudomembranous colitis with systemic signs of severe infection. Also complications have been seen from a perforation of the colon to a toxic megacolon, which usually needs surgical intervention (*i.e.* subtotal hemicolectomy).

*C. difficile* infections and the distribution of strains and types are described in recent pan-European study [10]. In this prevalence study it was demonstrated that there is not one prevalent strain all over Europe but nation by nation specific clusters of strains exist. The strains were classified by PCR-ribotyping, one of the most popular typing method in Europe [11]. It still remains unclear, what is the real potential of virulence of the so called hypervirulent strains like PCR-ribotype 027 or 078 or 001. It may be somewhat overestimated.

## Transmission of *C. difficile* in Health Care Facilities

*Clostridium difficile* is shed in feces. Any surface, device, or material (*e.g.* toilets, bathing tubs and electronic rectal thermometers) that becomes contaminated with feces may serve as a reservoir for the *C. difficile* spores. *C. difficile* spores are transferred to patients mainly via the hands of healthcare personnel who have touched a contaminated surface or item. On surfaces *C. difficile* can survive for long periods (upto 6 months).

## Prevention of *C. difficile*

In order to properly understand the nature of the CDI, the hospital hygiene personnel should be able to get first hand information on the problem. Research facilities should be provided for every larger hospital unit in order to facilitate early warning and effective countermeasures in time.

People who are hospitalized or on antibiotics are most likely to become ill. For safety precautions you may do the following to reduce the chance of acquiring *C. difficile* or spreading to others:

- Wash hands with soap and water, especially after using the restroom and before eating.
- Clean surfaces in bathrooms, kitchens, and other areas on a regular basis with household detergent/disinfectants.
- By thoroughly cleaning of hands and cleaning of surfaces adjacent to the patient with spore-active or non-active detergents/disinfectants transmission of spores can be prevented. The most important concept is the mechanical removal of the bacterial spores.

Transmission of *C. difficile* within hospitals has been observed through time-space clustering of new cases with identical strains and a greater risk of acquisition of *C. difficile* from exposure to roommates or other patients in close proximity who have positive cultures *C. difficile* spores have been found to contaminate the hands of healthcare workers and the hospital environment frequently.

Because alcohol-based hand sanitizers do not inactivate the spores of *C. difficile*, concern over their role in transmission of *C. difficile* have been raised. However, no increase of CDI incidence has been seen in hospitals using alcohol-based hand rubs as their primary means of hand hygiene after their introduction. Due to the theoretical advantage of hand washing over alcohol-based hand sanitizers, hand washing with a non-antimicrobial soap or antimicrobial soap and water should be considered after removing gloves in the setting of a CDI outbreak or if ongoing transmission cannot be controlled by other measures.

Patients with CDI should be placed on contact precautions and housed in single rooms with private bathrooms or, if unavailable, cohorted in rooms with other patients with CDI. Single-use disposables and patient-dedicated noncritical equipment should be used. Wearing gloves is one measure that has been proven to reduce the spread of *C. difficile* in hospitals.

Gowns and gloves should be donned prior to entering the room of a patient with CDI and removed followed by hand hygiene before leaving the room.

Although all hospital cleaning agents can inhibit the growth of *C. difficile* in culture, only chlorine-containing agents inactivate *C. difficile* spores. In the most definitive study evaluating environmental cleaning, the use of a 1:10 dilution of a 6% hypochlorite solution for daily room cleaning of CDI patients in a bone marrow transplant unit decreased the CDI rate significantly but had no effect on units with lower baseline CDI rates. Therefore, the use of hypochlorite might be most effective in units where CDI is highly endemic. The drawbacks of hypochlorite solutions are that most of them must be prepared fresh daily and they can be caustic and damaging to hospital equipment.

Antimicrobial use restrictions are another potential mechanism of controlling and preventing *C. difficile*. As with environmental cleaning, the exact role of antimicrobial restrictions is undefined due to the presence of confounding factors in most studies. However, several studies support the use of formulary restrictions promoting the use of narrow-spectrum antibiotics to reduce the incidence of CDI. Formulary substitutions of 8-methoxyfluoroquinolones for levofloxacin have also been proposed to control CDI outbreaks caused by the BI/NAP1/027 strain. While this appeared to be effective in one study it was ineffective in another, most likely because the overall use of fluoroquinolones in the hospital was not controlled.

Since resistance of the BI/NAP1/027 strain to fluoroquinolones is a class effect resulting in higher minimum inhibitory concentrations (MICs) to all fluoroquinolones the incidence of disease caused by such resistant strains is not likely to be reduced without controlling fluoroquinolone use in general.

In conclusion the increasing incidence and severity of CDI in North America and Europe present major challenges for control and management of this disease. Continued gathering of data on the epidemiology of *C. difficile* through disease surveillance both within and outside of healthcare facilities, and on the efficacy of prevention and treatment strategies is essential to reduce the burden of this disease. Meanwhile, all clinicians and especially critical care physicians should maintain awareness of the changing epidemiology of CDI and undertake measures to reduce the risk of the disease in their patients [12].

Therefore specific guidelines/position papers have been developed like the following European guideline [13]:

## Infection Control Measures to Limit the Spread of *Clostridium difficile*

The “TEN commandments” of Good practice in Infection Control for CDI:

1. Early diagnosis
2. Surveillance: get an idea about what is going in your institution?
3. Training of health care personnel: How to convince your colleagues to adhere to the rules?
4. Isolation precautions: How to implement isolation techniques under limited patient room resources?
5. Hand hygiene: alcoholic handrubs do not work against *C. difficile* spores!
6. Protective clothing: How much is safe enough?

7. Environmental cleaning: cleaning with mechanical removal of spores or sophisticated sporicidal surface disinfection?
8. Medical devices: bedpans and flexible endoscopes and their reprocessing a risk for transmission of *C. difficile*?
9. Antibiotics: restriction of AB = the single most efficient measure to reduce CDI!
10. Specific measures in outbreaks: do as much as you can.

### *1. Early Diagnosis*

The main purpose of screening cultures is to identify carriers of pathogens at the most early stage before inter-patient-spread can occur. The prevalence of CD in stool cultures of asymptomatic and otherwise healthy adults is less than 5%. In contrast, the rate of CD carriers among hospitalized patients varies significantly and may be as high as 25%. More than half of the CD strains from symptom-free individuals are toxinogenic.

Although screening cultures of CD have been performed in some nosocomial epidemics, there is no data showing that active screening of non-diarrheal patients in order to identify CD carriers would contribute to the reduction of the endemic baseline rate of CDI. Instead, one should focus on diarrheal patients as they represent the major reservoir for CD transmission. As recurrence of CDI after a symptom-free interval is not uncommon (up to 25-30% of cases), diagnostic testing for CD should also be performed at a new onset of diarrhea.

Besides reducing the risk of pathogen spread, the second rationale of screening cultures is to identify carriers who are at risk to develop endogenous nosocomial CDI. However, in a prospective observational study on long-stay patients Johnson *et al.* could show that symptom-free excretors of CD had even a slightly decreased risk of subsequent CDI (0 of 51 patients) compared to 7 of 229 (3.1%) cases of CDI in patients who had been initially culture negative. Shim *et al.* also observed that there were 22 cases of CDI in 618 previously non-colonized patients (3.6%) compared with 2 of 192 (1.0%) symptom-free carriers (p=.021). Additionally, CD eradication treatment of asymptomatic carriers has been shown ineffective. Hence, symptom-free CD colonization may actually be protective for subsequent symptomatic disease, but it is possible that asymptomatic carriers may still contribute to transmission of the organism. Screening of the environment is not generally recommended, but can be used to document environmental contamination or poor cleaning and disinfection.

### *2. Surveillance*

Generally active surveillance of CDI has been recommended. Surveillance is useful to detect an increase in the CDI incidence, or its severity, at an early stage, or to identify risk factors for CDI acquisition, and should include the identification of deaths to which CDI is either the primary cause or contributory. The significance of surveillance is not limited to outbreaks. In the endemic setting it may reveal high baseline rates, or significant variations between locations that require interventions. At least stool testing for CD toxins should be performed in the case of nosocomial diarrhea that cannot be unexplained otherwise, and for all patients that have been admitted to the hospital for severe non-nosocomial diarrhea. Microbiological laboratories should consider testing for CD systematically in all stool specimens from patients hospitalized for more than 3 days ("3-day testing rule"). If possible, culture and typing of CD toxin positive samples should be performed. In practice this is best achieved by storing (aliquots of) toxin positive fecal samples, so that these can be examined retrospectively to aid CD cluster/outbreak management, if necessary by a reference laboratory.

A threshold incidence or prevalence of CDI should be defined locally that would trigger implementation of additional control interventions. The alert level should be based on the incidence, severity of disease, institutional priorities, whether the patient population has conditions that would facilitate transmission or is at increased risk of adverse outcomes following CDI acquisition, and whether there is suspected or proven transmission within the facility.

Besides the surveillance of the number of CDI cases by infection control staff in a hospital notification system, surveillance of the antibiotic use in the hospital by the hospital's pharmacist in close cooperation with the medical microbiologist is also recommended.

### *3. Training of Health Care Personnel*

Education of staff has been shown to be one of the most effective measures to limit CD spread. This should include information about the basic pathogenetic mechanism of CD, its potential reservoirs, the route of transmission, contamination of the environment, and possible substance for decontamination of hands and surfaces. Training of staff should include not only the medical personnel (*e.g.* nurses or physicians), but also non-medical personnel – especially the staff involved in cleaning.

Education of visitors of patients with CDI is also necessary to prevent contamination of the environment by CD spores. Visitors should be encouraged to obey basic infection control measures with emphasis on appropriate hand hygiene. Visitors should not meet other patients after they have visited a patient with CDI. Individuals suffering from acute diarrhea themselves should generally be discouraged to visit patients in a hospital.

### *4. Isolation Precautions*

Isolation of patients colonized or infected with infectious agents in single rooms or in cohorts is one of the basic hygiene measures to limit pathogen spread. With few exceptions isolation of symptomatic patients with CDI is a key measure to control nosocomial CD outbreaks. Occasionally even closure of a complete ward or department is necessary to control an outbreak. Zafar *et al.* reported a 60% decrease from a mean of 155 cases per year to 67 per year by a combination of isolation enforcement, CDI surveillance, education of staff, phenolic disinfection of the environment, hand washing with soap containing 0.03% triclosan, and centralized re-processing of devices. Struelens *et al.* observed a CDI incidence decline from 1.5 to 0.3 per 1,000 admissions (73% reduction) by early isolation precautions, active initial CDI surveillance, sufficient environmental surface disinfection, and early therapy of patients with CDI. Additionally, re-isolation of patients presenting with diarrhea at subsequent re-admission who were previously known to suffer from CDI may reduce attack rate of new nosocomial CDI cases, reducing the overall healthcare cost.

If daily clinical practice does not allow isolating symptomatic patients in single rooms on a regular ward, cohorting several patients or even on a separate cohort/isolation ward may be implemented. On CDI cohort wards the staff may be better experienced in caring for such cases, cleaning protocols for CDI are easier to put into operation in separate areas, materials used on a cohort ward are usually not used elsewhere, and there are less people entering a cohort ward unnecessarily. Another positive effect of cohorting is to localize environmental contamination to a small part of the hospital environment. This is different from having isolation rooms used for CDI dispersed throughout a hospital in numerous locations. Each failure of hygiene in these locations represents a high risk of extended local contamination and secondary cases. The overall effect may be a significant reduction of the burden of environmental contamination to a single focus (the cohort ward), where it is recognized and containable.

Apart from isolation procedures it is essential that patients suffering from any form of diarrhea have a dedicated toilet or commode (*i.e.* should not be allowed to use general toilet facilities). Diarrheal stool samples should be processed as soon as possible to diagnose CDI. For practical purposes, patients with CDI can be sub-divided into risk groups, *e.g.* whether they are requiring treatment for CDI. This differentiation can be applied on both “regular wards” and on “CDI cohort wards”.

One critical issue is how long isolation and other control measures need to be continued. Few data are available on the excretion of vegetative cells and spores during CDI. The consensus is that nosocomial outbreaks can be terminated if precautions are kept in place until bowel motion has returned to a normal level for at least 48 hours. The environment of symptomatic patients with CDI is more frequently contaminated than that of asymptomatic CD carriers (49% vs. 29%; $p<.05$). However, even after adequate therapy of CDI and return to normal bowel movements, patients may still show detectable CD toxins in feces and continue to excrete CD. It is possible that 3% to 30% of stool samples remain toxin positive in patients treated with vancomycin or metronidazole and a correlation between stool cytotoxin levels and severity of gastrointestinal symptoms may not be present in all cases. There is, however, no clinical value in retesting CDI cases once symptoms resolve, and therefore the carriage status of patients will usually be unknown. Thus, the exact role of asymptomatic carriers for the spread of CD is currently unknown. Basic hygiene measures must therefore be an integral part of normal practice. Furthermore, for most other pathogens alcohol-based hand hygiene is recommended, unless a major contamination of hand has occurred, and thus guidance must be provided locally when this method should be recommenced (see below).

## 5. Hand Hygiene

Standard hand hygiene practice is based on alcohol-based products, unless hands are in contact with body fluids or are visually contaminated. When hands are clearly contaminated, decontamination by soap-based washing has to be performed (possibly prior to hand disinfection). It is clear that the hands of health care workers (HCWs) are likely to become contaminated when caring for patients with positive CD cultures. Unfortunately, bacterial spores are not killed by alcohol, and indeed alcohol is used in the laboratory setting to select for CD spores. In an experimental study using hands of volunteers contaminated by CD, Barbut *et al.* showed that 4% polyvidone soap was significantly more effective in reducing CD count than chlorhexidine or non-medicated soap, and these products were also more effective than alcohol-based products. In a recent observational study alcohol-based hand rub did not decrease the incidence of nosocomial CDI (mean 3-year incidence per 10,000 patient days before implementation of alcohol-based had rub: 3.24; mean 3-year incidence after: 3.38; $p=.78$), but one has to keep in mind that multiple factors are important for the successful prevention of CD transmission. In addition, Boyce *et al.* showed that a 10-fold increase in the use of alcohol-based had rub (3 to 30 L per 1,000 patient days; $p<.001$) within 4 years did not alter the incidence of CDI (1.74, 2.33, 1.14, and 1.18 per 1,000 patient days) in the corresponding time period[40]. Bacterial spores can be removed from hands by the physical action of washing and rinsing hands, using either non-antimicrobial liquid soap or antiseptic substances such as chlorhexidine gluconate. In a cross-over study no differences in residual counts of CD on bare hands of health care workers were observed comparing these the two agents. Those who used soap containing chlorhexidine gluconate found a significantly improved removal of spores in 6/7 (86%) in comparison with 2/16 (13%) using non-disinfectant soap ($p<.01$). Leischner *et al.* recently found that alcohol gels were significantly less effective at removing CD spores from the hands of volunteers when compared with hand washing with chlorhexidine ($p<.009$). However, there was a higher than expected reduction of spore counts following use of alcohol gels.

The role of the patient remains uncertain in the transmission of CD, although direct person-to-person transmission has been proposed. Endogenous infections can also occur in principle, even though there are some data suggesting that CD carriage is relatively protective for CDI. Thus hand washing by patients should be encouraged as usual especially after a toilet visit and before eating.

## *6. Protective Clothing*

The use of gloves to protect the hands of health care workers from contamination is generally recommended by the Healthcare Infection Control Practices Advisory Committee (HICPAC) as a part of infection control standard precautions. In a prospective controlled trial of vinyl-glove-use to prevent CD spread, the incidence of CDI decreased significantly from 7.7 to 1.5 per 1,000 patient discharges within a six month intervention period. In another observational study, hands of all 4 members of staff using gloves remained free of CD spores. This contrasts with a hand contamination rate of 7/15 (47%) for staff that did not use gloves and did not perform further hand hygiene measures (although the overall number of screened individuals was too low to reach statistical significance in this study; $p=.13$). Of course, contaminated gloves need to be removed prior touching non-contaminated surfaces. Contamination of hands may occur during removal of contaminated gloves. Therefore, hand washing and drying remains important regardless of previous glove use.

The use of gowns or aprons represents an additional measure of infection control standard precautions to prevent contamination of the regular working cloths by infectious agents, and should therefore be used when caring for known CDI cases. Only few data exist on the use of gowns to specifically prevent inter-patient spread of CD. Perry *et al.* compared the level of contamination of nurses′ uniforms before and after duty in 57 different staff on different ward areas and working shifts. Although the authors recommend wearing appropriate plastic aprons to reduce the levels of contamination on a uniform in principle, it remains unclear to what extend this was performed during the study time period. Because a professional laundry service was unavailable in that hospital, uniforms were laundered by staff at home. Unfortunately, there is no further information on storage conditions of cleaned uniforms or on possible ironing of uniforms when dry. CD was cultured from uniforms of 4 nurses after work that had been CD negative before duty. However, clothes of 7 additional nurses were CD positive before duty already. If such difficulties arise, one may also consider using disposable clothing instead of using an external professional hospital-laundry service. However, it should be emphasised that recovery of CD from uniforms could simply represent either direct or indirect contamination (*e.g.* for the environment) and does not necessarily implicate such fomites in transmission. Optimal hand hygiene and barrier precautions will counteract any risk that may be associated with CD contamination of uniforms.

### 7. Environmental Cleaning

It is well-documented that environmental contamination occurs during CDI especially if patients suffer from large amounts of liquid stool or stool incontinence. Remarkably heavy contamination takes place on floors, commodes, toilets, bed pans, and bed frames. The actual degree of spore recovery from environmental swabs may directly correlate with the incidence of CDI, although a recent molecular epidemiological study was unable to determine whether environmental contamination is the consequence of CDI rather than the source of infection, primarily because of the often clonal nature of nosocomial CDI. Once released in the environment, CD spores may persist for long periods (months or years) due to their resistance to drying, heat, and disinfection substances.

There is good evidence that environmental contamination can play a role in CD transmission. Wet cleaning using detergents only may be insufficient for decontamination purpose, and there is a need for effective and user-friendly sporocidal products. Various disinfection substances are available to inactivate CD spores but the choice of substance may influence formation of spores *e.g.* non-chlorine based products may enhance sporulation. Sporulation levels of outbreak strains such as type 027 or type 001 may also exceed those of other CD strains. In the current CDC (Centers for Disease Control and Prevention) / HICPAC guidelines no specific disinfection agent for standard environmental control of CD is recommended (*i.e.* in the absence of known CDI cases).

Hypochlorite-based disinfectants are recommended by the HICPAC in patient-care areas where CDI surveillance indicates ongoing CD transmission and are frequently used in many hospitals. Of potential importance, chlorine-based products are significantly less likely to enhance sporulation of CD strains *in vitro*. Compared to cleaning with a detergent only, hypochlorite use at a concentration of 1,000 parts per million (ppm) was associated with a significant reduction in the incidence of CDI. The use of phosphate-buffered hypochlorite (1,600 ppm available chlorine) may be more effective against CD spores than unbuffered hypochlorite solution (500 ppm chlorine). A possible disadvantage that needs to be considered in the choice of disinfection substance is the corrosive nature of hypochlorite on metal surfaces. Also, hypochlorite alone products are not suitable for removing organic matter. Products containing a combination of hypochlorite and a detergent may overcome this problem.

Quaternary ammonium (QA) solutions have also been used for environmental CD decontamination. However, while no differences in the CDI incidence were observed in patient care areas with low CDI incidences (neurosurgical intensive care unit: 3.0 per 1,000 patient days and general medicine ward: 1.3, respectively), the change from QA to unbuffered 1:10 hypochlorite led to a significant reduction in the CDI incidence (8.6 to 3.3 per 1,000 patient days) in bone marrow transplant patients; the incidence of CDI increased to 8.1 per 1,000 patient days after reverting back to QA cleaning.

Hydrogen peroxide vapour recently showed efficacy for environmental CD eradication. However, this method is expensive and involves having to vacate and seal clinical areas.

Glutaraldehyde is a disinfection substance known to be effective in inactivation of CD spores and has been used for that purpose in nosocomial CD outbreaks. However, due to its risks to human health and for environmental safety grounds it should not be used freely for environmental decontamination. Peracetic acid 0.2% (Perasafe®) is more active *in vitro* than chlorine-releasing agents such as sodium dichloroisocyanurate (NaDCC) at 1,000 ppm available chlorine. This high-level disinfection substance may also be a substitute for glutaraldehyde and has also shown better effectiveness than tertiary amines although long contact times of 15 to 20 minutes are required. Peracetic acid has not been used for environmental decontamination. When performing environmental decontamination of CD spores, one has also to keep in mind that the maximum permissible concentration of certain disinfection substances (*e.g.* chlorine) for environmental disinfection may differ depending upon diverse national regulations. Each organization responsible for cleaning of the hospital should have specific protocols for the treatment of rooms of CDI patients. At least once a day all objects with frequent contact to hands of patients and staff, such as tables, chairs, or telephones, should become disinfected. In clinical practice it has been shown to be essential to educate the cleaning personnel on regular basis; especially emphasizing the difference in cleaning and disinfection of areas used by CDI patients, in contrast to those for cases colonised/infected by MRSA or other multi drug resistant pathogens.

### *8. Medical Devices*

Use of disposable equipment whenever possible to limit spread of pathogens that lead to heavy environmental contamination such as CD has proven effective to control CDI outbreaks. Due to the intestinal habitat of CD, rectal thermometers seem likely to play an important role in bacterial transmission. Although electronic thermometers do not necessarily need to become CD contaminated, there are numerous investigations in which positive CD cultures were obtained from these devices. For example, Samore *et al.* showed that 3 of 38 (7.9%) thermometers became contaminated by CD during use. Shared rectal thermometers should therefore be replaced by individual thermometers, or alternatively consider a change to tympanic thermometers by which a 40% risk reduction for CDI has been achieved. An even greater risk reduction (56%; p=.26) was accomplished by the use of disposable thermometers instead of shared electronic ones in a randomized crossover study.

There have been no reports of endoscopes transmitting CD in the hospital setting. However, Hughes *et al.* found that 10 of 15 (67%) of endoscopes were CD culture positive immediately after use. Since single use is not an option for such expensive equipment, endoscopes need to be re-processed adequately before further use. Disinfection of endoscopes by alkaline glutaraldehyde solution 2% or by peracetic acid was capable of CD spores inactivation at an exposure time of 5 to 10 minutes after thorough cleaning. Therefore endoscopes have to be properly cleaned with adequate sporocidal disinfectants or by automated chemo-thermal reprocessing, as it should be the standard in most hospitals nowadays.

Additional devices that have been found CD positive may comprise blood pressure cuffs or oximeters. Although tested negative in one investigation, equipment for enemas may also count as a critical medical device in this context. Generally, instruments and equipment including stethoscopes and blood pressure cuffs should be used patient specific and cleaned carefully.

### *9. Antibiotics*

Antibiotic therapy or prophylaxis alters the colonic flora and allows *C. difficile* to flourish and cause diarrhea. Antibiotics are the most important predisposing factor for CDI. However, a systematic review of the studies that have examined the risk of CDI associated with different antibiotics concluded that most are flawed because of failure to control for potential confounding factors. Exposure to CD, antimicrobial polypharmacy and duration of antibiotics are frequently not addressed as causes of bias. Also, host humoral immunity to CD toxin(s) and the antibiotic susceptibility of the CD strain are likely to influence the risk of CDI development. It is not surprising therefore that there are conflicting studies in respect of the risk of CDI associated with specific antibiotics or classes.

Almost any antibiotic may caused CDI, but broad spectrum cephalosporines (in particular third spectrum cephalosporins), broad spectrum penicillins and clindamycin are most frequently implicated. Since 2000, fluoroquinolones have also been identified as possible risk factors for CDI, including CDI caused by the new hypervirulent fluoroquinolones resistant PCR ribotype 027. Conversely, ureidopenicillins (with or without beta lactamase inhibitors) appear to have low propensity to induce CDI. In many hospital patients receive combinations of antibiotics or multiple antibiotic courses. A Cochrane analysis indicated that antibiotic use is unnecessary or inappropriate in as many as 50% of cases and that interventions to improve antibiotic prescribing can reduce hospital acquired infections, and most notably CDI. The exact duration of risk to develop CDI after antibiotic exposure still needs to be determined, but it is has been demonstrated that the duration of therapy also influences this risk. Aggressive antibiotic restriction of high risk antibiotics, reducing polypharmacy, prevention of long duration of therapy and avoiding inappropriate prescribing are the first steps to decrease a high incidence rate of CDI. Some measures to achieve this goal comprise automatic stop dates, electronic prescribing, banning of certain antibiotic classes, prescriber education and production of guidelines or policies. There is also a need for continuous training of medical staff on appropriate antimicrobial use and feedback of success, since systematic reviews have been shown to have positive effect of audit and feedback in helping healthcare workers implement evidence based practice. In a controlled interrupted prospective trial on a geriatric ward Fowler *et al.* showed that feedback on improved antibiotic prescribing can be successful in reducing the use of broad spectrum antibiotics (amoxicillin/clavulanic acid and cephalosporins) in favour of more narrow spectrum antibiotics (benzyl penicillin, trimethoprime and amoxicillin). In consequence, this changed antimicrobial treatment regimen led to a significant decrease in CDI (p=.009). A definition of so-called “alert antimicrobials”, drugs that need a patient-specific feedback by an authorized person (*e.g.* the hospital’s pharmacist), may also be useful to further improve antibiotic use in the hospital.

Use of antibiotics is believed to lead to disturbance of the normal gastrointestinal flora. In consequence, this status may predispose for diarrhea after bacterial selection of CD. Probiotics (bacteria and yeasts) are thought to restore the balance of the gut flora when delivered orally to the patients. At present, only few data exists on the interference of probiotics and CDI. There are some randomized controlled trials that have shown a significant reduction in the risk of antibiotic-associated diarrhea by the use of *Saccharomyces boulardii*, but newer studies using *S. boulardii* or *Lactobacillus* GG failed to confirm these findings. Two recently published systematic reviews did not show sufficient evidence for the general recommendation of the use of probiotics for CDI prevention or treatment.

### *10. Specific Measures in Outbreaks*

The key to reducing infection risk is prevention of transmission of CD. When an increased number of CD cases are identified, infection control strategy should be introduced by risk assessment that takes into account the background epidemiological pattern and the risk category of the patients involved. Outbreak situations require immediate action to control the situation. Usually, this is associated with a combination of different infection control measures. Hence, the effectiveness of each individual measure is hard to determine. First of all, the adherence of all recommendations that apply in the endemic setting needs to be strengthened. These measures include strict separation of symptomatic patients, education of staff, increased awareness of CDI, and restricting the use of risk-antibiotics such as broad spectrum penicillins, third generation of cephalosporines, clindamycin and fluoroquinolones. In addition, there are specific measures that may be helpful in a nosocomial CD epidemic. Cohort nursing of confirmed CDI patients, and isolation of suspected CDI cases before laboratory test results are available have been shown to be effective during a CD outbreak. In a recently published nosocomial epidemic of CD on a geriatric ward Cherifi *et al.* reported that cohorting of infected patients on one ward with a single medical team was a key way of limiting the spread of infection. The potential to establish a dedicated nursing team for CDI patients need to be considered alongside competing pressures locally. Similarly, the measures used in outbreaks may depend on the wards involved and on the clinical severity of cases. However, a written local protocol must exist so that early adoption and/or review of control measures occur when CDI cases are identified.

## Treatment of *C. difficile* Infection: Current and Future Strategies

### *Recommendations for the Treatment of CDI [14a]*

In the case of mild CDI (stool frequency < 4 times daily; no signs of severe CDI), clearly induced by the use of antibiotics, it is acceptable to stop the inducing antibiotic and observe the clinical response, but patients must be followed very closely for any signs of clinical deterioration and placed on therapy immediately if this occurs. Antiperistaltic and opiate agents should be avoided, especially in the acute setting. There is no evidence that switching to 'low-risk' antibiotics, when the antibiotic treatment that triggered the episode of CDI cannot be stopped or stopping gastric acid suppressants, is effective. It seems rational, however, to always strive to use antibiotics covering a spectrum no broader than necessary. When the inciting antibiotic cannot be stopped, antibiotic treatment for CDI cannot be avoided.

In all other cases than mild CDI medical treatment for CDI should be started. Antibiotics may be started awaiting diagnostics when there is sufficient clinical suspicion. We recommend treatment of an initial episode of CDI with the following antibiotics, according to disease severity (implementation category between brackets), when oral therapy is possible:

- non-severe: metronidazole 500 mg tid orally for 10 days (A-I)
- severe: vancomycin 125 mg qid orally for 10 days (A-I)

CDI is judged to be severe when at least one of the markers of severe disease mentioned under 'definitions' is present .

When oral therapy is impossible, we recommend the following antibiotics, according to disease severity (implementation category between brackets):

- non-severe: metronidazole 500 mg tid intravenously for 10 days (A-III)
- severe: metronidazole 500 mg tid intravenously for 10 days + (A-III)
- retention enema: vancomycin 500 mg in 100 ml of normal saline every 4–12 h
  or
  vancomycin 500 mg qid by nasogastric tube (C-III)

Oral vancomycin may be replaced by teicoplanin 100 mg bid, if available.

There is no evidence that various genotypes of *C. difficile* should be treated differently if disease severity does not differ.

### *Recommendations for Surgical Treatment of CDI*

Colectomy must be performed to treat CDI in any of the following situations:

- perforation of the colon
- systemic inflammation and deteriorating clinical condition not responding to antibiotic therapy; this includes the clinical diagnoses of toxic megacolon and severe ileus

Since mortality from colectomy in patients with advanced disease is high, it is recommendable to operate in less severe stage. No strict recommendations as to the timing of colectomy can be given. Serum lactate may, inter alia, serve as a marker for severity, where one should attempt to operate before the threshold of 5.0 mmol/l.

### *Recommendations for Medical Treatment of Recurrent CDI*

Observational data [14b] suggests that the incidence of a second recurrence after treatment of a first recurrence with oral metronidazole or vancomycin is similar. Therefore, we recommend treating a first recurrence of CDI as a first episode, unless disease has progressed from non-severe to severe.

We recommend treatment of recurrent CDI with the following antibiotics (implementation category between brackets):

*First recurrence:*

See Recommendations for medical treatment of initial CDI.

Second recurrence and subsequent recurrences:

*If oral therapy is possible:*

- vancomycin 125 mg qid orally for at least 10 days (B-II)
  consider a taper/ pulse strategy (B-II)

*If oral therapy is impossible:*

- metronidazole 500 mg tid intravenously for 10 – 14 days + (A-III)
- retention enema of vancomycin 500 mg in 100 ml of normal saline every 4 – 12 h (C-III)
  or vancomycin 500 mg qid by nasogastric tube (C-III)

### *Recommendation for Prophylaxis of CDI*

Currently, there is no evidence that medical prophylaxis for CDI is efficacious and therefore we do not recommend prophylactic antibiotics. Of course, other preventive measures should be taken, such as hand hygiene of hospital personnel, prompt isolation of patients suspected of having CDI and prudent use of antibiotics.

**Table 1. Strength of recommendation and quality of evidence according to the Canadian Task Force on the Periodic Health Examination**

| | |
|---|---|
| *Strength of recommendation:* | |
| A: | good evidence to support a recommendation |
| B: | moderate evidence to support a recommendation |
| C: | poor evidence to support a recommendation |
| *Quality of evidence:* | |
| I: | evidence from ≥ 1 properly randomized, controlled trial |
| II: | evidence from ≥ 1 well-designed clinical trial, without randomization; from cohort or case-controlled analytic studies (preferably from ≥ centre); from multiple time-series; or from dramatic results from uncontrolled experiments |
| III: | evidence from opinions of respected authorities, based on clinical experience, descriptive studies, or reports of expert committees |

# Future Strategies and Trends

## New Pharmacological Agents

Due to the emergence of the highly virulent NAP1/BI/027 strain over the past few years, the response to metronidazole has been decreased [15,16] leading to an interest in new treatments. The greatest promise is with agents used to interrupt relapses.

In this category *ramoplanin, rifaximin, nitazoxanide, fidaxomicin*, toxin-binding agents (tolevamer), *probiotics* (*Saccharomyces boulardii* and *Lactobacillus ramosus*), *immune agents* (toxoid vaccine and hyperimmune globulin) and *fecal transplantation* have all shown activity against CDI.

## Ramoplanin

Ramoplanin is a lipoglycodepsipeptide antibiotic manufactured by Oscient Pharmaceuticals that acts by blocking peptidoglycan synthesis by binding to lipid II. It has shown bactericidal activity against aerobic and anaerobic gram-positive bacteria such as *Enterococcus* sp. and *C. difficile* [17,18]. Freeman *et al.* [19] observed similar efficacy between vancomycin and ramoplanin but with more efficacy in killing spores both in citron and *in vivo* in hamster gut. In a phase II multi centric trial the authors were able to establish a reasonable efficacy with limited side effects but the trial had insufficient power to establish non-inferiority to vancomycin [16]. The used doses were 200mg or 400 mg ramoplanin twice a day or vancomycin 125 mg four times a day for 10 days. The rate of cure at the end of therapy was 83-85% in the ramoplanin group versus 86% in the vancomycin group. The rate of relapses were 26.1% (ramoplanin 200mg), 21.7% (ramoplanin 400mg) and 20.8% (vancomycin 125mg). Mortality rates were similar in all groups and not attributable to the medicines. Larger trials are needed to gather evidence for establishing the role of this antibiotic in treatment of CDI.

## Rifaximin

Rifaximin is a nonabsorbed (bioavailability <0.4%) rifamycin that concentrates in the intestinal tract and acts by inhibiting bacterial RNA synthesis [20]. The drug appears to be particularly useful in treating and preventing recurrent CDI related to in-vitro activity of the drug against strains of *C. difficile* or the high drug concentrations available in the colon or both. In an open-label study, Boero *et al.* [21] compared rifaximin 200 mg 3 times daily for 10 days with oral vancomycin 500 mg twice daily for 10 days in 20 patients. Rifaximin was found to be effective in 90%, while vancomycin was successful in all (100%) patients who received it.

In order to prevent the recurrence of CDI, rifaximin can be given 2 weeks after completion of standard therapy [22,23]. Johnson *et al.* [24] later reported post-vancomycin strategies with rifaximin treatment in six patients with multiple recurrences of CDI. Four of the six patients (67%) had no further diarrhea episodes, but two patients failed shortly after or during the rifaximin treatment. *C. difficile* isolates from one of the two patients who failed treatment had an MIC of >256 ug/ml to rifaximin. Authors concluded that serial therapy with vancomycin, followed by rifaximin remains an option for some patients with multiple CDI recurrences. Polymorphous forms of rifaximin, processes for their production and use in medicinal preparations are successfully described in US Patent No: 7045620 [25]. However, emergence of resistance may be a problem.

Rifalazil, a newer benzoxazinorifamycin, is another potent rifamycin with fewer clinical data reported. The $MIC_{50}$ and $MIC_{90}$ for a total of 141 isolates range from 0.002 to 0.0075 µg/ml and 0.004 to 0.03 µg/ml, respectively [26]. The drug has a prolonged half-life permitting once-daily administration.

## Nitazoxanide

Nitazoxanide is a new thiazolide antiparasitic agent that has excellent activity in treating both interstitial protozoal and helminthic infections. It interferes with pyruvate-ferredoxin oxidoreductase (PFOR) enzyme dependent electron transport reaction, which is necessary for anaerobic metabolism. The drug acts by blocking anaerobic metabolism. *In vitro*, nitazoxanide has shown excellent activity against *C. difficile*. In one study, the 90% MIC ($MIC_{90}$) of the toxigenic strains of *C. difficile* was 0.5 μg/mL for nitazoxanide, which was identical to the $MIC_{90}$ of metronidazole and vancomycin [27]. Other investigators have reported lower MICs, ranging from 0.06 to 0.125 μg/mL [28].

In a larger, prospective, randomized, double-blind study, nitazoxanide was compared with metronidazole in treatment of hospitalized patients with *C. difficile* colitis [29]. In this study 44 patients received metronidazole 250 mg 4 times daily for 10 days, 49 received nitazoxanide 500 mg twice daily for 7 days, and 49 received nitazoxanide 500 mg twice daily for 10 days. The per-protocol results showed an 82.4% response rate in the metronidazole group compared with 89.5% in the nitazoxanide 7- and 10-day groups combined ($p = .20$). This study also evaluated the recurrence rate for 31 days after the start of treatment. A sustained response rate was demonstrated in 19 of 33 (57.6%) patients who received metronidazole for 10 days compared with 25 of 38 (65.8%) patients who received 7 days of nitazoxanide and 26 of 35 (74.3%) patients who received 10 days of nitazoxanide.

In another randomized double blind study, nitazoxanide was compared against vancomycin in CDI. Sustained response rates were 78% (18 of 23 patients) for the vancomycin group, and 89% (16 of 18 patients) for the nitazoxanide group [30].

These studies support nitazoxanide compared with metronidazole and vancomycin as a superior drug for both initial and sustained treatment of CDI in hospitalized patients. However, the efficacy of nitazoxanide used specifically in chemotherapy related CDI is unknown. US Patent Application No: 20080254010 describes the above mentioned methods including nitazoxanide in treating CDI associated diarrhea [31].

## Fidaxomicin (Tiacumicin B, OPT-80, PAR-101)

Fidaxomicin (OPT-80) is an 18-membered macrolide antibiotic. This antibiotic has demonstrated *in vitro* and *in vivo* activity against *C. difficile* and has a favorable pharmacokinetic profile for its potential use in CDI as demonstrated in a hamster model. Swanson *et al.* [32] found that MICs against 15 strains of *C. difficile* were 0.12 to 0.25 μg/mL for fidaxomicin and 0.5 to 1 μg/mL for vancomycin. They detected high concentrations of fidaxomicin in the cecum, but not the serum, of hamsters after oral administration).

Fidaxomicin, unlike metronidazole and vancomycin, does not affect the normal colonic flora which provides colonization resistance against *C. difficile*. The drug is well tolerated with oral administration and showed promising results in a phase 1 trial with a dose of 400 mg/d. Furthermore, the risk of relapse with fidaxomicin therapy is far less (only 5%) compared to either metronidazole or vancomycin [33]. Louie *et al.* [34] reported the results of the phase III trial of the drug, funded by Optimer pharmaceutical in February, 2011. A total of 629 patients were evaluated in a randomized trial treated for CDI with either fidaxomicin 200mg twice daily for 10 days or vancomycin 125mg four times a day for 10 days. The rates of cure with fidaxomicin were non-inferior to vancomycin in both modified intention-to-treat analysis (88.2% vs 85.8%) and the pre-protocol analysis (13.3% vs 24.0% p=0.004). There was lower rate of recurrence seen in patients with non-NAP1/027 strains when treated with fidaxomicin (7.8% vs 25.5%). Furthermore significantly higher rates of resolution of diarrhea in a shorter time span was seen with fidaxomicin as opposed to vancomycin (74.6% vs 64.1% p=0.006; 58 hours for resolution of symptoms for fidaxomicin versus 78 hours in vancomycin). The side effect profile was similar for the two agents.

## Toxin Binding Agents - Tolevamer

Tolevamer is a high molecular weight, styrene sulfonate polymer. It is given orally for the treatment of CDI. Tolevamer covalently binds and neutralizes toxins A and B. By this mechanism, it may prevent CDI-associated injury to the gastrointestinal tract without disrupting the reestablishment of normal bacterial growth [35]. The efficacy of Tolevamir was evaluated in a multicenter, multinational, double-blind phase 2 study [36] in patients with mild to moderately severe CDI associated diarrhea. Patients were randomized to 1 of 3 arms: tolevamer 3 g/day (n = 97), tolevamer 6 g/day (n = 95), or oral vancomycin 500 mg/day (n = 97) for 14 days. In the per-protocol analysis, with the primary endpoint being resolution of diarrhea, tolevamer 6 g/day was not inferior to vancomycin, with 58 of 70 (83%) patients achieving the primary endpoint versus 73 of 80 (91%) patients in the vancomycin-treated group. However, the 3 g/day dose of tolevamer was found to be inferior to vancomycin treatment. Interestingly, Louie *et al.* also found a trend toward lower recurrence rate in patients treated with tolevamer 6 g/day versus those who received vancomycin, 10% vs 19%, respectively (p = .19). In patients with metastatic disease where immune function is already compromised, tolevamir may be a much-needed alternative to antibiotic treatment. However, in two subsequent phase 3 trials, tolevamer was inferior to vancomycin and metronidazole for initial therapy [37]. In early 2008, a non-inferiority study versus vancomycin or metronidazole found that about half of the patient in the tolevamer group did not complete the treatment, versus 25% in the vancomycin and 29% in the metronidazole groups. *C. difficile* infection recurrence in patients reaching clinical success was reduced significantly by tolevamer (6% recurrence rate), vancomycin (18%) and metronidazole (19%). However, the significant reduction in recurrence rate seen in the tolevamer group may in part be attributed to its high drop-out rate. The development of the agent was halted since tolevamer did not reach its primary endpoint in this study.

## Immunotherapy

In *C. difficile* infections, patients who develop serum antitoxin A immunoglobulin G (IgG) titers in response to exposure tend to be 48 times less likely to develop diarrhea than those who do not mount a response [38]. In immunodeficient patients, intravenous immunoglobulins (IVIG) introduce a level of passive immunity. In a study of 5 children with relapsing *C. difficile* colitis and low baseline serum levels of antitoxin A IgG, Leung *et al.* [39] found that administration of IVIG 400 mg/kg every 3 weeks for 4-6 months was associated with a marked increase in serum antitoxin antibody and resolution of recurrent diarrhea. McPherson *et al.* [40] published a retrospective review of 14 patients given adjunctive IVIG (150-400 mg/kg once or twice, with the second dose given 3 weeks later) in addition to conventional treatment and found a response to treatment by stool normalization in 64% (n = 9) of patients with severe, refractory, or recurrent disease. Some isolated reports have suggested a benefit of IVIG therapy in refractory cases or in severe cases in patients who are unable to develop an immune response alone or as adjunct to traditional therapy. Less consistent results have been reported regarding the use of IVIG to treat patients with severe, refractory infection who had not had a response to standard therapy and for whom colectomy had been considered [41]. A recent phase II trial comprising 200 screened patients showed favorable outcomes with use of monoclonal antibodies in addition to metronidazole or vancomycin in patients with CDI. There was a reduced risk of recurrence seen with combined administration of two monoclonal antibodies against toxin A (CD A1) and and toxin B (CD B1) (7% in study group vs. 25% in placebo group, $p < .001$).

The combined administration of CDA1 and CDB1 human monoclonal antibodies in addition to antibiotics significantly reduced the recurrence of *C. difficile* infection. Larger studies are needed to confirm the findings of this phase II trial according to the authors [42]. US Patent No: 6680168 describes the method of passive immunization against *C. difficile* infection [43]

The published data regarding the efficacy of active immunization against *C.* difficile are even more sparse. Clinical trials using a toxoid vaccine (Acambis; Phase I) and monoclonal antibodies to toxin A and toxin B (Medarex; Phase II) are ongoing for prevention of *C. difficile* colitis. The vaccine is given in 3 doses (days 1, 8, and 30) and has a favorable side effect profile. Previous studies were done among healthy volunteers, and much of the data available supporting the use of intravenous immune globulin is from limited to uncontrolled studies and therefore at present it is not clear whether the vaccine will be effective in elderly, debilitated patients. [44]

In addition to the toxoid vaccine against toxins A and B, research is on-going in order to develop a carbohydrate conjugate vaccine against the NAP1/027 strains as well as immunization with recombinant *Bacillus subtilis* spores[45]. An US Patent Application described active and passive immunizations methods for CDI in detail [46]. More recently, a vaccine for the treatment or prophylaxis of *C. difficile* associated disease comprises a *C. difficile* gene or a *C. difficile* peptide/polypeptide or a derivative or fragment or mutant or variant thereof which is immunogenic in humans has been made for prophylaxis. The gene encodes a *C. difficile* surface layer protein, SlpA or variant or homologue thereof. The peptide/polypeptide is a *C. difficile* surface layer protein, SlpA or variant or homologue thereof. The vaccine may comprise a chimeric nucleic acid sequence. US Patent Application 20030054009 describes application of *C. difficile* vaccine for prophylaxis [47].

**Table 2. Summary of new treatments for *Clostridium difficile* infection**

| | Mechanism of Action | Clinical Activity | Results |
|---|---|---|---|
| • Ramoplanin | • Inhibits peptidoglycan synthesis by binding to Lipid II, an intermediate required for membrane formation. | • Produces bactericidal effect against aerobic and anaerobic gram positive bacteria | Current trial has had insufficient power to establish non-inferiority to vancomycin [67]<br><br>*Rate of Cure (at the end of therapy):*<br>• 200/400 mg ramoplanin: 83-85%<br>• 125 mg vancomycin: 86%<br><br>*Rate of Relapse :*<br>• 200 mg Ramoplanin: 26.1%<br>• 125 mg Vancomycin: 20.8%<br><br>*Mortality Rate:*<br>• Similar in all groups<br>• Larger trials need to be conducted in order to establish effectiveness of this drug. |
| • Rifaximin | • Inhibits bacterial RNA synthesis | • Treats and prevents recurrent CDI related to in-vitro activity of the drug against strains of *C. difficile* | • Rifaximin (200 mg, 3 times daily, 10 days)<br>• 90% effective [69]<br>• Vancomycin (500 mg, 2 times daily, 10 days)<br>• successful in all (100%) patients who received it |
| • Nitazoxanide | • Inhibits pyruvate-ferredoxin oxidoreductase, an enzyme (necessary for anaerobic metabolism) | • *In vitro*, nitazoxanide has shown excellent activity against *C. difficile* | • Metronidazole (250 mg, 4 times daily, 10 days) 82.4% response rate and Nitazoxanide (500 mg, 2 times daily, 7/10 days) 89.5% response rate (p = 0.20)<br><br>*Recurrance Rate (31 days):*<br>• Metronidazole (10 days):<br>• 57.6% sustained response rate<br><br>*Nitazoxanide (10 days):*<br>• 74.3% (26/35) sustained response rate<br>• 89% (16/18) sustained response<br>• Vancomycin |

| | Mechanism of Action | Clinical Activity | Results |
|---|---|---|---|
| | | | • Results support nitazoxanide as a superior drug for both initial and sustained treatment of CDI in hospitalized patients. |
| • Fidoxomicin (Tiacumicin B, OPT-80, PAR-101) | • Inhibits bacterial RNA synthesis by RNA polymerase | • Demonstrates *in vitro/ in vivo* activity specific to *C. difficile*<br>• Does not affect normal colonic flora | • Oral administration tolerated well<br>• Randomized trial [n=629]<br>• Fidaxomicin 200mg twice daily versus Vancomycin 125mg orally four times daily<br>• Rate of cure non-inferior to vancomycin<br>• Lower recurrence rate in patients with non-NAP1/027 strains, treated with fidaxomicin (7.8% vs. 25.5%)<br>• Higher rate of resolution of diarrhea in shorter time span<br>• Side effects similar to vancomycin |
| • Tolevamer | • Tolevamer covalently binds and neutralizes toxins A and B | • Prevent CDI-associated injury to the gastrointestinal tract without disrupting the reestablishment of normal bacterial growth | • Phase II study showed Tolevamer 6 g/day was not inferior to vancomycin.<br>• Patients with metastatic disease where immune function is already compromised, Tolevamir may be an alternative to antibiotic treatment.<br>• However, Since tolevamer did not reach its primary endpoint the development was halted. |
| • Immunotherapy | • Intravenous immunoglobulins (IVIG) introduce a level of passive immunity. | • Binds to toxin A and Toxin B to neutralize toxin effects.<br>• Additive advantage when use with antibiotics | • IVIG 400 mg/kg every 3 weeks for 4-6 months was associated with a marked increase in serum antitoxin antibody and resolution of recurrent diarrhea<br>• Favorable outcomes with use of monoclonal antibodies in addition to metronidazole or vancomycin in patients with CDI.<br>• There was a reduced risk of recurrence seen with combined administration of CD A1 and CD B1 monoclonal human antibodies (7% in study group vs. 25% in placebo group, $p < .001$). |

## Probiotics as Adjunctive Therapy

Adjunctive therapies for refractory disease include efforts to replenish colonic flora with the use of orally administered probiotics, usually *Lactobacillus* species or *Saccharomyces boulardii*. In Chapter 12 of this book the use of probiotics in CDI is also included.

A recent systematic review of randomized controlled trials to evaluate the efficacy of probiotic therapies in CDI identified only two treatment studies that showed some benefit of *S. boulardii*, although the benefit was restricted to subgroups of patients with severe or recurrent CDI. A more recent randomized, controlled study found some benefit of a yogurt containing *Lactobacillus* sp. and *Streptococcus thermophilus* in the prevention of antibiotic-associated diarrhea and CDI in patients over 50 years of age, although the applicability of the study has been questioned due to highly selective exclusion and inclusion criteria.

There is a concern over the safety of probiotics in severely ill or immunocompromised patients with several reports of *S. boulardii* fungemia and less frequent reports of sepsis due to *Lactobacillus* sp. In general, there is insufficient evidence to support the routine use of probiotics to prevent or treat CDI. Finally, case reports and case series have shown success with administration of donor stool or 'synthetic stool' (bacterial mixtures), either by nasogastric tube or colonoscopy.

## Summary: Future Strategies in Prevention

1. antibiotic resistance and *C. difficile* epidemic strains often show a typical resistance pattern but are in general not resistant against the most important drugs used for treatment like oral vancomycin and metronidazol.
2. controlling the usage of antibiotics with continuous CDI monitoring: in general no ban against specific antibiotics but against the general amount of all antibiotics consumed in hospitals; in some specific situation like CDI outbreaks with one hypervirulent strain like the PCR Ribotype 027 it makes sense to ban the quinolones for a certain time period.
3. hygienic measures: Infection control measures in the hospitals still have to be developed but the most important is keeping to the simple rules of thoroughly cleaning hands after direct and indirect contact with symptomatic patients and thoroughly cleaning the surfaces adjacent to the patient twice a day. If the hospital staff is sticking to these rules in most circumstances it is not necessary to use routine or targeted sporicidal disinfection of the surfaces around the patients.

## References

[1] 1a. Hakalehto E, Hell M, Bernhofer C, Heitto A, Pesola J, Humppi T, Paakkanen H. Growth and gaseous emissions of pure and mixed small intestinal bacterial cultures: Effects of bile and vancomycin. *Pathophysiology,* 2010; 17, 45-53.

1b. Hakalehto E. Simulation of enhanced growth and metabolism of intestinal *Escherichia coli* in the Portable Microbe Enrichment Unit (PMEU). In: Rogers MC, Peterson ND, editors. *E. coli infections: causes, treatment and prevention.* New York, USA: Nova Publishers; 2011.

[2] Hakalehto E, Hänninen O. Lactobacillic $CO_2$ signal initiate growth of butyric acid bacteria in mixed PMEU cultures. Manuscript in preparation. 2012.

[3] Hell M, Permoser M, Chmelizek G, Kern JM, Maass M, Huhulescu S, Indra A, Allerberger F. *Clostridium difficile* infection: monoclonal or polyclonal genesis? *Infection*, 2011; 39, 461-465.

[4] Hell M, Sickau K, Chmelizek G, Kern JM, Huhulescu S, Allerberger F. Absence of *Clostridium difficile* in asymptomatic hospital staff. *Am. J. Infect. Control.*, 2011. In press.

[5] Spigaglia P, Barbanti F, Mastrantonio P; European Study Group on *Clostridium difficile* (ESGCD). Multidrug resistance in European *Clostridium difficile* clinical isolates. *J. Antimicrob. Chemother.*, 2011; 66, 2227-2234.

[6] Kuijper EJ, Coignard B, Tüll P, the ESCMID Study Group for Clostridium difficile (ESGCD), EU Member States, and the European Centre for Disease Prevention and Control (ECDC). Emergence of *Clostridium difficile* -associated disease in North America and Europe. *Clin. Microbiol. Infect.*, 2006; 12 (Suppl 6), 2-18.

[7] Rodriguez-Palacios A, Staempfli HR, Duffield T, Weese JS. *Clostridium difficile* in retail ground meat, Canada. *Emerg. Infect. Dis.*, 2007; 13, 485-487.

[8] Aldape MJ, Bryant AE, Stevens DL. *Clostridium sordellii* infection: epidemiology, clinical findings, and current perspectives on diagnosis and treatment. *Clin. Infect. Dis.*, 2006; 43, 1436-1446.

[9] Rupnik M. Is *Clostridium difficile* -associated infection a potentially zoonotic and foodborne disease? *Clin. Microbiol. Infect.*, 2007; 13, 457-459.

[10] Bauer MP, Notermans DW, van Benthem BH, Brazier JS, Wilcox MH, Rupnik M, Monnet DL, van Dissel JT, Kuijper EJ. ECDIS Study Group. *Clostridium difficile* infection in Europe: a hospital-based survey. *Lancet*, 2011; 377, 63-73.

[11] Bidet P, Lalande V, Salauze B, Burghoffer B, Avesani V, Delmée M, Rossier A, Barbut F, Petit JC. Comparison of PCR-ribotyping, arbitrarily primed PCR, and pulsed-field gel electrophoresis for typing *Clostridium difficile*. *J. Clin. Microbiol.*, 2000; 38, 2484-2487.

[12] Gould CV, McDonald LC. Bench-to-bedside review: *Clostridium difficile* colitis. *Crit. Care.,* 2008; 12, 203.

[13] Vonberg RP, Kuijper EJ, Wilcox MH, Barbut F, Tüll P, Gastmeier P; European *C. difficile*-Infection Control Group; European Centre for Disease Prevention and Control (ECDC), van den Broek PJ, Colville A, Coignard B, Daha T, Debast S, Duerden BI, van den Hof S, van der Kooi T, Maarleveld HJ, Nagy E, Notermans DW, O'Driscoll J, Patel B, Stone S, Wiuff C. Infection control measures to limit the spread of Clostridium difficile. *Clin. Microbiol. Infect.*, 2008; 14 (Suppl. 5), 2-20.

[14] 14a. Bauer MP, Kuijper EJ, van Dissel JT, European Society of Clinical Microbiology and Infectious Diseases. European Society of Clinical Microbiology and Infectious Diseases (ESCMID). Treatment guidance document for *Clostridium difficile* infection (CDI). *Clin. Microbiol. Infect.*, 2009; 15, 1067-79.

14b. Pépin J, Routhier S, Gagnon S, Brazeau I. Management and outcomes of a first recurrence of *Clostridium difficile*-associated disease in Quebec, Canada. *Clin. Infect. Dis.,* 2006; 42, 758-764.

[15] Louie T, Gerson M, Grimard D, *et al.* Results of a phase III trial comparing tolevamer, vancomycin and metronidazole in patients with *Clostridium difficile* -associated diarrhea (CDI). Abstract presented at the 47th *International Conference on Antimicrobial Agents and Chemotherapy Meet*ing at Chicago IL September 17-20, 2007.

[16] Musher DM, Aslam S, Logan N, Nallacheru S, Bhaila I, Borchert F, Hamill RJ. Relatively poor outcome after treatment of *Clostridium difficile* colitis with metronidazole. *Clin. Infect. Dis.,* 2005; 40, 1586–1590.

[17] Farver DK, Hedge DK, Lee SC. Ramoplanin a lipoglycodepsipeptide antibiotic. *Ann. Pharmacother.*, 2005; 39, 863-868.

[18] Fulco P, Wenzel RP. Ramoplanin - a topical lipoglycodepsipeptide antibacterial agent. *Expert. Rev. Anti. Infect.,* 2006; 4, 939-945.

[19] Freeman J, Baines SD, Jabes D, Wilcox MH. Comparison of ramoplanin and vancomycin *in vitro* and *in vivo* models of clindamycin induced *Clostridium difficile* infection. *J. Antimicrob. Chemother.*, 2005; 56, 717-725.

[20] Jiang ZD, Ke S, Palazzini E, Riopel L, Dupont H. *In vitro* activity and fecal concentration of rifaximin after oral administration. *Antimicrob. Agents Chemother.,* 2000; 44, 2205–2206.

[21] Boero M, Berti E, Morgando A, Verme G. Treatment for *Clostridium difficile*: Results of a randomized open study with rifaximine vs. vancomycin. *Microbiol. Medica,* 1990; 5, 74-77.

[22] Garey KW, Jiang ZD, Bellard A, DuPont HL. Rifaximin in treatment of recurrent *Clostridium difficile*-associated diarrhea: an uncontrolled pilot study. *J. Clin. Gastroenterol.,* 2009; 43, 91-93.

[23] Johnson S, Schriever C, Galang M, Kelly CP, Gerding DN. Interruption of recurrent *Clostridium difficile* -associated diarrhea episodes by serial therapy with vancomycin and rifaximin. *Clin. Infect. Dis.,* 2007; 44, 846–848.

[24] Johnson S, Schriever C, Patel U, Patel T, Hecht DW, Gerding DN. Rifaximin Redux: Treatment of recurrent *Clostridium difficile* infections with Rifaximin immediately post-vancomycin treatment. Anaerobe doi:10.1016/j.anaerobe.2009.08.004.

[25] Viscomi GC, Campana M, Braga D, Confortini D, Cannata V, Severini D, Righi P, Rosini G. Polymorphous forms of rifaximin, processes for their production and use thereof in medicinal preparations. US Patent 7045620. 2006.

[26] Hecht DW, Galang MA, Sambol SP, Osmolski JR, Johnson S, Gerding DN. *In vitro* activities of 15 antimicrobial agents against 110 toxigenic *Clostridium difficile* clinical isolates collected from 1983 to 2004. *Antimicrob. Agents Chemother.,* 2007; 51, 2716–2719.

[27] McVay CS, Rolfe RD. *In vitro* and *in vivo* activities of nitazoxanide agents against *Clostridium difficile. Antimicrob. Agents Chemother.,* 2000; 44, 2254-2258.

[28] Dubreuil L, Houcke I, Mouton Y, Rossignol JF. *In vitro* evaluation of activities of nitazoxanide and tizoxanide against anaerobes and aerobic organisms. *Antimicrob. Agents Chemother.,* 1996; 40, 2266-2270.

[29] Musher DM, Logan N, Hamill RJ, Dupont HL, Lentnek A, Gupta A, Rossignol JF. Nitazoxanide for the treatment of *Clostridium difficile* colitis. *Clin. Infect. Dis.,* 2006; 43, 421-427.

[30] Musher DM, Logan N, Bressler AM, Johnson DP, Rossignol JF. Nitazoxanide versus vancomycin in *Clostridium difficile* infection: a randomized, double-blind study. *Clin. Infect. Dis.,* 2009; 48, e41-46.

[31] Sasser JM, Cousin KS. Controlling *Clostridium difficile*-associated disease in the gastrointestinal tract. US Patent Application No: 20080254010. 2008.

[32] Swanson RN, Hardy DJ, Shipkowitz NL, Freiberg LA, Lartey PA, Clement JJ. *In vitro* and *in vivo* evaluation of tiacumicins B and C against *Clostridium difficile. Antimicrob. Agents Chemother.,* 1991; 35, 1108-1111.

[33] Braunlin W, Xu Q, Hook P, Fitzpatrick R, Klinger JD, Burrier R, Kurtz CB. Toxin binding of tolevamer, a polyanionic drug that protects against antibiotic-associated diarrhea. *Biophysical. J.,* 2004; 87, 534-539.

[34] Louie TJ, Peppe J, Watt CK, Johnson D, Mohammed R, Dow G, Weiss K, Simon S, John JF Jr, Garber G, Chasan-Taber S, Davidson DM. Tolevamer, a novel nonantibiotic polymer compared with vancomycin in the treatment of mild to moderately severe *Clostridium difficile* -associated diarrhea. *Clin. Infect. Dis.* 2006; 43, 411-420.

[35] Louie TJ, Miller M, Donskey C, *et al.* Safety, pharmacokinetics and outcomes of PAR-101 in healthy subjects and patients with *Clostridium difficile*-associated diarrhea [abstract LB2–29], *Program and abstracts of the 45th Interscience Conference on Antimicrobial Agents and Chemotherapy (Washington DC)* volume. 226, American Society for Microbiology, Washington, DC (2005)

[36] Louie TJ, Miller MA, Mullane KM, Weiss K, Lentnek A, Golan Y, Gorbach S, Sears P, Shue YK OPT-80-003 Clinical Study group. Fidaxomicin versus vancomycin for *Clostridium difficile* infection. *NEJM,* 2011; 364-365.

[37] Chiu Y-h, Che TM, Romero A, Ichikawa Y, Shue Y-k. Polymorphic crystalline forms of tiacumicin B. US Patent No: 7378508. 2008.

[38] Owens RC. *Clostridium difficile*-associated disease: an emerging threat to patient safety. *Pharmacotherapy,* 2006; 26, 299-311.

[39] Leung DY, Kelly CP, Boguniewicz, Pothoulakis C, LaMont JT, Flores A. Treatment with intravenously administered gamma globulin of chronic relapsing colitis induced by *Clostridium difficile* toxin. *J. Pediatr.,* 1991; 118, 633-637.

[40] McPherson S, Rees CJ, Ellis R, Soo S, Panter SJ. Intravenous immunoglobulin for the treatment of severe, refractory, and recurrent *Clostridium difficile* diarrhea. *Dis. Colon. Rectum,* 2006; 49, 1-6.

[41] Juang P, Skledar SJ, Zgheib NK, Paterson DL, Vergis EN, Shannon WD, Ansani NT, Branch RA. Clinical outcomes of intravenous immune globulin in severe *Clostridium difficile*-associated diarrhea. *Am. J. Infect. Control.,* 2007; 35, 131-137.

[42] Lowy I, Molrine DC, Leav BA, Blair BM, Baxter R, Gerding DN, Nichol G, Thomas WD Jr, Leney M, Sloan S, Hay CA, Ambrosino DM. Treatment with monoclonal antibodies against *Clostridium difficile* toxins. *NEJM,* 2010; 362, 197-205.

[43] Thomas Jr. WD, Giannasca PJ, Zhang Z, Lei W, Monath TP. Passive immunization against *Clostridium difficile* disease. US Patent No: 6680168. 2004.

[44] Martin CE, Weishaupt MW, Seeberger PH. Progress toward developing a carbohydrate-conjugate vaccine against *Clostridium difficile* ribotype 027: synthesis of cell-surface polysaccharide PS-I repeating unit. *Chem. Commun.,* 2011; 47, 10260-10262.

[45] Permpoonpattana P, Hong H, Hetcharaburanin J, Huang JM, Cook J, Fairweather NF, Cutting SM. Immunization with *Bacillus* spores expressing toxin A peptide repeats protects against infection with *Clostridium difficile* strains producing toxins A and B. *Infect. Immun.,* 2011; 79, 2295-2302.

[46] Thomas WD, Giannasca PJ, Zhang Z, Lei W, Monath TP. Immunization against *Clostridium difficile* disease. US Patent Application No: 20060029608. 2006.

[47] Windle HJ, Doyle R, Kelleher D, Walsh JB, Ni Eidhin D. *Clostridium difficile* vaccine. US Patent Application No: 20030065466. 2007.

In: Alimentary Microbiome: A PMEU Approach
Editor: Elias Hakalehto
ISBN: 978-1-61942-692-4
© 2012 Nova Science Publishers, Inc.

*Chapter XI*

# Environmental Monitoring Using the Enrichment of Hygienic Indicators

***Lauri Heitto[1], Anneli Heitto[2], and Elias Hakalehto[3]***
[1]Environmental Reseach of Savo-Karjala Oy, FI, Kuopio, Finland
[2]Finnoflag Oy, FI, Kuopio, Finland
[3]Department of Biosciences, University of Eastern Finland, FI, Kuopio, Finland

## Abstract

Environmental monitoring of hygienic problems is based mainly on so called indicator bacteria. The most common bacteria for fecal indication are *E. coli* and fecal enterococci. Interestingly, they are among the most resistant microbes against bile acids, and therefore most frequent members of the duodenal flora and in the biliary tract. Wastewater treatment removes usually very efficiently pathogenic bacteria, but still purified waste waters entering natural waters can possess a serious hygienic risk for human health.

A special group is antibiotic resistant bacteria, which enter the environment via *e.g.* hospital waste waters. Environmental monitoring of the microbiological loads in natural waters is thus of great importance for human health. Using sample enrichment in the PMEU gives many advantages in environmental monitoring. The results can be obtained much faster, which has a special importance in the case of waste water spills. PMEU enrichment has also shown to give bacterial counts in situations where cultivation without enrichment has failed. For the same reason using PMEU in source tracking studies gives more reliable results.

The other important field in water research is the monitoring of household waters. The need for automated internet- based screening of microbes enriched from clean water systems can be fulfilled with PMEU-Coliline version.

# Introduction

The environmental monitoring is a wide concept. The concept could consist of all sensing outside our bodies. Microbiologically, the environment is the microenvironment outside the cells, or in many cases more precisely: the space between the cells.

In this Chapter we concentrate on the fate of the intestinal microbes in the waste waters and in nature and household waters. Their follow up is a necessary part of the environmental health disciplines. The major approach in determining fecal pollution is the detection and identification of indicator bacteria.

# Indicator Bacteria

Human feces contain huge amount of microbes, of which some may be pathogenic, such as *Salmonella*, different viruses, protozoa or parasites. These pathogens may also be found in the feces of warm-blooded animal. That is why it is very important for human health to know if there is any risk for fecal contamination in the environment.

Since pathogens need all their own methodology for identification, the concept of indicator bacteria has been widely accepted in environmental monitoring. Ashbolt *et al.* [2001] have described the early history of the concept:

> "The use of bacteria as indicators of the sanitary quality of water probably dates back to 1880 when Von Fritsch described *Klebsiella pneumoniae* and *K.rhinoscleromatis* as micro-organisms characteristically found in human feces [Geldreich, 1978]. In 1885, Percy and Grace Frankland started the first routine bacteriological examination of water in London, using Robert Koch's solid gelatine media to count bacteria [Hutchinson and Ridgway, 1977].
>
> Also in 1885, Escherich described *Bacillus coli* [Escherich, 1885] (renamed *Escherichia coli* by Castellani and Chalmers [Castellani and Chalmers, 1919]) from the feces of breast-fed infants. In 1891, the Franklands came up with the concept that organisms characteristic of sewage must be identified to provide evidence of potentially dangerous pollution [Hutchinson and Ridgway, 1977].
>
> By 1893, the 'Wurtz method' of enumerating *B. coli* by direct plating of water samples on litmus lactose agar was being used by sanitary bacteriologists, using the concept of acid from lactose as a diagnostic feature. This was followed by gas production, with the introduction of the Durham tube [Durham, 1893]. The concept of 'coliform' bacteria, those bacteria resembling *B. coli*, was in use in Britain in 1901 [Horrocks. 1901]. The colony count for bacteria in water, however, was not formally introduced until the first Report 71 [HMSO, 1934].

Therefore, the sanitary significance of finding various coliforms along with streptococci and *C. perfringens* was recognised by bacteriologists by the start of the twentieth century [Hutchinson and Ridgway, 1977]. It was not until 1905, however, that MacConkey [MacConkey, 1905] described his now famous MacConkey's broth, which was diagnostic for lactose-fermenting bacteria tolerant of bile salts. Nonetheless, *coli-forms* were still considered to be a heterogeneous group of organisms, many of which were not of fecal origin. The origins of the critical observation that *B. coli* was largely fecal in origin while other coliforms were not, could be claimed by Winslow and Walker [Winslow and Walker, 1907].

Parallel to the work on coliforms, a group of Gram-positive coccoid bacteria known as fecal streptococci (FS) were being investigated as important pollution indicator bacteria [Houston. 1900; Winslow and Hunnewell, 1902; Winslow and Walker, 1907]. Problems in differentiating fecal from non-fecal streptococci, however, initially impeded their use [Kenner. 1978]. It was not until 1957, however, with the availability of the selective medium of Slanetz and Bartley [Slanetz and Bartley, 1957] that enumeration of FS became popular. "

According to Ashbolt *et al.* [2001] indicator bacteria can be divided in three groups, process indicators, fecal indicators, and index and model organisms. Process indicators demonstrate the efficiency of a process, e.g, total coliforms for chlorine disinfection.

Fecal indicators indicate the presence of fecal contamination. Nowadays the most common used bacteria for fecal indicators are *E. coli* and fecal enterococci (earlier fecal streptococci). Interestingly, they are among the most resistant microbes against bile acids, and therefore most frequent members of the duodenal flora and in the biliary tract [Hakalehto. 2010; Hakalehto *et al.*, 2010]. Index and model organisms are species that indicate pathogen presence and behavior respectively, *e.g. E coli* is an index for *Salmonella*.

By definition, fecal indicator bacteria

1. are part of the natural flora of human and warm blooded animal intestine,
2. indicate fecal contamination but are not pathogenic themselves
3. do not multiply outside the intestine

Because of the wide variety of pathogens, not any bacterium is a universal indicator. The presence of *E. coli* is most often an indication that there is fecal contamination [Dufour. 1977], but absence is not a reliable indication that there are no pathogens. *E.g. Salmonella* has been shown to survive better than *E.* coli in nonhost environments in certain circumstances [Winfield and Groisman, 2003].

Tropical freshwaters can contain *E. coli* without fecal contamination [Jimenez *et al.*, 1989]. Also many viruses are more resistant *e.g.* water treatment management than *E. coli* [Muniesa *et al.*, 1999] and that is why bacteriophages have been used for fecal indication in some cases.

Other microbes applied for indication purpose, or for source-tracking the origin of the contamination, include *Bacteroides* sp. and its phages [Grabow, 2001], other human viruses, *Pseudomonas* sp., clostridial strains [Barrell *et al.*, 2000], and *Aeromonas* sp [Araujo *et al.*, 1991]. *E.g.* in developing countries the common aetiological agents are more likely to be viruses and parasitic protozoa than bacteria [Levy *et al.*, 1998].

Some strains of fecal indicators may also be pathogens. *E.g.* the toxigenic *E. coli* strains [Ohno *et al.*, 1997]. *E. coli* O157:H7 has been responsible for illness to recreational swimmers [Ackman *et al.*, 1997] and several deaths have been documented through food- and waterborne outbreaks [Jones and Roworth, 1996]. The detection of the various *E.coli* strains by the PMEU technologies has been reviewed earlier [Hakalehto, 2011].

## Survival of Fecal Indicators and Pathogens in Waste Water Treatment Processes

Once excreted from human or animal host, microbes like *Salmonella* and *E. coli* find themselves battling for survival, facing limited nutrient availability, osmotic stress, large variations in temperature' pH, and predation [Marshall. 1980; Rozen and Belkin, 2001; Savageau, 1983]. It has been suggested that one-half of the total *E. coli* population resides in the primary habitat of the host and one-half is in the external environment (i.e., the secondary habitat) [Savageau, 1983]. In urban areas most of the human feces is treated in wastewater plants, where various strains from all over are collected into one sludge. This, in turn, is giving an opportunity for the intestinal ecosystem to have again new balances sought after.

The history of waste water treatment is actually quite short. Until the early 1970s treatment *e.g.* in Europe and in the U.S. mostly consisted of primary treatment, which had removal of suspended and floating material, treatment of biodegradable organics, and elimination of pathogenic organisms by disinfection. In 1980's secondary treatment with biological activated sludge method improved the efficiency of organic matter and nutrient removal. Nowadays also a tertiary treatment has taken place with advanced treatment including *e.g.* microfiltration, carbon adsorption, evaporation /distillation, or chemical precipitation.

The activated sludge process is quite efficient in removing indicator bacteria from urban waste-waters. *E.g.* in one wastewater plant of a moderate size Finnish city (about 200 000 inhabitants) the activated sludge process removed about 94-96 % of the bacteria [Leino, 2008]. The concentration of *E. coli* in outgoing water varied between 1 – 840 000 CFU/ 100 ml and the ratio between fecal coliforms:fecal enterococci was about 7;1.

Because primary, secondary and tertiary waste water treatments do not eliminate all the pathogenic bacteria in many cases disinfection of waste water is needed. Disinfection methods can be either physical (*e.g.* heating, UV-light) or chemical (*e.g.* chlorine, ozone). Chlorine has been earlier widely used because of its relatively low cost and good disinfection efficiency, but it has harmful effects for other biota, *e.g.* fish. Chlorine and UV-light have been given good results in removing *E. coli*, but not so efficient *e.g.* in virus removal [Blatchley *et al.*, 2007; Koivunen and Heinonen-Tanski, 2005].

Viruses and other pathogens are not part of the normal fecal microbiota, but are only excreted by infected individuals. Therefore, the higher the number of people contributing to sewage or fecal contamination, the more likely is the presence of a range of pathogens. Hospital waste waters have a special role in the water treatment.

Besides many pathogens included in the waste water, the hospitals are a major site of antibiotics usage worldwide [Hakalehto, 2012]. It had been demonstrated that antibiotics are generally poorly absorbed by the animal and human body, and thus are extracted unchanged or transformed, via urine and feces [Gulkowska *et al.*, 2007]. It is estimated that in Sweden alone, 10 to 20 tons of active antibiotics are released each year to the environment via the urine of people and animals treated for infection [Gullberg *et al.*, 2011].

Hospital waste waters are usually treated mainly in the municipal plants. This means a significant environmental burden in the form of eg. antimicrobial substances, and also microbial strains resistant to these medicines. Therefore, they are increasingly also spreading the causative agents of the so called hospital-acquired (HA) infections [Hakalehto, 2006]. The

MRSA (methicillin-resistant *Staphylococcus aureus*) and VRE (vancomycin-resistant enterococci) rose into the headlines some 20 years ago. These Gram-positive bacteria are still contributing to the microbiological risks in the societies. Nowadays Gram-negative pathogens possessing the antibiotic resistance genes in plasmids have the opportunity to spread in wastewater treatment plots, or in animal farming, where uncontrolled use of antibiotics could also create remarkable community health risks. Actually, waste water plants can serve as selection machines for resistant bacteria. The major hazards could be, for example, ESBL strains (extended-spectrum beta-lactamase producing coliformic strains), multiresistant pseudomonads etc. [Yong *et al.*, 2009]. If these bacteria manage to continue their accelerated distribution, it could exhibit severe consequences, not only for the diseased individuals, but also for the entire healthcare system. For example, in India a total of 77% of the neonates had ESBL-producing *K. pneumoniae* or *E. coli* in their stool [Roy *et al.*, 2011]. The locally originating epidemics of novel multiresistant strains of intestinal origin could form a worldwide hazard [Rolain *et al.*, 2010]. In the case of one individual boy child we were able to demonstrate that the balance in the intestinal microbiota could be returned in approximately 3-6 months after the antibacterial medication [Pesola and Hakalehto, 2011]. This positive development was associated with a decrease of β-lactamase positive strains in the feces in this case.

The household waste waters contain also the medicines, cosmetics, washing powders and other detergents, wastes from food preparation and consumption etc. In the city of Stockholm these food and other organic materials from homes are collected by the water exhaust into biogas production units [Swedish EPA, 2009]. This approach makes the city waste management system to function in a more ecological way besides giving energy to the community.

We have also made preliminary work on biorefinery of waste waters. The Finnoflag Oy´s project entity for outlining the biorefinery treatment of municipal wastes and waste waters has been chosen the best project idea in the Baltic sea region countries´ REMOWE consortium in the meeting in Tallinn, Estonia, in October 2011. The idea behind the suggested and experimented project work was to convert the waste materials, sludges, and other recyclables into energy compounds and chemicals [Hakalehto *et al.*,2012]. By doing so the environmental burden of the wastes could be diminished in a remarkable fashion, and the economics are improved in the waste management. The products from these processes include methane, hydrogen, ethanol, butanol, acetone, 2,3-butanediol, organic acids, and several other molecules which originally are the products of the metabolism of intestinal microbes.

## Freshwater Ecosystems

After the treatment the purified waste water and indicator bacteria included are lead to the natural waterways like rivers and lakes. The die-off of these bacteria depends very much of the environmental conditions. Temperature, exposure to sunlight, availability of nutrients, predation, sedimentation and osmotic pressure are factors that affect the die-off of indicator bacteria [Marshall, 1980; Rozen and Belkin, 2001]. Sedimentation to the bottom sediments allows the indictor bacteria to live much longer than in free water [Jeng *et al.*, 2005].

In a harsh environment bacteria may enter to the so called VBNC (viable but not culturable) –state. The concept was introduced by Byrd and Colwell in the 1980's [Byrd *et al.*, 1991], and after that there is increasing evidence for the existence of such a state in microbes, particularly in the aquatic environment [Dunaev *et al.*, 2008; Sawaya *et al.*, 2008]. A bacterium in the VBNC state is defined as "a cell which is metabolically active, which being incapable of undergoing the cellular division required for growth in or on a medium normally supporting grown of that cell" [Oliver, 1993].

In the PMEU the environment for the bacteria (temperature, culture media, nutrient availability, oxygen concentration) can be optimized [Hakalehto, 2006] so that it would be possible to leave the VBNC-state. In many of our studies in the lake environment we have managed to get *E. coli* or enterococci strains from PMEU enriched samples, whereas cultivation with standard plate methods gave zero results.

As discussed earlier, antibiotics and antibiotic resistant bacteria are connected with waste waters entering natural waters. Although concentrations of antibiotics are low, they have showed to have direct toxic effect to aquatic organisms, mainly bacteria [Batt *et al.*, 2006; Gulkowska *et al.*, 2007; Kummerer *et al.*, 2000; Tamtam *et al.*, 2008]. Antibiotic substances eliminate sensitive bacteria, which might play an important ecological role [Chelossi *et al.*, 2003; Costanzo *et al.*, 2005]. On the other hand the number of antibiotic resistant bacteria in aquatic environments has increased. We carried out some experimentation outside Stockholm together with the researchers from the Karolinska Institutet in summers 2005 and 2006 (Figure 1 and 2). Then it was possible to isolate antibiotic resistant bacterial strains form the bathing waters in low concentrations. According to other studies, from lakes that receive effluent from hospitals, antibiotic-resistant bacteria are up to 70 percent more common than in uncontaminated environments [Shah, 2010]. The number of coliformic antibiotic resistant hospital strains has especially been increasing (Figure 3). Hygienic indicator bacteria can be used not only for evaluating of hygienic risk but also for source-tracking of different types of waste waters [Kühn, 1985]. In Lake Kallavesi, Finland, we could demonstrate the area affected by forest industry and municipal waste waters with a help of fecal enterococci phenotyping (Figure 4) [Heitto *et al.*, 2009].

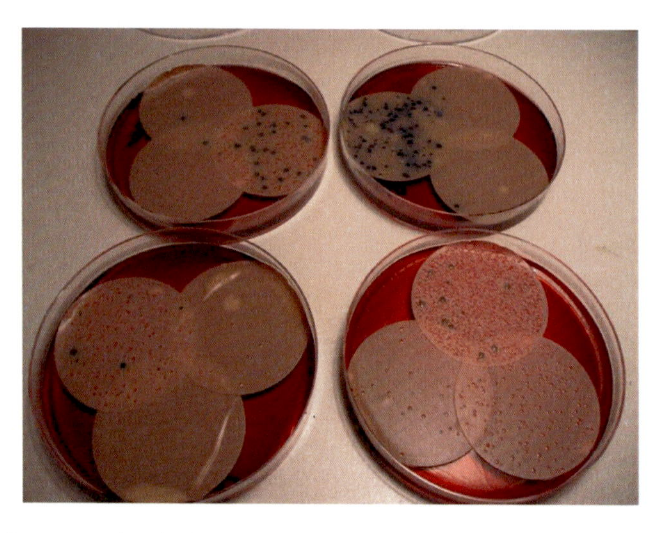

Figure 1. Colonies of enterococci and other waterborne bacteria membrane filtered and cultivated on plates at the laboratorium of Karolinska Institutet in Stockholm, Sweden.

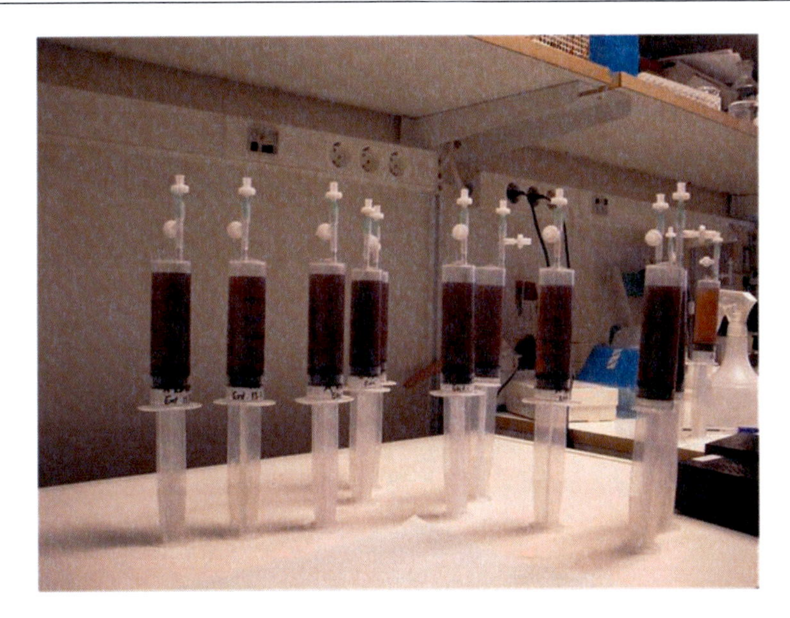

Figure 2. PMEU sampling and enrichment syringes after selecting water enterococci. The cultivation served as a pre-enrichment for the plate studies presented in Fig.1.

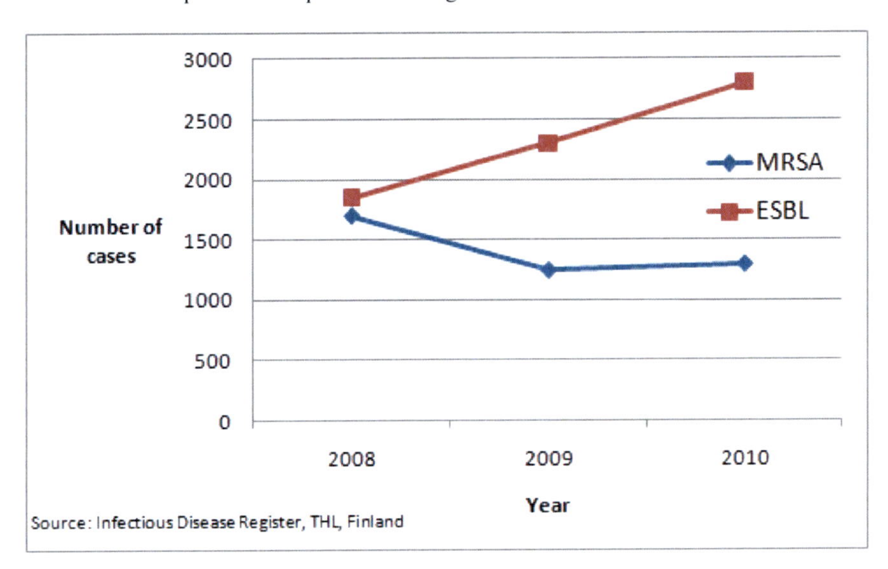


Figure 3. Number of MRSA and ESBL cases in Finland 2008-2010. Figure from an article in "Science against microbial pathogens: communicating current research and technological advances" (Microbiology Book Series) Formatex Research Center, Spain [Hakalehto, 2012].

Near the outlet of forest industry the most common enterococci group was so called C-G-F -group (*Enterococcus casseliflavus, E. gallinarum* and *E. flavescens*), which is not of fecal origin. This has been noticed earlier by Niemi *et al.* [1993]. Fecal enterococci (mainly *E. faecalis* and *E. faecium*) were dominant near urban waste water outlets. PMEU enrichment of samples improved the analysis because of the increased number of phenotypes (Figure 5). This combination of Pheneplate$^{TM}$ –method and the PMEU is a fast, simple and cost-effective tool for source-tracking.

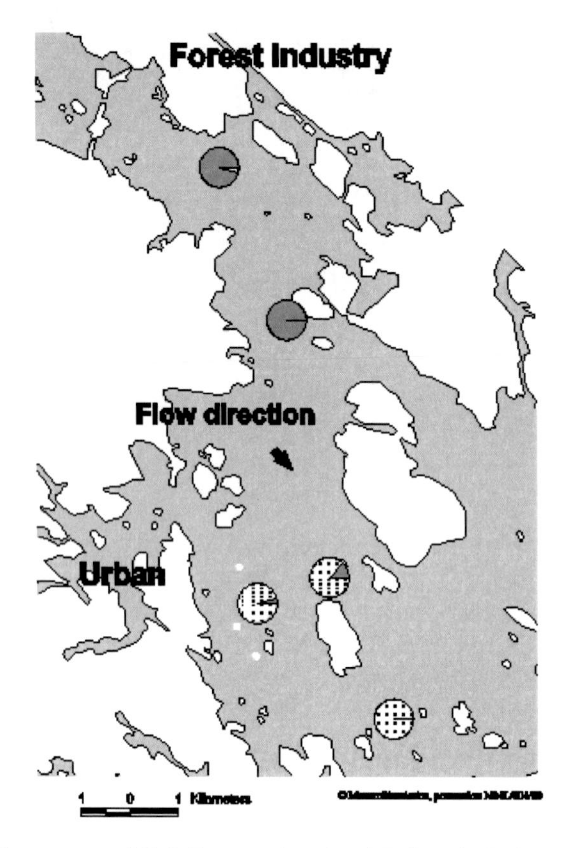


Figure 4. Proportion of the enterococci C-G-F -group (grey) and *E. faecalis/E. faecium* (black and white) at Lake Kallavesi near forest industry and urban sewage treatment plants.

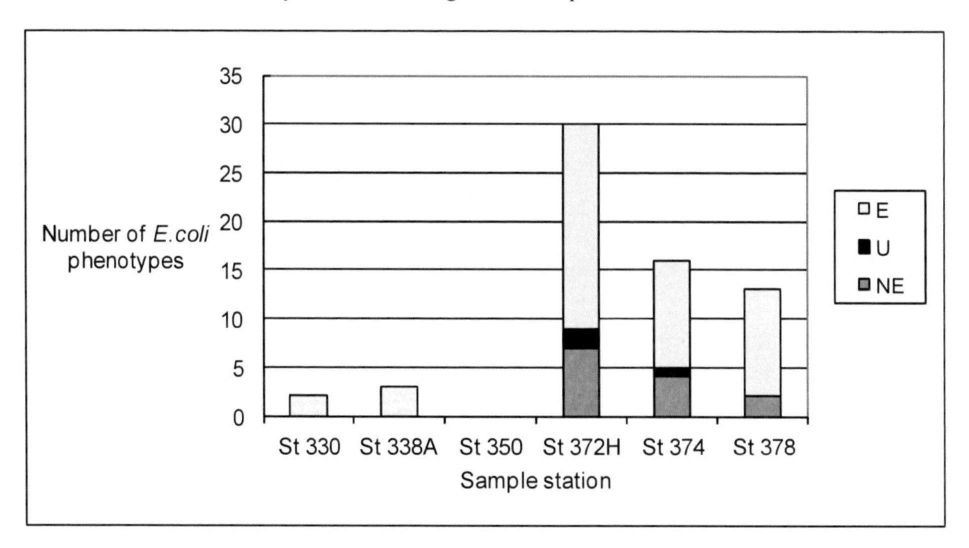


Figure 5. Number of different *E.* coli phenotypes in different sampling sites. E = number of phenotypes with PMEU enrichment only, NE = number of phenotypes without PMEU enrichment, U=number of phenotypes, that were found both with and without PMEU enrichment. Note, that in sample stations 330 and 338 culturable *E. coli* strains could be found only in samples with PMEU enrichment.

PMEU is an analysis device, which can be used in the field conditions (Figure 6). Because of this, results can be obtained much faster as from the samples with which the analysis starts not until in the laboratory. *E.g.* in the case of waste water spills in a lake environment it is very often difficult to know, what is the affected area because of internal currents. Samples incubated in the PMEU just after sampling show which samples are contaminated with bacteria in a few hours (Figure 7). With this information operations can be directed to infected areas during the day of the waste water spill.

The PMEU method has been used for the hospital hygiene monitoring [Hakalehto, 2006; Hakalehto *et al.*, 2010], follow up of *Salmonella* sp. [Hakalehto, 2011; Hakalehto *et al.*, 2007], microbiological screening of *Campylobacter* sp. in natural and household waters [Pitkänen *et al.*, 2009], for monitoring pulp and paper industry process and waste waters [Mentu *et al.*, 2009], environmental typing of enterococci [Heitto *et al.*, 2006; Heitto *et al.*, 2009], coliform analysis [Hakalehto, 2011; Wirtanen and Salo, 2010] and for control in water departments [Hakalehto *et al.*, 2011].

After entering the recipient water, waste waters contribute to the cycles of many substances in the environment. It is our track in the nature, and gives a starting point to many microbiological nutrient chains in the seas, lakes and rivers. Being subjected to the microbe load of our human origin, the nature has to return the valuable components back to the circulation of matter. This is necessary for keeping up the waterways functioning. Together with the microbial pollution the water ecosystems have to cope with the all forms of chemicals resulting from human activities. The water of river Rhein is said to be exploited for 16 times before it arrives to the sea. During its course it receives not only the loads from the municipal water treatment plants in different countries, but also various industrial waste waters.

(a)

(b)

Figure 6. Sampling (a) and sample enrichment (b) with the PMEU in field conditions.

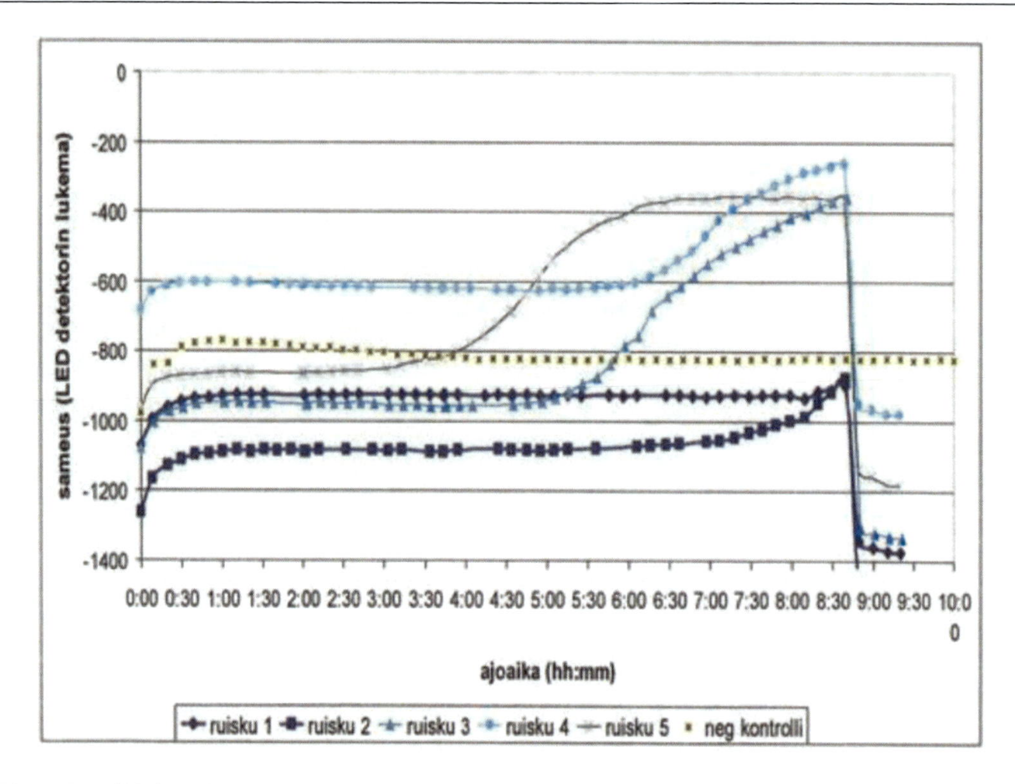


Figure 7. Validation of *E. coli* cultivation in PMEU Spectrion®. The sample with grey curve (ruisku 5), which showed the existence of *E coli* already after 4 hours, had a concentration of $10^5$ cell/ml in the beginning and the sample with a concentration of 2-4 cells/ml (ruisku 1) in the beginning after 8,5 hours [Wirtanen and Salo, 2010]. This experiment proved out that the PMEU Spectrion® method in the hygiene control could verify the presence of *E. coli* in about 10 hours at least in the laboratory conditions starting from singular cells. The same rate was observed also for the detection of salmonellas in tap water [Hakalehto *et al.*, 2011].

# Household Waters

Lakes and rivers serve as sources of water for drinking, recreation and irrigation. The same hygiene indicators are used for verifying the hygienic quality in various samples. Thus the quality of waste waters, natural waters and household waters are followed up using largely the same measurements. Monitoring of household waters is utmost importance because of its immediate consequences for human health.

In Finland, a major outbreak of bacterial, viral and protozoan diseases occurred in 2007 in the city of Nokia, where about 10 000 people were consuming for 2-3 days water contaminated with waste water [Hulkko *et al.*, 2010]. As a consequence, several health problems and epidemics occurred among population. For example, after a month around 30 cases of Giardia infections were reported. Other pathogens that were included in connection with the outbreak were salmonellas, campylobacteria, yersinias, norovirus and many other common pathogens of the gastrointestinal tract. This incidence pointed out the importance of keeping apart the household waters and the sewage drains. In reality, both tubes often locate in the very same pipeline supporting structures beneath the earth in the world cities. This is increasing the risks of leakages besides sudden breakages and sudden outbreaks.

In the USA and other industrialized countries there are huge amounts of leaking tubes underground [Feeney *et al.*, 2009]. The most shocking water crisis in the USA was in Milwaukee in 1993 with about 100 deaths and 400000 consumers getting sick because of *Cryptospodium* [Mac Kenzie *et al.*, 1994]. The resulting horror was a reminder of the seriousness of the water hygiene and protection projects.

We have tested the PMEU method for the surveillance of the clean water distribution system of the city of Kuopio, Finland, with 100000 inhabitants and about 1000 km of pipeline [Hakalehto *et al.*, 2011]. Samples from 10 sampling sites around the network were studied weekly in order to monitor the water quality. The PMEU approach produced nearly 30 early warnings during one month time experimental period in comparison with the couple of alarms by the control method detecting only coliforms by the Colilert TM system (IDEXX Inc. USA). It is worth mentioning that the Colilert media have been tested in different PMEU versions where they have been shown to function well together with the equipment. This provides the water control authorities an option for realtime warning system.

The PMEU technology has been participating in the national Finnish Polaris project intended for improvement of water quality, and the control methods for safequarding it [Hakalehto, 2010] (www.samplion.com website, PMEU in hygiene control). The Polaris study comprises such authorities as the National Institute of Health and Welfare, and the Geological Survey of Finland, as well as Finnish companies, water departments, and universities. There have been several field tests during this project taking place in three different water departments in Finnish cities, as well as in a couple of industrial facilities. The PMEU techniques were operating on spot as stand-alone systems monitoring either automatically (PMEU Coli-line) or by manually taken samples in PMEU Spectrion® system. The PMEU method has been studied also in the Savonia water plant pilot unit in Kuopio (Figure 8).

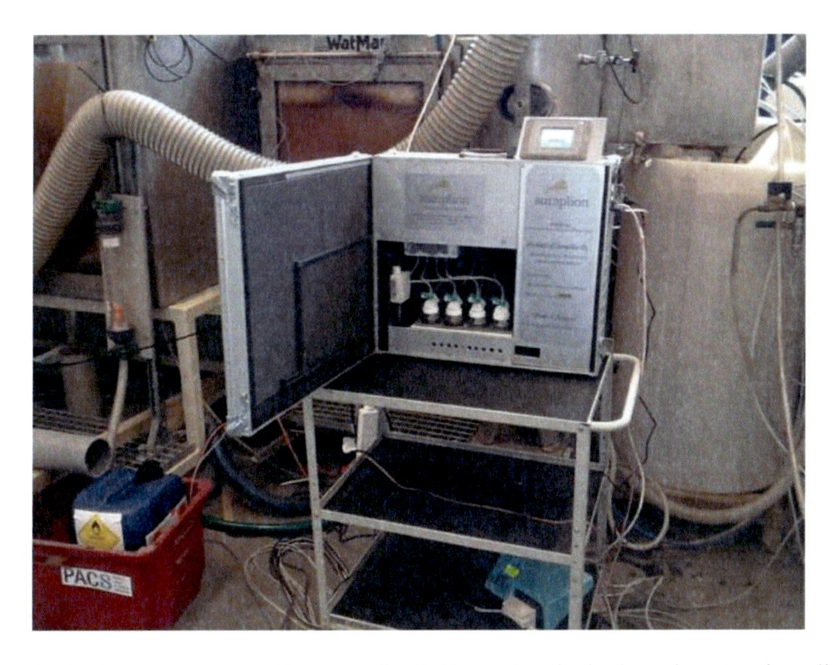


Figure 8. PMEU-ColiLine connected to the water distribution system in the Savonia water plant pilot unit in Kuopio, Finland.

The control units can distribute information on the bacterial growth in control samples via the Ethernet. Many common indicator organisms have been studied with this system [Heitto *et al.*, 2009; Hakalehto, 2011]. In November 2012, the PMEU manufacturer Samplion Oy in consortium with the developer of the PMEU technologies, Finnoflag Oy, are joining an extensive EU project.

Besides the Polaris field tests, and tests in Kuopio, the PMEU technologies have been experimented in the water departments of such cities as Tampere and Turku. During the trials in the water distribution networks of these cities with about 200- 250.000 inhabitants, the various water storage tanks and their pipelines were extensively tested using the PMEU Spectrion® technologies. Some results from one water reservoir indicate the growth with one peak in the incoming water, whereas the outgoing water flow produced two peaks (Figure 9 a and b.).

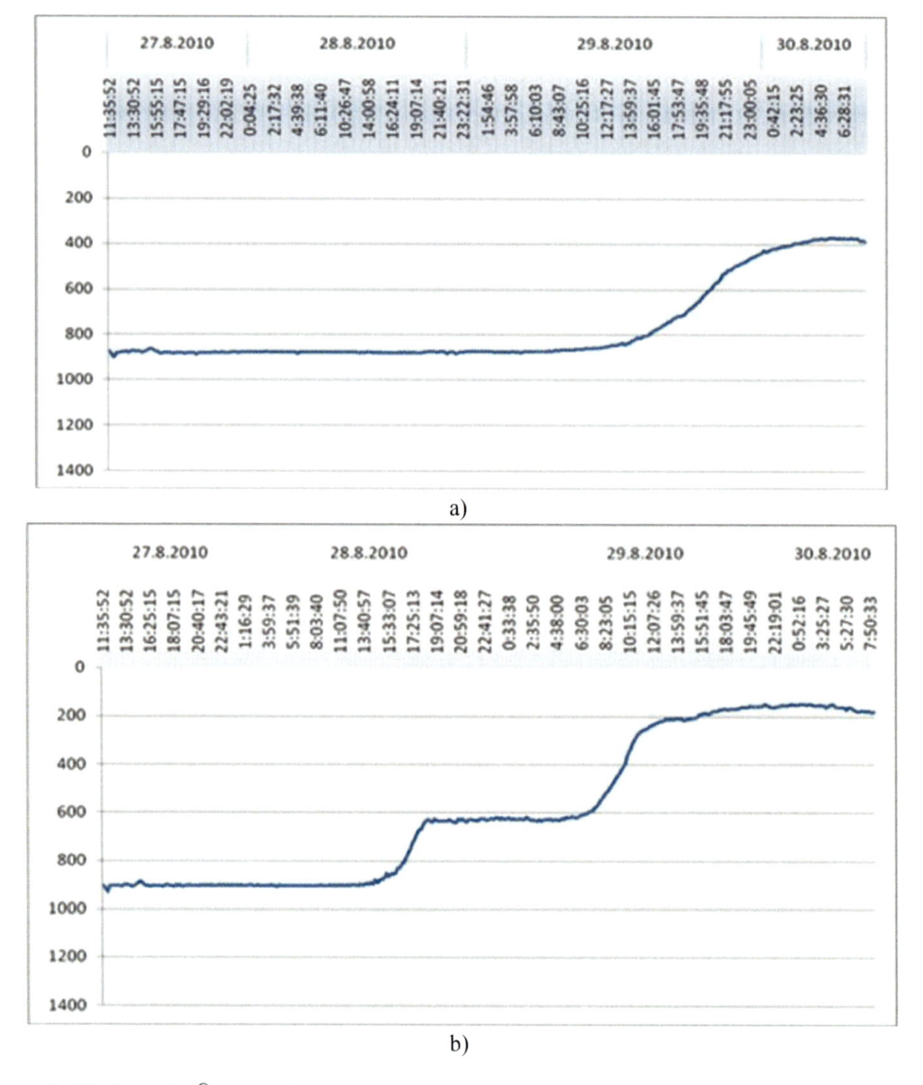


a)

b)

Figure 9. PMEU Spectrion® curves for incoming (a) and outcoming (b) water of a reservoir. These curves strongly imply to an additional microbial growth inside the reservoir tank.

The problems with water management in developing countries could reach huge magnitudes. During the year 2011, the PMEU equipment has been in field trials in Burkina Faso, Africa, where irrigation water was contaminated with *Yersinia* and *Campylobacter* from the municipal waste water pool where it originated from [Heitto *et al.*, 2011]. Prior to the field trials, preliminary method development was carried out in Finnoflag Oy's laboratory in Siilinjärvi, Finland., in order to set up procedures, where the screening of pathogenic bacteria would be facilitated. In this task, the outnumbering other microbes had to be selected out, and the PMEU assisted in the method development. The field method for the microaerobic *Campylobacter* screening included the use of sacs filled with an appropriate gas mixture. It was possible to figure out the real propositions of the threat of contamination by the dirty irrigation water. A couple of years earlier, the PMEU system was in field use in Mosambik, where coliforms and salmonellas were screened from river water in a development project including Finnish, South African, Kenyan and Mosambique students.

The PMEU Scentrion® prototype device has been used for the rapid screening of *Salmonella* sp. from tap water [Hakalehto *et al.*, 2011]. In this survey, the contaminations strating from single cells of *Salmonella enterica* Serovars *typhimurium* and *S. enteritidis* were verified after about ten hours of enhanced enrichment in this PMEU version, which can monitore volatile emissions (see Figure 9). These small amounts of gaseous compounds liberated from the hygiene monitoring cultivations reveal and indicate the presence of the pathogens in such a concise period of time. The shortening of the pathogen detection was achieve also in the field tests of natural and contaminated household waters for the presence of contaminating *Campylobacter* sp. [Pitkänen *et al.*, 2009]. The study was carried out in cooperation with the National Institute of Health and Welfare in Finland. According to its results, the PMEU method enhanced the detection of campylobacteria remarkably prior to detection by real time PCR techniques. This improvement was remarkable when compared with the international standard method, ISO17995 (2005), for the cultivation of *Campylobacter* sp. In the Finnish patient material *Campylobacter jejuni* was constituting the majority with *C. coli* also detected in the specimens [Nakari, 2011]. The former species was of domestic origin for 41%, whereas the latter originated from Finland only in 17% of the cases. The verification of both species could be intensified and enhanced with the PMEU remarkably, and the final concentrations of the bacterial cells in the enrichment broth were 10-100 fold when compared with the normal cultivation [Pitkänen *et al.*, 2009].

# References

Ackman, D; Marks, S; Mack, P; Caldwell, M; Root, T; Birkhead, G. Swimming-associated haemorrhagic colitis due to *Escherichia coli* O157:H7 infection:evidence of prolonged contamination of a freshwater lake. *Epidemiol. Infect*, 1997; 119, 1-8.

Araujo, RM; Arribas, RM; Pares, R. Distribution of *Aeromonas* species in waterswith different levels of pollution. *J. Appl. Bacteriol.*, 1991; 71, 182-186.

Ashbolt, NJ; Grabow, WOK; Snozzi, M. Indicators of microbial water quality. In: Fewtrell L, Bartram J, editors. World Health Organization (WHO). *Water Quality: Guidelines, Standards and Health.* London, UK.: IWA Publishing; 2001; 289-315.

Barrell, RAE; Hunter, PR; Nichols, G. Microbiological standards for water and theirrelationship to health risk. *Commun Dis. Public Health*, 2000; 3, 8-13.

Batt, AL; Bruce, LB; Aga, DS. Evaluating the vulnerability of surface waters to antibioticcontamination from varying wastewater treatment plant discharges. *Environ. Poll*, 2006; 141, 295-302.

Blatchley, ER3; Gong, WL; Alleman, JE; Rose, JB; Huffman, DE; Otaki, M; Lisle, JT. Effects of wastewater disinfection on waterborne bacteria and viruses. *Water Environ. Res.*, 2007; 79, 81-92.

Byrd, JJ; Xu, H-; Colwell, RR. "Viable but nonculturable bacteria in drinking water" *Appl. Environ. Microbiol.*, 1991; 57, 875-878.

Castellani, A; Chalmers, AJ. *Manual Tropical Medicine*. 3rd ed. London: Bailliere,; 1919.

Chelossi, EC; Vezzulli, L; Milano, A; Branzoni, M; Fabiano, M; Riccardi, G; Banat, IM. Antibiotic resistance of benthic bacteria in fish-farm and control sediments of theWestern Mediterranean. *Aquaculture*, 2003; 219, 83-97.

Costanzo, SD; Murby, J; Bates, J. Ecosytem response to antibiotic entering the aquaticenvironment. *Mar. Poll. Bull.,* 2005 218-223.

Dufour, AP. *Escherichia coli*: the fecal coliform. In: Hoadley AW, Dutka BJ, editors. *Bacterial Indicators/health Hazards Associated with Water*. PA: American Society for Testing and Materials; 1977; 48-58.

Dunaev, T; Alanya, S; Duran, M. Use of RNA-based genotypic approaches for quantification of viable but non-culturable *Salmonella* sp. in biosolids. *Water Sci. Technol.,* 2008; 58, 1823-1828.

Durham, HE. A simple method for demonstrating the production of gas by bacteria. *BMJ*, 1893; 1, 1387.

Escherich, T. Die Darmbakterien des Neugeborenen und Säuglings. *Fortschr Med*, 1885; 3, 515-522-547-554.

Feeney CS, Thayer S, Bonomo M, Martel K. *Condition Assessment of Wastewater Collection Systems*. 2009;EPA/600/R-09/049.

Geldreich, EE. Bacterial populations and indicator concepts in feces, sewage,stormwater and solid wastes. In: Berg G, editor. *Indicators of Viruses in Water and Food*. MI: Ann Arbor; 1978; 51-97.

Grabow, WOK. Bacteriophages:Update on application as models for viruses in water. *Water SA*, 2001; 27, 251-268.

Gulkowska, A; He, Y; So, MK; Leo, WY,; Young, HW; Leung, HW; Giesy, JP; Paul, KS; Martin, M; Richardson, J. The occurrence of selected antibiotics in Hong Kongcoastal waters. *Mar. Poll. Bull.*, 2007; 54, 1287-1293.

Gullberg, E; Cao, S; Berg, OG; Ilbäck, C; Sandegren, L; Hughes, D; Andersson, DI. Selection of resistant bacteria at very low antibiotic concentrations. *PLoS Pathog.,* 2011; 7, e1002158.Hakalehto, E. Antibiotic resistant traits of facultative *Enterobacter cloacae* strain studied with the PMEU (Portable Microbe Enrichment Unit). In Press. In: Méndez-Vilas A, editor. *Science against microbial pathogens: communicating current research and technological advances. Microbiology book series Nr. 3.* Badajoz, Spain: Formatex Research Center; 2012.

Hakalehto, E. Hygiene monitoring with the Portable Microbe Enrichment Unit (PMEU). 41st R3 -Nordic Symposium. Cleanroom technology, contamination control and cleaning. VTT Publications 266. Espoo, Finland: VTT (State Research Centre of Finland); 2010.

Hakalehto, E. Semmelweis' present day follow-up: Updating bacterial sampling and enrichment in clinical hygiene. *Pathophysiology,* 2006; 13, 257-267.

Hakalehto, E. Simulation of enhanced growth and metabolism of intestinal *Escherichia coli* in the Portable Microbe Enrichment Unit (PMEU). In: Rogers MC, Peterson ND, editors. *E. coli infections: causes, treatment and prevention.* New York, USA: Nova Publishers; 2011.

Hakalehto, E; Heitto,L; Heitto,A; Humppi,T; Rissanen, K; Jääskeläinen, A; Paakkanen, H; Hänninen, O. Fast monitoring of water distribution system with portable enrichment unit – Measurement of volatile compounds of coliforms and *Salmonella* sp. in tap water. *JTEHS,* 2011; Vol 3(8), 223-233.

Hakalehto, E; Hell, M; Bernhofer, C; Heitto, A; Pesola, J; Humppi, T; Paakkanen, H. Growth and gaseous emissions of pure and mixed small intestinal bacterial cultures: Effects of bile and vancomycin. *Pathophysiology*, 2010; 17, 45-53.

Hakalehto, E; Jääskeläinen, A; Humppi, T; Heitto, L. Production of energy and chemicals from biomasses by micro-organisms. Manuscript in preparation, 2012.

Hakalehto, E; Pesola, J; Heitto, L; Närvänen, A; Heitto, A. Aerobic and anaerobic growth modes and expression of type 1 fimbriae in *Salmonella. Pathophysiology*, 2007; 14, 61-69.

Heitto, A.; Haukka, K.;Martikainen, O.; Miskala-Jaakkonen, M.; Rissanen, K.; King, K.; Heitto, L.; Hänninen, O.; Hakalehto, E. Minute pathogen levels detectable from fresh water flow using the PMEU technology. *19th International Conference on Environmental Indicators*. 11-14.9.2011, 2011, Haifa, Israel. Poster presented.

Heitto, L; Heitto, A; Hakalehto, E. Tracing wastewaters with faecal coliforms and enterococci In: Simola H, editor. *Seminar on Large Lakes 2006*. Joensuu, Finland; 2006; 101-106.

Heitto, L; Heitto, A; Hakalehto, E. *Tracing wastewaters with faecal enterococci*. (Poster). Second European Large Lakes Symposium. Norrtälje, Sweden; 2009.

HMSO. The bacteriological examination of water supplies. Reports on Public Health and Medical Subjects 1934;71.

Horrocks, WH. An introduction to the bacteriological examination of water. London: J and C. Churchill; 1901.

Houston, AC. On the value of examination of water for *Streptococci* and *Staphylococci* with a view to detection of its recent contamination with animal organicmatter. Sup. *29th Ann. Report of the Local Government Board containing the Reportof the Medical Officer for 1899–1900*. London: London City Council; 1900; 548.

Hulkko, T; Lyytikäinen, O; Kuusi, M; Seppälä, S; Ruutu, P. *Infectious Diseases in Finland 1995–2009*. THL, Report 2010;28/2010.

Hutchinson, M; Ridgway, JW. *Microbiological Aspects of Drinking Water Supplies*. London: Academic Press; 1977.

Jeng, HC; Sinclair, R; Daniels, R; Englande, AJ. Survival of *Enterococci faecalis* in estuarine sediments. *Int. J. of Environ. Stud*, 2005; 62, 283-291.

Jimenez, L; Muniz, I; Toranzos, GA; Hazen, TC. Survival and activity of *Salmonella typhimurium* and *Escherichia coli* in tropical freshwater. *J. Appl. Bacteriol.*, 1989; 67, 61-69.

Jones, IG; Roworth, M. An outbreak of *Escherichia coli* O157 andcampylobacteriosis associated with contamination of a drinking water supply. *Pub. Health,* 1996; 110, 277-282.

Kenner, BA. Fecal streptococcal indicators. In: Berg G, editor. Indicators of Viruses in Water andFood. *Ann Arbor, MI: Ann. Arbor. Science;* 1978; 147-169.

Koivunen, J; Heinonen-Tanski, H. Inactivation of enteric micro organisms with peracetic acid, hydrogen peroxide, chlorine, UV-irradiation and combined chemical/UV treatments. *Water Research*, 2005; 39, 1519-1526.

Kühn, I. Biochemical fingerprinting of *Escherichia coli*: a simple system method for epidemiological investigations. *J. Microbiol. Methods*., 1985; 3, 159-170.

Kummerer, K; Al-Ahmad, A; Mersch-Sundermann, V. Biodegradability of some antibiotics, elimination of the genotoxicity and affection of wastewater bacteria in a simple test. *Chemosphere*, 2000; 40, 701-710.

Leino, N. *Pathogens in treated effluent and the need for disinfection* (in Finnish, abstract in English). Lappeenranta, Finland: Lappeenranta University of Technology; 2008.

Levy, DA; Bens, MS; Craun, GF; Calderon, RL; Herwaldt, BL. Surveillance for waterborne-disease outbreaks: United States, 1995–6. Morbid Mortal Weekly Rep, 1998; 47, 1-34.

Mac Kenzie, WR; Hoxie, NJ; Proctor, ME; Gradus, MS; Blair, KA; Peterson, DE; Kazmierczak, JJ; Addiss, DG; Fox, KR; Rose, JB; Davis, JP. A massive outbreak in Milwaukee of *Cryptosporidium* infection transmitted through the public water supply. *NEJM..*, 1994; 331, 161-167.

MacConkey, AT. Lactose-fermenting bacteria in faeces. *J. Hyg,* 1905; 5, 333.

Marshall, KC. Adsorption of microorganisms to soils and sediments.. In: Britton G, Marshall KC, editors. *Adsorption of microorganisms to surfaces*.. New York, N.Y.: Wiley; 1980; 317-329.

Mentu, JV; Heitto, L.; Keitel, HV; Hakalehto, E. *Rapid Microbiological Control of Paper Machines with PMEU Method.* Paperi ja Puu / Paper and Timber, 2009, 91, 7-8.

Muniesa, M; Lucena, F; Jofre, J. Comparative survival of free shiga toxin 2-encoding phages and *Escherichia coli* strains outside the gut. *Appl. Environ. Microbiol.*, 1999; 65, 5615-5618.

Nakari, UM. Identification and epidemiological typing of *Campylobacter* strains isolated from patients in Finland. Helsinki, Finland 2011: National Institute for Health and Welfare (THL); 2011.

Niemi, RM; Niemelä, SI; Bamford, DH; Hantula, J; Hyvärinen, T; Forsten, T; Raateland, A. Presumptive fecal streptococci in environmental samples characterized by one-dimensional sodium dodecyl sulfate-polyacrylamide gel electrophoresis. *Appl. Environ. Microbiol.,* 1993; July, 2190-2196.

Ohno, A; Marui, A; Castrol, ES; Reyes, AA; Elio-Calvo, D; Kasitani, H; Ishii, Y; Yamaguchi, K. Enteropathogenic bacteria in the La Paz River of Bolivia. *Amer. J. Trop. Med. Hyg.,* 1997; 57, 438-444.

Oliver, D. Formation of viable but nonculturable cells. In: Kjelleberg S, editor. *Starvation in Bacteria.* New York, NY, USA: Plenum Press; 1993.

Pesola, J; Hakalehto, E. Enterobacterial microflora in infancy - a case study with enhanced enrichment. *Indian J. Pediatr.*, 2011; 78, 562-568.

Pitkänen, T; Bräcker, J; Miettinen, I; Heitto, A; Pesola, J; Hakalehto, E. Enhanced enrichment and detection of thermotolerant *Campylobacter* species from water using the Portable Microbe Enrichment Unit (PMEU) and realtime PCR. *Can. J. Microbiol.*, 2009; 55, 849-858.

Rolain, JM; Parola, P; Cornaglia, G. New Delhi metallo-beta-lactamase (NDM-1): towards a new pandemia? *Clin. Microbiol. Infect*, 2010; 16, 1699-1701.

Roy, S; Mukherjee, S; Singh, AK; Basu, S. CTX-M-9 group extended-spectrum β-lactamases in neonatal stool isolates: Emergence in India. *Indian J.Med. Microb.*, 2011; 29, 305-308.

Rozen, Y; Belkin, S. Survival of enteric bacteria in seawater. *FEMS Microbiol Rev,* 2001; 25, 513-529.

Savageau, MA. Escherichia coli habitats, cell types, and molecular mechanisms of gene control. *Am. Nat*, 1983; 122, 732-744.

Sawaya, K; Kaneko, N; Fukushi, K; Yaguchi, J. Behaviors of physiologically active bacteria in water environment and chlorine disinfection. *Water Sci. Technol.*, 2008; 58, 1343-1348.

Shah S. As pharmaceutical use soars,drugs taint water and wildlife. *Environment* 360. 2010; 15.4.2010.

Slanetz, LW; Bartley, CH. Numbers of enterococci in water, sewage and fecesdetermined by the membrane filter technique with an improved medium. *J. Bacteriol.*, 1957; 74, 591-595.

Swedish EPA. *Better air quality in Stockholmthanks to sewage sludge*. 2009.

Tamtam, F; Mercier, F; Le Bot, B; Eurin, J; Dinh, TQ; Clement, M; Chevreuil, M. Occurrence and fate of antibiotics in the Seine River in various hydrological conditions. *Sci. Total. Environ.*, 2008; 393, 84-95.

Winfield, MD; Groisman, EA. Role of nonhost environments in the lifestyles of *Salmonella* and *Escherichia coli*. *Appl. Environ. Microbiol.,* 2003; 69, 3687-3694.

Winslow, CEA; Hunnewell, MP. *Streptococci* characteristic of sewage andsewage-polluted waters. *Science*, 1902; 15, 827.

Winslow, CEA; Walker, LT. Note on the fermentation of the *B. coli* group. *Science,* 1907; 26, 797.

Wirtanen G, Salo S. *PMEU-laitteen validointi koliformeilla* (Validation of the PMEU equipment with coliforms). 2010; Report VTT-S-01705-10, Statement VTT-S-02231-10.

Yong, D; Toleman, MA; Giske, CG; Cho, HS; Sundman, K; Lee, K; Walsh, TR. Characterization of a new metallo-beta-lactamase gene, bla(NDM-1), and a novel erythromycin esterase gene carried on a unique genetic structure in *Klebsiella pneumonia* sequence type 14 from India. *Antimicrob. Agents Chemother.*, 2009; 53, 5046-5054.

In: Alimentary Microbiome: A PMEU Approach
Editor: Elias Hakalehto
ISBN: 978-1-61942-692-4
© 2012 Nova Science Publishers, Inc.

Chapter XII

# What Should a Future Probiotic Be Like?

***Elias Hakalehto*[1]*, Markus Hell*[2]*, and Osmo Hänninen*[3]**

[1]Department of Biosciences, University of Eastern Finland, FI, Kuopio, Finland

[2]Department of Hospital Epidemiology and Infection Control, University Hospital Salzburg, Salzburg, Austria

[3]Department of Pathology, University of Eastern Finland, FI, Kuopio, Finland

## Abstract

Intestine provides several niches for a big variety of micro-organisms. These strains are generally striving for balanced collaboration due to the fact that the alimentary microbiome is an entity rather than a track of competition. The survival of each strain depends on its ability to maintain the balance by the particular micro-organism. In exceptional conditions, due to health or nutritional problems of the host, the microbiota may get severely disturbed. This gives the opportunity to some microbes to turn into pathogens, or any incoming true pathogen may get a chance for invasion.

In order to avoid such situations, the implementation of probiotic strains is often needed. Practises of probiotic treatment need modernization. More tools are required for the prevention of imbalances of the body and in its microbiome. Additional research is required, as well as vigorous testing of ideas; compatibilities of particular strains, relations between them etc.

Such studies have been conducted with the PMEU when a realistic view on the microflora is acquired. It is extremely important for our health to learn to take care of the microbial floras in our alimentary tract.

# Introduction

## Why the Probiotics?

The idea of supplementing foods with probiotic microbial strains was invented and designed to promote positive health-related interactions between microbes and man. Firstly such contribution of any strain or mixture of strains had to be documented. Nowadays the common belief is that the lower gastrointestinal tract, especially the colon, is the target site of the probiotic supplementation [Farnworth and Champagne, 2010]. However, as discussed earlier in this book, entire gastrointestinal tract is subjected to microbial influences and often the surfaces of mucous membranes are covered with microbial communities. Also, encountering with the microbial strains in any part of the body system means that host responses will develop for them. For example, in the proximal parts of the intestinal area, the bile acid secretion together with extensive degradative enzyme functions, and such body defenses as like the antibodies, defensins and gastric hydrochloric acid, keep the internal surfaces of the alimentary canal free from excessive microbial colonization. *e.g.*The immunological reactions define the interaction in the long run, and then potential neurological and hormonal stimulation has to be taken into account [Lyte 2010; Trakhtenberg *et al.*, 2011].

One important guideline in using the probiotics should be to maintain or restore the balance within the alimentary tract microflora, and also between the man and his microbes. This objective drives the entire field forward more individualized treatment, and toward the personalized medicine. For the maximal benefit of any bacterial or other supplementation we need to understand the behavior of the added strains in the target ecosystem. What could be the tools for that monitoring? – In the case of treating or preventing *e.g.* diarrhea, the consequences could be observed [Salvatore and Vandenplas, 2010]. These effects dependon the nutritional conditions of the host, that has also important contribution to the normal flora. For example, breast milk contains more than 130 different oligosaccharides with bifidogenic and anti-infective properties. The influences of various infant food formulas on the intestinal enterobacterial flora have been documented by the PMEU (Portable Microbe Enrichment Unit) [Pesola *et al.*, 2009]. Therefore, any such well-planned addition of microbial strains in early childhood can have dramatic life-protecting impacts on the individual diseases. Necrotizing enterocolitis (NEC) is a disease that affects about 0.5% of the newborn infants [Finegold *et al.*, 1986]. For example, concrete results on reduction of the infection rate of fatal and non-fatal neonatal sepsis associated with NEC have been achieved by probiotic treatments [personal communication with NICU's of Vienna University Hospital and Salzburg University Hospital].

## What Kinds of Probiotic Additions?

The present understanding regarding the probiotics is largely based on the use of LAB (lactic acid bacteria) as the target organisms [Goldin and Gorbach, 2008]. This bacterial group belongs to our normal flora, and its members have been shown to have numerous positive effects on our health. Perhaps even more importantly, this group is not containing any pathogenic strains, which makes the usage of the strains relatively safe. It has also been documented that the LAB strains survive relatively well in the GI tract, including the acidic gastric regions of healthy individuals [Hakalehto *et al.*, 2011]. Many LAB strains have been selected for commercial use, but also new LAB strains should be studied in order to find the optimal ones for different individual needs.

However, if we wish to extend our understanding on the alimentary microbiology, as well as to increase the potential to cure and prevent the diseases using this knowledge, also other groups of microbes should be considered in treatment. Non-lactic acid probiotics often work by increasing acidity in the intestine, and inhibiting the proliferation of bacteria, destroying undesirable microorganisms, and suppressing inflammation. In the future it should be a goal to produce such probiotics or their mixtures, which take into account the entire microbial community of the gut, and its interactions with the host. In fact, it should be kept in mind that one part of the health effects of the lactic acid bacteria are related to the lactic acid itself. This was documented by Metchnikoff already more than hundred years ago [Metchnikoff , 1907].

The regulatory mechanism on the probiotics usage should be flexible enough for achieving the goals of the supplementations. At the same time, the individuals have to be protected from any side effects. Keeping this baseline in mind the restrictions of the current policies should be wisely simplified by the following ways:

1. The probiotic treatments with several simultaneous additions (by multi-species-preparations) should be made more easily achievable, and more research should be directed on them. The correct combinations of species and strains should be searched for.
2. The absolute amounts of the probiotic cells should be increased for the best effect.
3. The selection of the probiotic organisms should not be restricted to lactic acid bacteria (LAB) only.
4. The interactions of chosen formulae should be simulated on various foods and nutritive factors by using such tools as the PMEU (Portable Microbe Enrichment Unit) for the monitoring of these interactions.
5. Novel means for following up the positive health effects both in human populations and in the individual alimentary systems have to be designed.
6. Also other sites within the alimentary tract, besides the colon, have to be targeted. For example, in the PMEU studies there has been shown to exist a balanced symbiotic community of different enterobacterial species in the duodenum [Hakalehto *et al.*, 2008; Pesola *et* al., 2009; Hakalehto *et al.*, 2010; Hakalehto, 2011].

In many studies, the joint effect of *e.g.* bifidobacteria with *Streptococcus thermophilus* has been documented [Saavedra *et al.*, 1994; Thibault *et al.*, 2004]. In additions of mixed cultures it could be possible to achieve a long-lasting effect, which is directed toward the various parts of the alimentary tract at the same time. For example, we isolated from a 16 years old patiênt with candidosis the co-culture of *Lactobacillus* and_*Streptococcus* strains from mucosal membranes in all parts of the mouth as predominant species. These strains had been used for the treatment of *Candida albicans* [Hakalehto E., unpublished results]. In the case of preventing antibiotic associated diarrhea (AAD) a mixture of species: LGG, *Bifidobacterium lactis* (former *B. bifidus*), *Streptococcus thermophilus* and *Saccharomyces boulardii* reduced the risk from 28.5% to 11.9% [Szajewska *et al.*, 2006]. *S. boulardii* yeast also significantly reduced the risk of diarrhea on days 3, 6 and 7 [Szajewska *et al.*, 2007]. Also the risk of diarrhea lasting for more than seven days was significantly reduced.

The probiotic addition could consist of several kinds of micro-organisms, not only of the LAB. For example, the use of viruses, bacteriophages has been suggested for the treatments in the gastrointestinal tract [Lederberg, 1996]. These components are numerous in the GI tract in any case. Being like parasites of the bacteria, they could be used to control or block out some undesired strains. It is also imaginable that phages could be used to maintain, manipulate, and carry out the formulation of the microbial ecosystem. Another option and trial for a new kind of implementation of the probiotics has been the use of yeasts for outgrowing *Salmonella typhimurium* or *Clostridium difficile* in the mice intestines [Martins *et al.*, 2005]. The *Saccharomyces cerevisiae* yeasts isolated from natural sources were able to decrease mortality in a group of conventional mice challenged with *S. typhimurium* and to rule out the effects of *C. difficile* in germ-free gnotobiotic mice. However, these probiotic additions in humans require caution because the yeasts as such may constitute a risk of microbial overgrowth, and could thus have detrimental effects on the microbiome as a whole. It has been stated that "one factor impeding the introduction of effective probiotics has been our very limited understanding of the composition of the human microbiome, as well as the biological requirements for these organisms" [Dominguez-Bello and Blaser, 2008]. For this purpose, it is not enough to study the metagenome of the micro-organisms in the various hosts, but to also increase knowledge of competition and cooperation between the microbes, and the host-microbe signalling. Only robust scientific basis makes it possible to develop further the probiotic strategies leading to clinical trials and commercialization of the novel products. The PMEU offers one tool for testing the microbial interactions, and the influences of the gut environmental factors on the specific strains and the mixed cultures [Hakalehto, 2010; Hakalehto, 2011; Pesola and Hakalehto, 2011]. This work should be extended as infection control to the nearest environments of the patients inside the hospital systems, and in the homes and working places [Hakalehto, 2006; Hakalehto, 2010]. For example, such strains in the hospital environment as the *Enterobacter cloacae* may develop more easily multiresistant forms than the actual intestinal bacteria [Hakalehto, 2012]. These antibiotic resistance markers then pose a risk for distribution and transfer to the latter strains, and to the pathogenic bacteria.

Many probiotics are lactic acid bacteria, which are believed to promote healing in the intestine by reducing gut permeability and enhancing intestinal immune responses. Lactic acid is the least absorbing organic acid molecule in the alimentary tract. Its production by the bacteria is taking part of the nutrients out of the body's reach. In our recent studies, one probable mechanism explaining this is the production of gaseous compounds such as the $CO_2$ by the LAB in the intestines [Hakalehto, 2011]. This substance is keeping *Clostridium* in an active growth phase, thus inhibiting the sporulation and spreading or penetration onto the gut walls. It was observed in our experiments also that the 100% $CO_2$ as the gassing agent in the PMEU was well tolerated by *Clostridium difficile* and other clostridia [Hell *et al.*, 2010]. Also other bacteria such as *E. coli* have been shown to be provoked by some $CO_2$ but they generally also have an upper limit for that gas [Hakalehto and Hänninen, 2012]. The requirement for small carbon dioxide concentration for promoting the growth among eubacteria was detected already a long time ago [Volley and Rettger, 1927]. This phenomenon has not been explained so far. In our studies with the PMEU it is possible to investigate the precise effects of various gases on the microbial populations. Their gas generation has also been studied in several projects with the PMEU Scentrion® [Hakalehto *et al.*, 2008; Hakalehto *et al.*, 2009; Hakalehto, 2010; Hakalehto, 2011; Hakalehto *et al.*, 2011].

## Positive Results Accumulate from Probiotic Trials and Usage

Randomized placebo controlled trials have shown promising results for probiotics in the treatment of certain childhood conditions, including gastroenteritis, or stomach flu, and diarrhea. *Lactobacillus rhamnosus* strain GG, for example, has been shown to reduce by 18 hours the duration of acute diarrhea, particularly that caused by rotavirus [Szajewska and Mrukowicz, 2001]. Prophylactic *Lactobacillus* GG reduced antibiotic-associated diarrhoea in children with respiratory infections [Arvola *et al.*, 1999]. In Austria, there have been in active clinical use different multi-species preparations, with 10 strains mainly in one supplement. These products have acutely helped in adjunctive therapy of severe CDI (*Clostridium difficile* infections) cases and reduced mortality significantly [Hell *et al.* Case series report submitted]. In an Israeli trial at 14 day-care centers, the children taking probiotics (*Bifidobacterium lactis* and *Lactobacillus reuteri*) had fewer episodes and shorter periods of diarrhea and less occurrence of fever [Van Niel, 2005].

## Metamorphosis of the Probiotic Functions

In order to better facilitate the prevention of contagious diseases, or GI dysfunctions, we need to develop new tools for exploiting the probiotic functions. These novel strategies may necessitate the usage of organisms that belong to such genera or groups which contain pathogenic strains. One example could be *E. coli* which is both a multivalent member of our normal microflora, and a potent disease causing bacterium [Hakalehto, 2011]. Therefore there is an urgent need to build up methods for testing different characteristics and capabilities of various strains [Hakalehto, 2011]. For instance, the toxicities of previously unknown endotoxins (lipopolysaccharides, LPS) can be tested with several biochemical tests [Helander *et al.*, 1984]. Antibiotic resistance needs to be evaluated [Hakalehto, 2012].

The microbial monitoring combined with potential treatment methods urgently requires new innovations. The decades of rapid developments in genetic techniques have produced plentiful of information about the genetic variation in the microbial ecosystems. In addition, proper understanding of these ecosystems would require more knowledge of the physiology and the interactions between various strains.

This approach could then give an idea on *e.g.* spread of pathogens and their epidemic outbreaks [Tullus *et al.*, 1984]. In the example of intestinal, P-fimbriated *E. coli* strains, clones of which were able to nosocomially spread and cause neonatal acute pyelonephritis in epidemic infections. During one of such recorded epidemics in Sweden 1975-1976 the majority of isolated *E. coli* strains (16 out of 23) in equally many children, a common antibiotic resistance pattern was observed [Tullus and Sjöberg, 1986].

These strains were resistant to ampicillin and cefalotin, and showed intermediary sensitivity to gentamycin, sulfonamide and carbenicillin. A similar strain was behind some septic cases associated with this outbreak. There is an urgent need for increasing the probiotic implementation in order to prevent the spread of antibiotic resistant bacterial strains.

Small children acquire fully mature microbiomic flora by the age of 3-5 years, by some estimates in 6 years [Eckburg *et al.*, 2005]. The microbiome needs to be completely developed and in balance in order to resist illnesses. Any imbalances may cause distortion in the flora composition, and consequently the development of disease. In studies with children suffering from inflammatory bowel diseases, ulcerative colitis or Crohn's disease of ages of 10-17 years, the bile acid secretion was 2-5 fold when compared to a group of healthy children [Ejderhamn *et al.*, 1991]. This included total bile acids, unconjugated bile acids, and glycine and taurine conjugates. It has been supposed that the colonization of anaerobic bacteria may be initiating and promoting cytokine responses mediating development of chronic colitis in rats [Hata *et al.*, 2001]. Also the antibody levels against some anaerobes could be elevated [Saitoh *et al.*, 2002].

From these rather scattered pieces of evidence one could deduce that the excessive colonization of anaerobic bacteria may be preventing the bile acid circulation, as well as cause inflammatory reactions within the host. The prevention of these consequences could be facilitated by active normal flora, especially the lactic acid bacteria with some other facultative bacteria actively adjusting the intestinal pH and providing the host with maximal energy availability [Hakalehto *et al.*, 2008; Hakalehto, 2010; Hakalehto et al., 2012]. The instant growth of the facultatives deprives the obligate anaerobes from rapidly exchangeable food materials. Therefore, these strict anaerobes are kept under control of the healthy intestinal flora. In fact, by favoring these facultatives with bile acid flushes and the peristaltic gut movements, for which the fimbriated coliformic strains are prepared in an excellent fashion, the host body is building up a natural barrier against infections. However, the balance of gut microflora should not be restricted to the avoidance of the infections of epithelial surfaces only.

The so called functional disturbances of the gastrointestinal tract can be associated with bacterial or other microbial overgrowth. This dysfunction is not necessarily causing changes in blood parameters or other laboratory values which are normally monitored. Therefore, the monitoring of microflora is divided into the follow up of its

1. composition
2. function
3. development, and
4. effects

All of these sections of the microbiological monitoring needs to be taken into account. This is needed in order to acquire a broader understanding on the actions of the various probiotic preparations in the human body, and in relation with the other microbes. There is a continuous dynamic process of microflora development taking place in our intestines, but on the contrary, the permanent members of the microbiome organ do have their reservoirs within our body system, and especially in the alimentary tract.

## How the Probiotics Are Delivered

The usage of dried probiotic products is generally preferred. It is beneficial to combine them with a prebiotic matrix, which makes the usage temperature-independent or indifferent with the storage temperatures (no problems are caused anymore by high temperature / hot climate zones). Better and quicker clinical effects are obtained by using high dosages per day and application twice a day.

Various vehicles for the transport of the probiotics in the GI tract have been developed [Chen *et al.*, 2011]. Tablets, capsules and sachets have been developed, and some of the techniques are presented in this book in Chapter 5. In a recent study with gastric endoscopic samples in the PMEU we have shown that some probiotic strains survive in the stomach of the test individuals overnight on the mucose membranes regardless of the acidity [Hakalehto *et al.,* 2011]. This finding indicates that the acidic pH in the gastric areas is not necessarily destroying the strains fully, but they may stay in these areas, and pass to the lower compartments. The tolerance for acidic environment could also be important in eradicating local and invasive yeast infections (*e.g.* prevention of invasives candidosis in neonates) [Brissaud *et al.*, 2011].

## Attachment on the Epithelial Surfaces

In order to fully exploit the different probiotics it is useful to learn about their surface structures. For example, many lactobacilli attach with their S-layer consisting of regularly shaped (RS) protein subunits. These proteins are quite easily removeable in the laboratory by relatively mild acid treatments [Hakalehto *et al.* 1984], but could also provide some protection against harsh environmental conditions. This adsorption tactics is quite different from that of many enteric pathogens which exploit rapidly synthesized surface organelles for primary anchoring onto the gut walls [Hakalehto *et al.*, 2000]. These attachment strategies have the drawback that the protrudements are often cut off by the bacteria in short time if the nutrients are exhausted, or the conditions get unfavorable. It is possible that the LAB is able to establish more stable colonies onto the walls at first hand. Of course, this is only a hypothesis so far.

The lactobacilli have, however, shown to contain also pili (fimbriae). The widely used probiotic. *Lactobacillus rhamnosus* GG was genetically compared with the similarly sized genome of *L. rhamnosus* LC705, an adjunct starter culture exhibiting reduced binding to mucus [Kankainen *et al.*, 2009]. Some pilin genes were detected only with the GG strain. Using anti-SpaC antibodies (against pilin protein), the physical presence of cell wall-bound pili was confirmed by immunoblotting. As the adherence of strain GG to human intestinal mucus was blocked by SpaC antiserum, it was concluded that the presence of SpaC is essential for the mucus interaction of *L. rhamnosus* GG and likely explains its ability to persist in the human intestinal tract longer than the ordinary starter strain.

The wall-attached probiotic strain could at best keep the tight junctions really tight, and is holding back of invasive, not wanted pathogens. A study showed that a *L. rhamnosus* strain GG was able to attach *in vivo* to colonic mucosal membranes and, to remain for more than a week after discontinuation of the GG administration in whey drinks [Alander *et al.*, 1999].

The results from this study also demonstrated according to the authors that the research on fecal samples alone is not sufficient in evaluation of the colonization by a probiotic strain. This was deduced from the difference between results from the fecal and colonoscopic samples.

## Probiotics in Combination with Nutritional Factors in Preventing Toxication and Disease

The reduction of meat-consumption and replacing it by more lacto-vegetarian food would be beneficial, as well as. replacing milk by fermented milk-products such as yoghurts. In Finnish tradition a product called "viili" is widely recognized. It contains specific streptococcal strains producing extensive polysaccharide (PS) capsules. These bring a characteristic composition to the food, which also has a *Geotrichum* sp. mould for producing the surface of the sour substance. In Kuopio, Finland, the effects of yoghurt containing viable *Lactobacillus* strain GG and/or fiber supplements on fecal enzyme activities (β-glucuronidase, nitroreductase, 0-glucosidase, glycocholic acid hydrolase, urease) and on bacterial metabolites in urine (phenol, p-cresol) were studied in 64 young adult females [Ling *et al.*, 1994]. The probiotic GG strain was proven to be the decisive factor in elevating most enzyme levels in the intestines. However, also the fibres in food have shown to possess many beneficial influencing into the GI tract [Kristensen and Jensen, 2011].

The gastrointestinal tract also forms the first line of defence in the body against the main load of xenobiotics [Hänninen *et al.*, 1987]. The monooxygenase activities of the gastrointestinal mucosa in most species are relatively low in the mucosa as compared to the liver, but conjugation, for example, glucuronides change this picture. UDP-glucuronosyl-transferase activities can exceed those in the liver. Glutathione S-transferase activity is also high. The biotransformation activities are readily inducible in the mucosa and are partially responsible for the oral-aboral gradient seen in enzyme activities. A similar gradient can also be found from the gut lumen, in both germ-free and specific pathogen-free rats. The cells in the middle of the villi appear to be most responsive under the influence of inducers.

In recent research the gut barrier function of the probiotics has been documented with respect to invading pathogens. They prevent and treat the disorders of the alimentary tract [Majamaa *et al.*, 1995; de Vrese and Marteau, 2007]. The cited conclusion from one study: "Specific probiotics may enhance gut barrier function and aid in the development of immune responses. Thus, specific probiotics may provide protection against offending macromolecules in the gut and control the future infections by accelerated immunological maturation (ClinicalTrials.gov ID NCT01148667)" [Nermes *et al.*, 2011]. Indication about the participation of the probiotics in immunomodulation has been obtained in several studies. For example, it was found that the preincubation of mononucleic white blood cells of allergy patients with lactobacilli but not with *E. coli*, could inhibit the production of some Th-2 interleucins and simultaneously increased the interferon production [Pochard *et al.*, 2002].

The IBS-associated bacterial alterations were reduced during multispecies probiotic intervention consisting of *Lactobacillus rhamnosus* GG, *L. rhamnosus* Lc705, *Propionibacterium freudenreichii* ssp. *shermanii* JS and *Bifidobacterium breve* Bb99 [Lyra *et al.*, 2010]. The intervention alleviated gastrointestinal symptoms of IBS, and it returned the microbiome condition into the direction which prevailed prior to the onset of disease. Certain probiotics have been found to maintain intestinal equilibrium by enhancing the gut mucosal barrier via manipulation of expression of several their own and the host's genes [Kalliomäki *et al.*, 2008]. All of the 12 mucus adhering probiotic strains tested were able to inhibit and displace ($P<0.05$) the adhesion of *Bacteroides*, *Clostridium*, *Staphylococcus* and *Enterobacter* strains [Collado *et al.*, 2007]. Specific probiotic bacteria have also been shown to stabilize the gut microbial environment and the intestine's permeability barrier, and to enhance systemic and mucosal IgA responses [Isolauri, 2007].

## Conclusion

Different bacteria act probiotically in different situations, depending on the composition of host's normal flora. This reflects the prevailing condition and composition of the flora as well as the host situation. According to recent studies the probiotic strains can have strong preventive impacts on the intrusion of pathogens onto the intestinal membranes. If the microbiota is severely disturbed, building up any kind of co-operation between increasing amount of strains could be beneficial. Especially neonates and small children with developing microbiota, as well as sick persons, may need a strong enough supplementation of the nutrition with microbial strains. It is also advisable to have the surroundings of the patients under surveillance with infection control means in order to maintain their microbiological health. We strongly recommend using multi-species preparations, and also individually designed treatments in order to improve the probiotic preventive care and in conjunction with all kinds of replacement therapies. In order to achieve the best influences of the probiotic therapies, it is also important to develop the field in such a way that all types of organisms in the alimentary microbiome are researched with this respect. Proper use of probiotics can maintain our health from that part, and secure our lives in situations when the body is struggling for survival. Therefore, the importance of this treatment cannot be underestimated. After all it is also an indication that we care for our relation with the alimentary microbiome we carry with us.

## References

Alander, M; Satokari, R; Korpela, R; Saxelin, M; Vilpponen-Salmela, T; Mattila-Sandholm, T; von Wright, A. Persistence of colonization of human colonic mucosa by a probiotic strain, *Lactobacillus rhamnosus* GG, after oral consumption. *Appl. Environ. Microbiol.,* 1999; 65, 351-354.

Arvola, T; Laiho, K; Torkkeli, S; Mykkanen, H; Salminen, S; Maunula, L; Isolauri, E. Prophylactic *Lactobacillus* GG reduces antibiotic-associated diarrhea in children with respiratory infections: a randomized study. *Pediatrics,* 1999; 104, e64.

Brissaud, O; Tandonnet, O; Guichoux, J. Invasive candidiasis in neonatal intensive care units. *Arch. Pediatr.,* 2011; 18 Suppl 1, S22-32.

Chen, S; Zhao, Q; Ferguson, LR; Shu, Q; Weir, I; Garg, S. Development of a novel probiotic delivery system based on microencapsulation with protectants. *Appl. Microbiol. Biotechnol.,* 2011.

Collado, MC; Meriluoto, J; Salminen, S. Role of commercial probiotic strains against human pathogen adhesion to intestinal mucus. *Lett. Appl. Microbiol.,* 2007; 45, 454-460.

de Vrese, M; Marteau, PR. Probiotics and prebiotics: effects on diarrhea. *J. Nutr.,* 2007; 137, 803S-11S.

Dominguez-Bello, MG; Blaser, MJ. Do you have a probiotic in your future? *Microbes Infect.,* 2008; 10, 1072-1076.

Eckburg, PB; Bik, EM; Bernstein, CN; Purdom, E; Dethlefsen, L; Sargent, M; Gill, SR; Nelson, KE; Relman, DA. Diversity of the human intestinal microbial flora. *Science,* 2005; 308, 1635-1638.

Ejderhamn, J; Rafter, JJ; Strandvik, B. Faecal bile acid excretion in children with inflammatory bowel disease. *Gut,* 1991; 32, 1346-1351.

Farnworth, ER; Champagne, CP. Production of probiotic cultures and their incorporation into foods. In: Watson RR, Preedy VR, editors. *Bioactive Foods promoting health: Probiotics and Prebiotics* : Academic Press/Elsevier Inc.; 2010; 3-17.

Finegold, SM; Lance-George, W; Mulligan, ME, editors. Anaerobic infections. A Disease-a-month classic. Year Book Medical Publishers. Chicago, USA, 1986.Goldin, BR; Gorbach, SL. Clinical indications for probiotics: an overview. *Clin. Infect. Dis.,* 2008; 46 Suppl 2, S96-100; discussion S144-51.

Hakalehto, E. Antibiotic resistant traits of facultative *Enterobacter cloacae* strain studied with the PMEU (Portable Microbe Enrichment Unit). In Press. In: Méndez-Vilas A, editor. *Science against microbial pathogens: communicating current research and technological advances. Microbiology book series Nr. 3.* Badajoz, Spain: Formatex Research Center; 2012.

Hakalehto, E. Hygiene monitoring with the Portable Microbe Enrichment Unit (PMEU). *41st R3 -Nordic Symposium. Cleanroom technology, contamination control and cleaning. VTT Publications 266.* Espoo, Finland: VTT (State Research Centre of Finland); 2010.

Hakalehto, E. Semmelweis' present day follow-up: Updating bacterial sampling and enrichment in clinical hygiene. *Pathophysiology,* 2006; 13, 257-267.

Hakalehto, E. Simulation of enhanced growth and metabolism of intestinal *Escherichia coli* in the Portable Microbe Enrichment Unit (PMEU). In: Rogers MC, Peterson ND, editors. *E. coli infections: causes, treatment and prevention.* New York, USA: Nova Publishers; 2011.

Hakalehto, E; Haikara, A; Enari, TM; Lounatmaa, K. Hydrochloric acid extractable protein patterns of *Pectinatus cerevisiophilus* strains. *Food Microbiol.,* 1984; 1, 209-216.

Hakalehto, E; Hänninen, O. Lactobacillic $CO_2$ signal initiate growth of butyric acid bacteria in mixed PMEU cultures. Manuscript in preparation. 2012.

Hakalehto, E; Hell, M; Bernhofer, C; Heitto, A; Pesola, J; Humppi, T; Paakkanen, H. Growth and gaseous emissions of pure and mixed small intestinal bacterial cultures: Effects of bile and vancomycin. *Pathophysiology,* 2010; 17, 45-53.

Hakalehto, E; Hujakka, H; Airaksinen, S; Ratilainen, J; Närvänen, A. Growth-phase limited expression and rapid detection of *Salmonella* type 1 fimbriae. In: Hakalehto E. *Characterization of Pectinatus cerevisiiphilus and P. frisingiensis surface components. Use of synthetic peptides in the detection fo some Gram-negative bacteria.* Kuopio,

Finland: Kuopio University Publications C. Natural and Environmental Sciences 112; 2000. Doctoral dissertation.

Hakalehto, E; Humppi, T; Paakkanen, H. Dualistic acidic and neutral glucose fermentation balance in small intestine: Simulation *in vitro*. *Pathophysiology,* 2008; 15, 211-220.

Hakalehto, E; Pesola, J; Heitto, A; Deo, BB; Rissanen, K; Sankilampi, U; Humppi, T; Paakkanen, H. Fast detection of bacterial growth by using Portable Microbe Enrichment Unit (PMEU) and ChemPro100i((R)) gas sensor. *Pathophysiology,* 2009; 16, 57-62.

Hakalehto, E; Vilpponen-Salmela, T; Kinnunen, K; von Wright, A. Lactic Acid bacteria enriched from human gastric biopsies. *ISRN Gastroenterol,* 2011; 109183.

Hakalehto, E;Tiainen, M; Laatikainen, R; Paakkanen, H; Humppi, T; Hänninen, O. Rapid bacterial metabolic activity by the production of ethanol and 2,3 -butanediol without population growth protects intestinal flora against osmotic stress. Manuscript in preparation, 2012.

Hänninen, O; Lindström-Seppä, P; Pelkonen, K. Role of gut in xenobiotic metabolism. *Arch. Toxicol.,* 1987; 60, 34-36.

Hata, K; Andoh, A; Sato, H; Araki, Y; Tanaka, M; Tsujikawa, T; Fujiyama, Y; Bamba, T. Sequential changes in luminal microflora and mucosal cytokine expression during developing of colitis in HLA-B27/beta2-microglobulin transgenic rats. *Scand. J. Gastroenterol.,* 2001; 36, 1185-1192.

Helander, HF; Sundell, GW. Ultrastructure of inhibited parietal cells in the rat. *Gastroenterology,* 1984; 87, 1064-1071.

Hell, M; Bernhofer, C; Huhulescu, S; Indra, A; Allerberger, F; Maass, M; Hakalehto, E. How safe is colonoscope-reprocessing regarding *Clostridium difficile* spores? *J. Hosp.Inf.,* 2010; 76, 21-22.

Isolauri, E. Probiotics in preterm infants: a controversial issue. *J. Pediatr. Gastroenterol. Nutr.,* 2007; 45 Suppl 3, S188-9.

Kalliomaki, M; Salminen, S; Isolauri, E. Positive interactions with the microbiota: probiotics. *Adv. Exp. Med. Biol.,* 2008; 635, 57-66.

Kankainen, M; Paulin, L; Tynkkynen, S; von Ossowski, I; Reunanen, J; Partanen, P; Satokari, R; Vesterlund, S; Hendrickx, AP; Lebeer, S; De Keersmaecker, SC; Vanderleyden, J; Hamalainen, T; Laukkanen, S; Salovuori, N; Ritari, J; Alatalo, E; Korpela, R; Mattila-Sandholm, T; Lassig, A; Hatakka, K; Kinnunen, KT; Karjalainen, H; Saxelin, M; Laakso, K; Surakka, A; Palva, A; Salusjarvi, T; Auvinen, P; de Vos, WM. Comparative genomic analysis of *Lactobacillus rhamnosus* GG reveals pili containing a human- mucus binding protein. *Proc. Natl. Acad. Sci. USA,* 2009; 106, 17193-17198.

Kristensen, M; Jensen, MG. Dietary fibres in the regulation of appetite and food intake. Importance of viscosity. *Appetite,* 2011; 56, 65-70.

Lederberg, J. Smaller fleas ... ad infinitum: therapeutic bacteriophage redux. *Proc. Natl. Acad. Sci. USA,* 1996; 93, 3167-3168.

Ling, WH; Korpela, R; Mykkanen, H; Salminen, S; Hanninen, O. *Lactobacillus* strain GG supplementation decreases colonic hydrolytic and reductive enzyme activities in healthy female adults. *J. Nutr.,* 1994; 124, 18-23.

Lyra, A; Lahtinen, S; Tiihonen, K; Ouwehand, AC. Intestinal microbiota and overweight. *Benef. Microbes,* 2010; 1, 407-421.

Lyte, M; Li, W; Opitz, N; Gaykema, RP; Goehler, LE. Induction of anxiety-like behavior in mice during the initial stages of infection with the agent of murine colonic hyperplasia *Citrobacter rodentium*. *Physiol. Behav.,* 2006; 89, 350-357.

Majamaa, H; Isolauri, E; Saxelin, M; Vesikari, T. Lactic acid bacteria in the treatment of acute rotavirus gastroenteritis. *J. Pediatr. Gastroenterol. Nutr.,* 1995; 20, 333-338.

Martins, FS; Nardi, RM; Arantes, RM; Rosa, CA; Neves, MJ; Nicoli, JR. Screening of yeasts as probiotic based on capacities to colonize the gastrointestinal tract and to protect against enteropathogen challenge in mice. *J. Gen. Appl. Microbiol.,* 2005; 51, 83-92.

Metchnikoff E. Lactic acid inhibiting intestinal putrefaction. In: In: The prolongation of life: optimistic studies. London. W. Heinemann. pp.161, editor. ; 1907.

Nermes, M; Kantele, JM; Atosuo, TJ; Salminen, S; Isolauri, E. Interaction of orally administered *Lactobacillus rhamnosus* GG with skin and gut microbiota and humoral immunity in infants with atopic dermatitis. *Clin. Exp. Allergy,* 2011; 41, 370-377.

Pesola, J; Hakalehto, E. Enterobacterial microflora in infancy - a case study with enhanced enrichment. *Indian J. Pediatr.,* 2011; 78, 562-568.

Pesola, J; Vaarala, O; Heitto, A; Hakalehto, E. Use of portable enrichment unit in rapid characterization of infantile intestinal enterobacterial microbiota. *Microb. Ecol. Health Dis.,* 2009; 21, 203-210.

Pochard, P; Gosset, P; Grangette, C; Andre, C; Tonnel, AB; Pestel, J; Mercenier, A. Lactic acid bacteria inhibit TH2 cytokine production by mononuclear cells from allergic patients. *J .Allergy Clin. Immunol.,* 2002; 110, 617-623.

Saavedra, JM; Bauman, NA; Oung, I; Perman, JA; Yolken, RH. Feeding of *Bifidobacterium bifidum* and *Streptococcus thermophilus* to infants in hospital for prevention of diarrhoea and shedding of rotavirus. *Lancet,* 1994; 344, 1046-1049.

Saitoh, S; Noda, S; Aiba, Y; Takagi, A; Sakamoto, M; Benno, Y; Koga, Y. *Bacteroides ovatus* as the predominant commensal intestinal microbe causing a systemic antibody response in inflammatory bowel disease. *Clin. Diagn. Lab. Immunol.,* 2002; 9, 54-59.

Salvatore, S; Vandenplas, Y. Prebiotics and Probiotics in therapy and prevention of gastrointestinal diseases in children In: Watson RR, Preedy VR, editors. *Bioactive Foods in Promoting Health: Probiotics and prebiotics.*: Elsevier Science; 2010; 181-204.

Szajewska, H; Mrukowicz, JZ. Probiotics in the treatment and prevention of acute infectious diarrhea in infants and children: a systematic review of published randomized, double-blind, placebo-controlled trials. *J. Pediatr. Gastroenterol. Nutr.,* 2001; 33 Suppl 2, S17-25.

Szajewska, H; Ruszczynski, M; Chmielewska, A; Wieczorek, J. Systematic review: racecadotril in the treatment of acute diarrhoea in children. *Aliment. Pharmacol. Ther.,* 2007; 26, 807-813.

Szajewska, H; Ruszczynski, M; Radzikowski, A. Probiotics in the prevention of antibiotic-associated diarrhea in children: a meta-analysis of randomized controlled trials. *J. Pediatr.,* 2006; 149, 367-372.

Thibault, H; Aubert-Jacquin, C; Goulet, O. Effects of long-term consumption of a fermented infant formula (with *Bifidobacterium breve* c50 and *Streptococcus thermophilus* 065) on acute diarrhea in healthy infants. *J. Pediatr. Gastroenterol. Nutr.,* 2004; 39, 147-152.

Trakhtenberg, EF; Goldberg, JL. Neuroimmune Communication. *Science, Immunology,* 2011; Vol 334, 47-48.

Tullus, K; Horlin, K; Svenson, SB; Kallenius, G. Epidemic outbreaks of acute pyelonephritis caused by nosocomial spread of P fimbriated *Escherichia coli* in children. *J. Infect. Dis.,* 1984; 150, 728-736.

Tullus, K; Sjöberg, P. Epidemiological aspects of P-fimbriated E. coli. II. Variations in incidence of *E. coli* infections in children attending a neonatal ward. *Acta Paediatr. Scand.,* 1986; 75, 205-210.

Van Niel, CW. Probiotics: not just for treatment anymore. *Pediatrics,* 2005; 115, 174-177.

Volley, G; Rettger, LF. The influence of carbon dioxide on bacteria. *J. Bacteriol.,* 1927; 14, 101-137.

# Index

## A

## B

## C

## D

## E

## F



## G

## H



## I

## J



## K

## L



## M

## N

## O



## P

## Q

## R



## S

## T



## U

## V



## W



## X



## Y



## Z